ADVANCES IN PRINTING SCIENCE AND TECHNOLOGY

ADVANCES IN PRINTING SCIENCE AND TECHNOLOGY

Proceedings of the 16th International Conference of Printing Research Institutes, Key Biscayne, Florida, USA, June 1981

Edited by W. H. Banks
for the International Association of Research Institutes for the Graphic Arts Industry

PENTECH PRESS
London : Plymouth

First published 1982 by
Pentech Press Limited
Estover Road
Plymouth
Devon PL6 7PZ

ISBN 0 7273 0108 X

British Library Cataloguing in Publication Data

International Conference of Printing Research
Institutes (16*th: 1981: Key Biscayne*)
Advances in printing science and technology.
1. Printing—Congresses
I. Title II. Banks, W. H.
686.2 Z120.5

ISBN 0-7273-0108-X

Printed in Great Britain by
Billing and Sons Limited
London, and Worcester

Preface

This book contains the papers and discussions of the 16th Research Conference of IARIGAI held in June 1981 in Key Biscayne, Florida. The papers describe original researches and development taking place in the principal graphic arts research institutes of the world. The topics are widely ranging but apart from a few of a general character, they can be grouped under Press and Paper, Colour Reproduction, Ink Transfer, and New Technology.

The success of the Conference owes much to the able direction of our Host Mr Harvey F. George, Executive Vice President and Research Director of Gravure Research Institute. Thanks are also due to Mr Kenneth Hoare for preliminary planning.

The Editor wishes to record his thanks to Mrs Mary Ovretveit for her meticulous typing of the final copy.

W. H. Banks
Editor

Contents

RESEARCH IN THE GRAPHIC ARTS - WHERE IS IT TAKING US?

Gordon O. F. Johnson, LogEtronics Inc., Springfield, Virginia, USA

INTRODUCTION

Printing as we know it today must be considered a mature industry. One characteristic of a mature industry is that each incremental improvement in technology offers less and less increased benefit to the customer but costs more and more for the innovator.

Printing is an industry where the practitioners of the trade do not do their own research. The printer's customers decide what medium they will use to convey their message, Print or Other, and his suppliers decide what new tools he the printer will have available to ply his trade. For printing, as for many other industries, the real technological breakthroughs will come from outside the circle of conventional suppliers to the industry, rather than inside.

In this presentation, I will try to review some of the ideas and opinions discussed in Rochester in May 1981 in a panel discussion on research in the graphic arts which was part of the annual conference of the Technical Association of the Graphic Arts.

STATE OF THE NATIONAL EFFORT

There were six members of our panel*, each bringing a different point of view to our subject. To set the stage for our discussion, we first looked at some overall 'facts of life' about Research and Development:

If we really want to know where the future lies, we should look at who is spending the R and D money in industry today. Business Week magazine each year makes an analysis of the amount of research and development reported by U.S. industrial firms in their annual reports to their stockholders. Table 1 shows data selected from among the 723 public U.S. companies which in 1979 had sales over $35 million and research over $1 million or over 1% of sales:

*The composition of the panel is given in the Appendix to this paper.

(1) The average R and D investment for all 723 R and D investing companies was 1.9% of sales or $1,553 per employee.

(2) IBM was one of the biggest spenders with $1.4 billion*, approximately 10 times the amount spent by Digital Equipment at $138 million. DEC, in turn, spent more than 10 times the $11 million spent by Compugraphic.

(3) Companies like Kodak, Du Pont and Xerox which are deeply involved with new imaging technologies are all major investors in R and D.

(4) Harris, at $40 million, spends significantly more than AM International, which in turn is more than some others under $10 million, which were not included in this list:

Sun Chemical - a little over $9 million
Itek - a little under $9 million
Hunt Chemical - a little over $5 million.

(5) Paper companies were omitted from this list because they are not big R and D spenders. The 14 paper companies reported on by Business Week spend a combined total of $167 million, representing 0.8% of sales and $651 per employee.

TABLE 1. INDUSTRIAL R AND D (1979)

($ millions)

Selected Companies	R and D	% Sales	$/Employee
IBM	$ 1,360	5.9%	$4,034
Kodak	459	5.7	3,635
Du Pont	415	3.3	3,093
Xerox	376	5.4	3,253
3M	238	4.4	2,711
DEC	138	7.7	3,142
Harris	40	4.0	2,036
Bell & Howell	27	5.0	2,372
AM	16	2.1	800
Compugraphic	11	4.4	2,108
723 Companies	23,836	1.9	1,553

Source: Business Week, 7 July 1980; (Public Companies: Sales over $35M; R and D over $1M.)

One obvious and important trend in the graphic arts today is the 'blurring together' of the functions which once made up separate and independent steps in the graphic arts process. This comes as a natural consequence of the continuing quest for increased productivity.

One of the basic rules to improve productivity in any manufacturing operation is to eliminate every possible stopping and collecting point in the manufacturing process, and to create as much as possible a contin-

* A billion in the U.S.A. is equal to 1000 million.

uous flow of work through the plant. This in turn requires a 'systems approach' to the process - a recognition that change in any one part of the process will almost always also impact, often unexpectedly, somewhere else in the process. For our purposes, let us look at the Graphic Arts System as made up of two basic sub-systems: an Image Generating Sub-System and a Paper Handling Sub-System.

(1) Tremendous technological change is taking place in the image generating sub-system as computers convert what was once a series of separate composing and photomechanical steps into one continuous flow of work.

(2) Paper handling systems have seen a major technology change in the shift from sheetfed to continuous web for quality offset color printing, as well as in the increasing number of in-line finishing systems.

(3) At the same time, new laser imaging technology, applied in the new intelligent copiers, has eliminated almost all of the post press finishing and bindery complexity in printing 'one book at a time', one color only, on cut sheet paper - and in the process it also eliminated entirely the imposition and platemaking steps which would have been required for large multiple-page signatures produced in conventional printing. Intelligent copiers are slow and the quality is inferior to conventional printing - commercial printers may therefore feel they do not have to worry about these machines competing for some of their business, but these new printing processes do offer the ultimate in high convenience and low labor cost where a large quantity of multiple copies may not be needed all at one time and place. For example, according to a study by Dataquest, the concumption of 'cut-size paper', (up to 11 x 17 in) was 1.9 million tons in the U.S. in 1979 - that is, 3.8 billion pounds of cut-size paper.

In our panel discussions, R. M. Montgomery did not feel that there is any 'replacement technology' on the horizon which would actually replace printing as a means to put images rapidly on paper. He did point out to us, however, that two forces at work in our economy today will impact significantly how we meet the expanding information needs of modern society: inflation and high cost energy.

How we distribute information becomes critical, and with increasing paper and distribution costs, accompanied by the rapidly decreasing cost of electronics, our society will look more and more to moving information in preference to moving people and things. For printing, this will mean shorter press runs, more electronic controls and software and satellite communications networks with direct imaging systems at the receiver output end.

A. J. Powers observed that enthusiasm for the new often tends to minimize the cost in effort and money necessary to bring products to a commercial stage and that it usually costs as much to go from 95 to 100% with a new venture as to move from 0 to 95%. His company has been actively working in several directions to produce film and paper products of camera or contact speed that are low in silver content or contain none at all. They also are combining camera technology with electrostatic platemakers for use with Dow Jones flatbed laser scanners which eventually will be located in eighteen different locations throughout the United States to publish what is at the moment the only truly national daily newspaper in America.

Microprocessors for camera controls, direct positive halftone paper and film products, new simplified single angle screens for newspaper color were examples of other work underway to simplify and smooth the flow of pre-press work from copy to press.

He agreed that many American companies do tend to focus on enhancing return on investment, with the observation that ROI sometimes seems to be an acronym for 'Restraint on Innovation'. Perhaps that accounted for the fact that the first new lightweight newspaper press called for by the American Newspaper Publishers Research Institute, to upgrade the flexographic Anilox inking system and reduce waste and energy consumption, was built by a Japanese company. As a result of examples like this, in a number of different industries, he felt that there is a new awareness in the U.S. today of the need to produce a commercial climate where taking risks for future growth is worthwhile.

H.T. Marcy observed that we are often loose in our use of the word research, and that one man's research may be another man's sales engineering. Change will occur, whether or not we involve ourselves, due to 'outside' forces. He talked about 'technical momentum' which involved not only the rate of change of a technology, but also the volume of usage - and the greater the volume of usage the greater also becomes the economy of learning.

Some of the important points in this check list in the development of strategy included:

(1) Get a sense of economic and time change which characterizes the Industry.

(2) Chart the areas of high momentum; that is, detail where technical advance is coupled with volume from related activities which are moving independently.

(3) Chart the areas which are important to your situation where momentum is low and uncertainty in achievement is high.

His discussion of the areas of high momentum, brought out a number of fields of new technology where rate of change and volume of usage is high:

Computer technology
Electro optics technology
Graphics and word processing
Data printing

Communications
Automatic control systems.

To optimize the cost effectiveness of our commercial printing man/machine systems, he focused on four particular areas which deserve our priority investment:

(1) A better means to provide proof copy of both text and art at remote sites.

(2) More cost effective digital data transfer to the printing press. Present film intensive technology will be less and less cost effective in the future.

(3) Device independence to allow multiple installations to effectively cooperate - without locking in on single suppliers or single generations of equipment.

(4) Specialized software to deal with the complexities of multiprocessing operations.

G. R. White covered a wide range of topics in his remarks. In evaluating the merits of a new technology he suggested four criteria:

(1) Its scientific merit.

(2) Its embodiment - remember that there is nothing totally new. New electronic or electromechanical devices still need casters and dial lights, for example.

(3) Its operational merit.

(4) Its potential revenues - i.e., its market merit.

Three areas of basic research deserve watching closely:

(1) Electronics - the cost reduction tide in this industry will wash over all of us.

(2) Software to control optics and mechanics which are not coming down in cost.

(3) Information systems.

Measuring the productivity of a research effort 'takes longer than you ever dare think or tell'. Xerox' first laser imaging experiment, for example, took place in August 1964, and commercial products such as the 5700 are only now coming into the marketplace - another example of the 10 - 20 years required to bring research to market.

When do you terminate a research project? When it doesn't make sense anymore. This question is akin to the question 'Who guards the guard?', and 'Who judges the judges?'. Who dares call the best researchers wrong if, and this is important, if they are tackling the right problem - if they are not working on the right problem, then management needs to look at itself, not the researchers.

The real vigor of an industry should be judged not so much based on whether it generates its own breakthroughs, as on its ability to usurp someone else's breakthroughs.

The fundamental problem controlling the entry of computers and electronics into the graphic arts is software. At Xerox they have adopted a radical computer structure, involving a unique programming language which divides the image of each page into an electronic bit map. Basically they are substituting an electronic description for a piece of paper as a means to describe an image. The system's options which follow from this decision are 'bewilderingly rich and large'.

G B Bassett urged that we look at the total lithographic system as a system. Research done by the suppliers tends to focus on individual components or subsections - press, ink, paper, film, plate, dryer, etc. The installation of GATF's new web offset color press, for example, involved 18 manufacturers for various phases of the total system. The task is challenging and much needed but it is doubtful anyone has the resources - technical and financial - to do a thorough job.

GATF resources for research are also heavily impacted by its necessary continuing involvement with government at all levels, particularly in connection with environmental regulations, both potential and actual. Like many other organizations, the 'today' problems manage to take priority over the longer range research problems because these are the concerns which generate the funds for ongoing operations.

C. Noon also called for more systems research to address the unique combination of craft/artistry and science/technology which constitutes the graphic arts. Data rates for a 38 in, 4-color web press, at over 10 billion Bits per second, still far outstrip the capabilities of electronic systems on any foreseeable cost effective basis, and our industry needs continuing research to improve our fundamental knowledge of the printing processes. Paper with its high energy content continues to increase in cost, and controlling that cost is vital to the industry's future. In a typical web plant, for example, where paper is 60% of its total cost, a 10% waste factor translates into almost 6% of the sales dollar - a key difference between profit or loss in a competitive business.

C. Noon summarized our thoughts for all of us when he said that print today is indeed alive and well, with challenges and opportunities for continuing growth and change. Offset, gravure and screen printing continue to grow, and even letterpress is an up-and-coming process again today, but we call it flexography.

In short, our panel concluded that images on paper will continue to grow in importance in the new information age in which we live. Machines and high technology sooner or later have to transfer their information to humans, for whom paper-based images are still the most effective, versatile and persuasive means of communications. United Telecom put it very well in a recent advertisement:

"The toughest part of communications is the last few inches. Modern technology has given us the means to move information farther, faster, and to more people than was ever dreamed possible. Communication, however, means more than just moving information. It also means moving minds. And technology, no matter how far it advances, will never be able to communicate for us. At United Telecom, we're in the business of moving information: across the street, around the world, even to the moon and back. Our job is maintaining and improving communication technology; the easy part. The hard part is shortening the distance between reception and understanding."

Images on paper are still the most effective means to cover the last few inches.

APPENDIX: Panel members

(1) Dr. Robert M. Montgomery, Chief Scientist for the six divisions which make up the Harris Printing Equipment Group, from Melbourne, Florida, opened our meeting to share his views on how it looks from the vantage point of a company which is still an important manufacturer of printing machinery but for whom electronic data handling and communication equipment and systems have today become the largest part of their business.

(2) Mr. A. J. Powers, Jr., President of Powers Chemco Inc. in Glen Cove, New York, spoke from the point of view of a supplier of both conventional and unconventional pre-press expendable supplies and equipment.

(3) Mr. H. Tyler Marcy, Vice President, Computer Technology at R. R. Donnelley & Sons Co. in Chicago, and formerly for many years with IBM, presented a view from a large commercial printer, one of the few large enough to be able to afford its own in-house R and D program.

(4) Dr. George R. White, Corporate Vice President for Research, Development and Engineering at Xerox in Stamford, Connecticut, followed with some views on how it looks to one not wedded to putting ink on paper if there is another alternative, such as electrostatic toners, to deposit images on paper.

(5) Mr. G. B. Bassett and Mr. Cyril Noon, Executive Director and President, respectively, of the Graphic Arts Technical Foundation in Pittsburgh and the RIT Research Corporation in Rochester, who spoke from the point of view of organizations which must generate funds to support graphic arts research not from sales of a product sold to the industry, but directly in the form of sponsored research with defined purposes and objectives paid for by one or a group of clients.

GRAPHIC ARTS RESEARCH & DEVELOPMENT IN PACKAGING

J. C. Votava, Gravure Research Institute, U.S.A.

INTRODUCTION

My comments are based on a total career in R and D in the packaging sector of the industry, and participation in many of the graphics organizations in this country. For your background, my employer, American Can Company, prints by most of the current printing processes: web and sheet lithography, flexography, rotogravure, dry offset, (web, sheet, and formed containers) ink jet, and modified forms of each. The only process we do not use to any great extent is silk screen and, even there, we do some pattern printing by this process. You can see from this background that our interests are broad and our needs varied.

No company today can afford to do all of their graphics R and D effort in-house to develop their graphics needs for current and future products. By this I mean press equipment, prepress, ink and varnishes, and systems development. With the rapid advance of technology in the graphic sector, the scientific disciplines required are increasing and the number that would be needed is economically difficult to justify. With the current trend to electronics and computer control in all aspects of graphics reproduction, more dependence is being placed on the suppliers and the basic work being done by R and D organizations, institutes, associations, and advanced educational institutions. The majority of the technical efforts by printing companies go into modifying currently available equipment, working with suppliers to develop their particular needs, and supplying the technical assistance required by the manufacturing arm of the company. Even this effort increases the technical demand on personnel. This does not mean some basic work is not done today for specialized systems by companies where proprietary technology is a necessity for profitability and marketplace position.

There are few scientists being trained in our educational institutions today for the graphics industry; in fact, very few of them even know the graphics industry exists. The few schools which do offer graphics training stress the manufacturing aspects rather than the technical end.

There is a trend to more technical and engineering education in our graphics students but much more is required to match the current need. At American Can we generally find it easier to teach graphics to engineers and chemists than the reverse.

R and D management today must justify its programs based on monetary return from new products or cost reduction or both or from a quality viewpoint. These programs are then supported by the manufacturing and marketing departments. It is also the R and D department's responsibility to recommend programs to the company that it feels are necessary based on existing or coming technology, not waiting for marketing or manufacturing to state their needs. R and D should be aware of technological change and is in the best position to foresee it.

Keeping up to date on graphics technological trends is done in many ways: through trade journals, seminars, trade associations, technical institutes, professional organizational meetings, and suppliers to the trade. This can be very time consuming but must be done by progressive companies to maintain their operations at the 'state of the art' and be competitive.

Due to the cost of R and D today, basic R and D is hard to justify. Work such as understanding the interactions of all the variables on a litho press is hard to put in terms of monetary return against the investment. Yes, we have to know these technical interactions but we turn to the suppliers and to the organizations and institutes such as those represented here today for the technical information. Work such as this can be justified by organizations because of the total industry effect rather than the individual effects on one company. Corporations can, however, justify membership and monetary support since this is less costly than doing it themselves. Industry standards are also a big justification for memberships in industry organizations since these have an economic impact on the cost of doing business.

With the above personnel and management aspects in mind, where is graphics in the packaging industry today and what are its technical needs? I would like to take the processes individually:

FLEXOGRAPHY

Flexo is becoming the dominant printing process for flexible packaging because of its cost for short run work and the technical advancement in recent years. Process work, using 120 or 133 lines per inch, can be done today through the development of doctor blade systems for ink film thickness control, photopolymer plates, and improved mechanical control. There are some problems remaining, however, such as Anilox roll wear, press speed on high quality work, and the cost of photopolymer plates. Photopolymer plates give a better reproduction of the separation negatives but do not give the cushion effect needed for substrate caliper variation or mechanical conditions across a wide common impression flexo press. Foam backed adhesives help rubber plates but their effect is negated by the polyester backing on the new photopolymer plates. The time required for remakes is also a detriment.

LITHOGRAPHY

The work horse in carton manufacturing is sheet fed litho, although rotogravure is becoming dominant in long run work. One of the major problems facing the industry is the cost of press replacement. Modern presses have press speeds of 10,000 sheets per hour, automated inking units, and some color control but the cost of these new presses is hard to justify when considering the gains made over current equipment. Many companies have opted for major press overhauls with some improvements. This trend will change.

Most of the development effort in litho will be directed at better color control and reduced makeready. The improvements in these areas give better payback than achieving higher press speeds. Quality will be pushed by better color density control and control of the ink-water balance. Development of Sensors to measure the relative amounts of water-ink is necessary and progress is being made in the industry. More effort is being given to the definition of color and color control through the use of spectrophotometry and computer aided color formulation systems (CACF). The CACF is potentially a powerful tool to reduce ink costs in packaging due to the many line colors which must be formulated. I do not feel there would be many in opposition if I said a simplified inking system for litho is required. The Dahlgren inker development may give the needed color control. Only time will tell.

The subject of litho cannot be left without discussing the use of radiation cured inks and varnishes. Inroads have been made into the speciality market due to the smooth finish achieved. In order to reduce costs, some development effort in the industry is being devoted to the application of uv curing varnishes over conventional oxidation curing ink systems. Some success has been achieved in the speciality markets; in fact, Americure uv technology, with its lower cost and current improvements, is being actively licensed by American Can. This technology is not being used in litho, but rather in gravure applications. Where the energy costs are high, such as metal decorating, some uv inroads have also been made. Successful applications have been made as well in the two piece can and plastic container market.

ROTOGRAVURE

A lot of effort is currently going into the automation of the gravure press. Much of this effort was stimulated by the basic work done by The Gravure Research Institute. The initial entrance will be in monitoring rather than control mode using the new systems developed by Hurletron and Crossfield which use the new mark-to-mark eye register control systems. Some good developments have been made in doctor blades, and their use during the past couple of years has helped with wiping problems. A new blade which has twice the milled thickness at the edges of the cylinder versus the middle has helped blade life.

One area requiring more research effort is the resolution of the cause of blushing and streaking. The reasons for these problems and methods used to resolve them are not thoroughly understood.

INK JET

One printing system that seems to be just over the horizon is ink jet or noncontact printing. The state of the art is progressing rapidly, primarily for alphanumeric character printing. Very little progress has been made in its application to packaging except for coding of products for traceability, shelf life, or recall purposes. This system will find its unique niche in the printing processes by being an adjunct to rather than a substitute for the basic systems. The major drawbacks to doing line and process reproduction are the current cost of proposed systems plus reliability.

PREPRESS

The prepress area is currently going through a tremendous change. The publication area is heading toward a total filmless system via the electronic route. This trend, however, will happen in packaging last for two reasons: cost of equipment versus workload and the lack of a hard copy proof. Packaging's main effort is in reruns and lacks the volume individually for new color work in contrast to publication and commercial work. Electronic color separation is making inreads and will eventually take over from the camera approach. Certainly more ink standardization must be achieved in all of the basic printing systems to make this a total reality.

Preproofing systems are slowly coming into their own for packaging. Du Pont's and 3M's systems are being used for sheet fed litho and some development effort has enabled chromalin to be used for rotogravure in special applications. These systems also show promise for flexo, particularly with the growing trend toward photopolymer plates.

In flexo and roto the hurdle of matching a preproof to a proof to a press still requires a lot of development. Progress toward this end seems more imminent for flexo than for roto. I do not know of a good system in development for roto short of building a stripped-down press. Much more effort is required in the industry to resolve this dilemma. For rotogravure to gain a larger share of the packaging printing market, the cylinder cost must come down. A lot of development effort is going into new automated engraving systems: however, I see no great cost reduction coming for it to compete on short runs with sheet fed litho and flexography. The Helio electromechanical engraving cylinders have been a gain in cylinder manufacture, but mostly from a quality viewpoint rather than cost.

GOVERNMENT REGULATION

Compliance with the new government regulations on air and water pollution is also consuming development effort. The deep etch litho plate is dead in packaging due to regulations, but the supplier industry has responded well with photopolymer plates of equal or better life. Rotogravure and flexography, however, present major problems. The solutions to air pollution are incineration, solvent recovery, or water-borne inks and coatings. Costwise, the packaging industry prefers the water-borne materials route. With the many different substrates used, particularly in flexible packaging, the job is enormous. Good progress is being made on inks for paper, paperboard, and polyethylene films. The difficult ones will be inks for propylene, nylon, polyester, and the new polymer blends and materials. American Can currently runs uv curable proprietary varnishes on three roto presses to help alleviate the solvent pollution problem.

CONCLUSIONS

What I have just gone through is certainly not all inclusive due to time limitation, but it does lead up to one final question: what should the graphics organizations be doing to help the progress of the printing industry? Certainly each of you have your specific charter or expertise and work within those guidelines. However, I urge you to attack those problems which need resolution where the answers can be put to immediate use. Research for research sake has little immediate impact or use. New projects or programs should have in their evaluations the question: 'How can my results be used by the industry I serve if I am successful?' If the answer is not positive, don't go forward.

DISCUSSION

M. Bruce Lyne Ink jet printing should not be discounted as a future four-color, full width, high speed printing process. We have cooperated with several large American corporations who are interested in producing wide ink jet arrays with much higher drop delivery rates than in current systems and which use inks and paper surfaces that are conducive to good print sharpness, color density, and gloss. These characteristics, together with the flexibility in imaging should make this form of impactless printing very attractive for packaging.

J. Votave Certainly the electronic hardware and software problems will eventually be resolved and the cost factor will be reduced. Jet reliability needs a lot of work to develop a reliable day to day press. The big factor will be ink development. High density color is required, particularly in the line work, which cannot be achieved today because pigments cannot be used. All the inks contain dyes and these are limited. When it is eventually successful it will be in the commercial and publishing arena before the packaging arena for high speed four color process work.

STATUS OF PRINTING IN THE U.S.A. - 1981

M. H. Bruno, Nashua, New Hampshire, U.S.A.

ABSTRACT

The printing and publishing industries in the U.S.A. continue to grow at about the same rate as the Gross National Product, or about 10% per year in 1980 dollars (2-3% real growth). Value of products produced are predicted to be over $100 billion* by 1984 and $110 billion by 1985. This growth is dependent on trends in the various segments of the printing and publishing industries; newspapers, magazines, book publishing and printing, business forms, commercial and other printing. This paper analyzes these trends and assesses the effects of new developments on them. The new developments covered and their effects are electronics, computers and software; typesetting and page composition; electronic pre-press systems; electronic scanners; electronic cameras; non-silver films; pre-press proofing; automated composition and imposition systems; platemaking; press control systems; ink drying and curing; automated bindery systems; computerized industrial robots; electronic printing systems; small business computers; and environmental considerations. The biggest impact on growth will be computer software to automate systems to improve productivity and reduce waste. Lithography will continue to dominate the printing processes, while flexography and gravure grow at the expense of letterpress.

PRINTING STATISTICS

The printing industry continues to grow at about the same rate as the Gross National Product (GNP), even in recessionary times as in 1980. This growth is shown in Table 1 along with estimates of predicted growth for the next five years[1]. The table has data only for the printing and publishing industries and does not include statistics for the packaging industry, which is over $60 billion, about a third of which represents

*A billion in the U.S.A. is equivalent to 1000 million.

TABLE 1

PRINTING AND PUBLISHING INDUSTRIES 1979 - 1985

Value of Shipments (millions of current U.S. Dollars)

	1979	% Change 78-79	1980	% Change 79-80	1981	% Change 80-81	1985*	% Annual Change 80-85
Printing & publishing	62,248	11.4	68,063	9.3	74,865	10.0	109,000	10.0
Newspapers	16,125	11.2	17,500	8.5	19,250	10.0	27,400	9.5
Magazine publishing	8,052	12.5	8,937	11.0	9,920	11.0	15,000	11.0
Book publishing & printing	7,975	10.0	8,815	10.6	9,790	11.0	14,300	10.0
Business forms	4,001	21.0	4,441	11.0	4,707	6.0	7,700	11.0
Commercial printing	18,200	10.3	19,700	8.2	21,800	10.7	31,500	10.5
Other printing & publishing**	7,895	10.9	8,670	9.8	9,575	10.4	14,000	10.0

* 1985 projections by author based on information in Industrial Outlook text.

** Includes engraving and plate printing, photoengraving, greeting card publishing, blankbooks, bookbinding, typesetting, electrotyping and stereotyping, lithographic plate services and miscellaneous printing.

printing or decorating, and quick and inplant printing which produce products estimated at $14 billion annually, much of which is duplicating and much of this could be considered printing.

As was demonstrated at the international printing equipment exhibits, PRINT 80 in Chicago and IPEX-80 in Birmingham, England, the hallmark of the decade of the 80's will be computer programs or software to increase cost effectiveness of operations and productivity of manpower by increasing the utilization rate of the digital computers which were the principal product of the 70's. Computers have been estimated to be less than 25% utilized. The new digital prepress systems and the spread of computers and automation to the press, bindery and other manufacturing processes in the graphic arts, so visible at PRINT 80 and IPEX, are concrete evidence of the many programs and software being developed to increase productivity by raising the utilization potential of the computers used. As these efforts accelerate, the 80's will become known as the 'Decade of Software'.

The new equipment, techniques and methods developed by the software will not only increase productivity, improve cost effectiveness and reduce waste and spoilage, which justify the often high investments in the new technology, but they will keep the processes competitive, conserve critical materials like paper and ink, and assure their continued supply to support the anticipated growth of processes like web offset. In some instances the new technology creates new information and communications processes which do not use conventional printing components such as films, plates and presses and some do not even use ink or paper. These developments are making many progressive printers and publishers aware that they are not just in the business of putting information on paper, but actually only a part of the huge information and communications industries - the business of disseminating information which now involves almost two-thirds the work force in the U.S.A. The decade of the 80's may well be remembered as a period when the printing and publishing industries underwent significant changes not only in processes, products and markets, but also in concepts.

This does not imply that conventional printed products like newspapers, magazines, books, catalogs, business forms, maps, advertising brochures, labels and packaging will be replaced by other processes or products. On the other hand, many of these products and their effectiveness will be complemented and enhanced by new products like the electronic newspaper, electronically composed and satellite-transmitted magazines, paperless books, electronic shopping, etc. Electronic transfer of funds, electronic publishing systems like Xerox 9700 and Mitel Corp. of Canada, and electronic telephone directories using video terminals to display public announcements of time, weather, horoscopes, sports, and emergency numbers can, however, pose threats to existing printing markets unless the printer or publisher recognizes them as new opportunities in the information and communications industry and participates in them. Many of these new electronic processes or products use video or soft displays in place of hard copy or paper so they can affect the markets for printing supplies and equipment such as papers, inks, plates, photographic films, cameras, presses, solvents, etc. Suppliers and manufacturers especially should be aware of the new technologies and how their applications will affect the use of their products.

TRENDS

A review of the current trends in the various segments of the printing and publishing industries, what conditions and constraints affect them at present, and how new technology and other factors can influence them in the future, will help the printer, publisher, supplier and manufacturer understand and evaluate the data in Table 1 and the timetable for the application, if any, of new technology.

Newspapers

The conversion of newspapers to web offset continues to be the dominant trend in newspaper printing since the use of phototypesetting accelerated with the introduction of the video display terminal (VDT) in 1970[2]. In 1980 there were 1763 daily newspapers and 7954 weeklies with a total circulation of 62.2 million for the dailies and 42.3 million for the weeklies. In 1979, the latest year for which data is available, over 90% of the weeklies and 75% of the dailies were printed by web offset. The data for daily newspapers is in Table 2. In 1977, the offset dailies represented 31.5% of the circulation. Since then many large dailies like the New York Times, Los Angeles Times, Chicago Tribune, Washington Post, Detroit News, Detroit Free Press and New York Newsday, have converted or ordered equipment to convert to web offset so that offset will soon command a much larger and representative share of the daily circulation.

TABLE 2 Conversion of Daily Newspapers to Offset, 1972 - 1979

	Daily Newspapers			Circulation (millions)		
	Total	Offset	Percent of total	Total	Offset	Percent of total
1972	1749	868	49.6	62.0	11.6	18.8
1973	1761	976	55.4	62.6	15.9	25.4
1974	1774	1110	62.6	63.1	17.1	27.1
1975	1768	1173	66.3	61.9	17.2	27.8
1976	1779	1217	68.4	61.3	18.3	29.8
1977	1786	1260	70.5	61.6	19.4	31.5
1978	1781	1272	72.5	62.1	21.2	36.6
1979	1779	1340	75.3	62.2	26.2	42.8

An important trend favoring the use of web offset, besides the conversion to photo and computer typesetting, has been the establishment of satellite printing plants by many of the large daily newspapers.

Most of these plants are equipped with new presses like the Goss Metroliner, the Harris, the TKS (Tokyo Kikai Seisakusho). The success of Di-Litho in converting letterpress presses to using lithographic plates, and the introduction of offset conversion units for letterpress presses have also contributed to this increased use of offset by newspapers in their central plants. Especially important in this trend are the offset conversion units for letterpress presses which are allowing large newspapers like the New York Times, Los Angeles Times and Detroit News to convert their undepreciated letterpress presses to offset at about one-third the cost of buying new web offset presses. Publishers Equipment Corp. (PEC) has converted 71 units of letterpress to offset at the New York Times; Los Angeles Times converted the letterpress presses in its Orange County plant to offset using Smith LOC units and planned to convert the presses in its central plant with units of its own design; the Detroit News converted 54 units of letterpress with TKS offset conversion units.

Three manufacturers, TKS, Faustel and Kidder-Stacey (a division of Motter Printing Press Co.), introduced flexographic printing units for newspaper presses at the ANPA-RI Production Management Conference and Exhibit in 1980 but it will take several years to test these units and assess the impact of flexographic printing on newspaper printing.

Newspapers have traditionally been at the leading edge of new technology, particularly in the pre-press area including photo- and electronic type composition, facsimile and satellite transmission of data, and direct computer to plate systems. The latest developments in this area have been page composition systems incorporating both text and graphics. A system which attracted considerable attention at the 1980 ANPA-RI Exhibit was the HASTECH (Hendrix Advanced Systems Technology) Page Pro. There are other systems which were predicted to make big reductions in the cost of pre-press operations. In 1970 total pre-press (type composition and engraving) costs were 20% of a total newspaper's costs; in 1980 they were 10%, and by 1990 they are predicted to be down to 3%[3].

Newspapers were the first to use satellite data transmission systems. The Wall Street Journal has used them for several years and last year added three new plants to its satellite hook-up. The New York Times announced it will become a national newspaper by printing editions in Chicago using satellite transmission. Both Associated Press (AP) and United Press International (UPI) are building nationwide satellite systems, and newspapers across the country are erecting dish antennas to receive the wire services information. The UPI will have a system with 3700 sites in operation by mid-1983 which will save the news service two thirds of its annual telephone bill for current distribution. AP will start with a 900 station system to offer satellite service to companies presently receiving its high speed data service. The ANPA is currently studying the feasibility of distributing national advertising by satellite in its project SATFAX.

While the main thrust of new technology is automating and computerizing post-press operations, particularly mail room equipment, to improve distribution and cope with the increase in newspaper inserts, newspapers have revealed a new area of innovative leadership in the birth of the electronic newspaper. Newspapers have become the first graphic arts group to realize that they are in the information and communications business beyond the boundaries of ink on paper. On July 1 1980 the Columbus (Ohio) Dispatch became the first newspaper in the U.S. to go electronic, using responsive CATV and home and business computers. American Telephone and Telegraph (AT&T) and Knight-Ridder Newspapers have begun a joint trial of an interactive home electronic information system called Viewtron in Coral Gables, Florida, which provides a combination of news, information and in-home shopping. The Courier Journal and Louisville Times Co. in Louisville, Kentucky, plans to program news and information from its two newspapers on three channels on each of four cable TV systems in the Louisville area. Phoenix Newspapers Inc., publishers of the Phoenix (Arizona) Gazette and Arizona Republic, has plans to develop a similar cable TV package by leasing three channels on a local cable network. The Eau Claire (Wisconsin) Leader Telegram began its version of an electronic newspaper using cable TV in January 1981. The success of the electronic newspaper is bound to encroach on newspaper circulation. On the other hand, it will extend the utilization of the newspaper's database and help the publisher take advantage of a new technology which otherwise could compete with and undermine his business.

Despite these trials with electronic newspapers, they are not expected to have much impact on circulation in the next five years[4]. Consumption of newsprint is expected to grow at the rate of about 2% per year. Consumption in 1980 was 11.5 million tons, and it is projected at 12 million tons for 1981. U.S. capacity of newsprint will reach 5 million tons in 1981 to which will be added 7 million tons, or 70% of Canada's capacity[1].

Magazine Publishing

The magazine industry which had incredible growth of 15% to 20% a year in the 1975-1978 period, has settled down to an annual growth of about 11% which is expected to continue to 1985. The industry publishes 10,665 active publications and is divided into general consumer magazines, business and professional or trade publications and farm periodicals. Consumer magazines account for three-fifths of the industry's receipts; trade publications are about one-fourth, and farm journals the rest. The top ten magazines, which are (1) TV Guide, (2) Time, (3) Reader's Digest, (4) Newsweek, (5) Sports Illustrated, (6) People, (7) Playboy, (8) Family Circle, (9) Better Homes and Gardens, and (10) Women's Day, represented 24% of all advertising and circulation revenues generated by all magazines in 1979. The largest 400, or 4% of the magazines published, represented 76% of all revenues. These statistics are from the Folio 400 Study of Magazine Performance printed in the September 1980 issue of

Folio, the Magazine for Magazine Management (New Canaan, Connecticut).

Magazine advertising revenues, which are a major source of revenue for magazines, have been increasing in the past several years. Since 1976, national advertisers have shifted a significant share of their advertising budgets from high priced network TV advertising to magazines. In 1976, 8.4 cents of every advertising dollar went to magazines; in 1980 it was 9.4 cents. This trend should continue as magazines can deliver selective audiences to advertisers, especially with the growing number of special interest magazines. For the same reason, cable TV can become a threat to magazine publishers since it too can target audiences for advertisers.

Production and postal costs are still chronic problems for magazine publishers as are high rates of return of unsold news-stand magazines. With paper costs rising 10% to 12% a year, ink costs increasing at over 20%, production expenses increasing 8%-9%, delivery costs for newsstand magazines up 9%, postal rates up 10% for mailings, and salaries up 8%, consumer and business magazines may have to increase advertising rates, cover prices and subscription renewal rates at least 10% a year to maintain current profit margins.

Magazines have been a beneficiary of the new electronic and computer composition technology used by newspapers and some of the weekly news magazines are using this with satellite transmission to reduce production costs. A number of magazines are doing their own phototypesetting and page make-up. Three of the leading weekly news magazines are using the sophisticated text and graphics composition and page make-up systems like the Atex front end systems and Information International Inc. (III) Automated Illustrated Documentation Systems (AIDS) to compose complete pages and transmit them by satellite to printing plants in the U.S. and Far East (Hong Kong) where the complete pages of text and graphics are composed on film. The films are then used directly for platemaking, cutting the time from transmission of copy to finished magazines to hours instead of days. These systems, used by Time, Newsweek and U.S. News and World Report, have not only shortened deadlines but they have saved costs by eliminating considerable overtime in the printers' production departments.

Another effective attack on magazine production costs has been the War on Waste (WOW)[2] programs waged by many magazine printers, mainly through prodding by the publishers and active guidance by the Graphic Communications Computer Association (GCCA) division of Printing Industries of America (PIA). These programs are based on information gathering systems which target the causes of waste and indicate the means for reducing it. Aggressive programs have resulted in paper waste reductions of over 5% (about 25 tons of paper per press per year), with corresponding savings in makeready, ink and other costs. WOW programs have been so effective they are spreading to other segments of the printing industry and can have a significant effect on paper and ink conservation.

The problem of high rates of return of unsold news-stand magazines, estimated as high as 50%, is being attacked in several ways. The large weekly news magazines and other popular magazines are printed in multiple plants. Newsweek is printed in 11 printing plants across the U.S. This serves two purposes. It reduces both distribution costs and returns by putting the manufacturer closer to the consumer. This, combined with computerization of distribution and the use of the Universal Product Code (UPC) to record news-stand patterns, sales and performance, makes distribution more efficient and should help reduce the number of unsold copies.

Book Publishing and Printing

The U.S. book publishing industry continues to produce 35,000-40,000 new titles yearly which are marketed by a vigorous distribution system consisting of over 13,000 retail bookstores, almost 500 book club and mail order operations, and over 100,000 retail outlets for paperback books. The industry serves a variety of highly specialized markets including textbooks, professional, reference, religious, and general (trade) books. This trade book category includes mass market paperbacks, book club, mail order, juvenile and adult books which cover fiction, non-fiction, biography, history, drama, art and other literature works. All categories are showing good growth except reference books which are affected by a lack of growth in library and institutional funding and the challenge of on-line computer terminals which are beginning to provide businesses, schools, libraries and homes with access to electronic data banks. Even text book sales show some increase despite decreases in enrollment in virtually all grade levels from elementary schools to college. Renewed emphasis on teaching basic skills and more highly structured learning programs have helped text book sales at the expense of supplementary educational materials like audio-visual packages.

As with news-stand magazines, high rates of returns of up to 50% continue to plague the mass paperback book market. Low U.S. production costs and high visibility in the over 100,000 retail outlets have justified large print orders but buyer resistance is surfacing as prices of paperbacks approach $3. To help control the problem of returns, runs of 100,000 or more are being limited to as few titles as possible. In fact, more than half of all book runs are under 10,000. The Universal Product Code (UPC) and computerized inventory tracking are helping to reduce returns for at least one distributor, which maintains a computerized warehouse of 4000 paperback titles, to between 30% and 35%.

While the revised U.S. Copyright law (1978) will provide some benefits like wider dispersion of printed materials, stronger protection for authors, and increased international copyright protection for publishers, the easing of the restrictions in the Manufacturing Clause, which required books to be printed and bound in the U.S. for U.S. Copyright protection, is expected to have a serious impact on U.S. book printers when it is eliminated in 1982[2]. In a speech before the International Association of Platemakers in Palm Beach, Florida, in November 1980, James R. McClintock, division vice president, 3M Printing Products, stated: 'Ironically the pressure to move book printing work out of the U.S. is not coming from foreign printers, but from U.S. book publishers who are actively touring areas such as South and Central America and the Far East, seeking out potential printers. Estimates are that from 10 to 40 percent of domestic book publishing business could be lost within a few years. In addition, some experts say that a domino effect could occur, causing increased foreign competition for non-copyrighted and pre-press materials. These effects of the elimination of the manufacturing clause were not considered in the projections of book publishing and printing growth in Table 1.

One answer to the possible effects of the elimination of the Manufacturing Clause is for book printers to reduce costs and improve productivity. During the 1970s book printers have undergone a technological transition from letterpress to offset and from cast metal to photo- and computer typesetting. From 40% in 1960, lithography accounted for over 90% of book production in the U.S.A. in 1980[5]. Annual expenditures for capital equipment have risen from $25 million a year in the early 70s to almost $100 million in 1980. Even allowing for inflation, these investments have resulted in notable gains in productivity in the U.S. book printing industry. For the period 1970-1980 value added per employee has almost doubled from 5.2% to 9.9% per year.

Another answer could be the revival of in-line printing systems like the Cameron belt press[2]. The introduction of new flexible relief plates has given the belt press the facility for producing higher quality halftone reproductions in finer screen rulings up to 150-line and this has revived interest in the in-line belt press concept.

As the quality of ink jet and electronic printing improves, these non-contact plateless printing systems will be developed into complete in-line processing systems to produce books on demand. Combining phototypesetting, digital graphics, computer storage and programs for page composition and imposition with ink jet or electronic printers will produce systems that can print signatures in the proper order for collating in-line into complete books. Prototypes of such systems exist in the Xerox 9700, IBM 6670 and the Wang Image Printer. A more sophisticated system has been assembled by Mitel Corp. in Canada using a Xerox 9700, an AM/ECRM Autokon 8400 laser camera, a PDP11/34 computer and a Kurzweil omni-font OCR to interface text from a Compugraphic 7700 typesetter. It will not be long before these or other similar systems which use no phototypesetting paper, photography or film, plate, printing press or ink, will be used to produce the hundred or so proof books needed for advance reviews and book club solicitation. Soon they will be suitable and cost effective for short run educational and other utility book products.

Newspaper publishing is not the only area facing an invasion of video based technologies. The first paperless book, The Computer Cookbook, has recently been published by a California publisher. The introduction of computers in the classroom has alerted text book publishers to the opportunities of offering computer based products for teaching at elementary and secondary levels[6]. Two text book publishers, Houghton Mifflin Co. and Scott, Foresman Co., are offering software to teach elementary reading and math, help the teacher to spot weak areas which need additional work and assist administrators with school records. Houghton Mifflin estimates that in five years one-third of its products for schools will be computer-related.

A professional book publisher, John Wiley & Sons, is trying to shorten publishing schedules for its technical and scientific publications by providing some of its authors with computer consoles to speed delivery of manuscripts to typesetting. In other trials it is transmitting articles and other information electronically to clients with video display terminals. Arete Publishing Co. has put the entire 10 million word content of its 21-volume Academic American Encyclopedia onto computer tape. World Book Encyclopedia is negotiating to sell computer tapes of its printed contents to several information retrieval sources; and Encyclopedia Britannica has agreed to supply some of its contents to NEXIS, a retrieval service of Mead Corporation, Data Central operation. This could be one way for publishers to compensate for the decrease in growth of sales of reference books.

Business Forms

Business forms usage parallels changes in computer markets. These have been dominated by sales of stock continuous forms for use on large general purpose computers. The increasing use of computer output microfilm (COM), non-impact printers and forms management is gradually eroding this market. Even though general purpose computers are increasing at a slower rate of 2% to 3% a year, the new computers are from 2 to 10 times more powerful than computers they are replacing so that even though other output devices are being used in place of forms, the market for stock continuous forms for these computers is still holding. It is being challenged, however, by smaller computers and distributed information processors which use custom continuous forms.

Business forms manufacturers are encountering marketing problems with these small business computers, terminals and minicomputers. Many of the users of these new computers are not large-volume users of forms and forms suppliers must develop new marketing approaches to sell to an increasing number of low-volume users. Several of the major forms suppliers are trying to broaden their base by marketing small business computers and others are developing specialized business software that can be used on small retail computers.

Another growth area for forms manufacturers is the increasing use of forms presses to produce multicolor forms or letters for direct mail advertising. These one part forms, many printed on both sides, must be in roll form as an increasing number are surprinted with computer generated variable information by a line printer or ink jet unit. The Reader's Digest and other sweepstakes mailers are examples of this type of direct mail piece printed on forms presses. The use of these commercial quality, one-part forms will continue to grow and become profitable products for forms manufacturers.

Like newspapers, forms printers are receptive to new technology. Because of the small amount of graphics, business forms will be among the first products besides newspapers to use direct imaging of plates by lasers. By 1985 several forms printers will be making plates directly from forms designs and composition computer-generated at a remote location. Forms printers will also be among the first to use ink jet variable printing as a primary printing process. Tests are underway to produce sets of continuous forms.

Commercial Printing

Lithography is the dominant process used for printing at least 70% of commercial printing which consists of five basic product groups: (1) publications, (2) advertising, (3) labels and rolls, (4) financial and legal, and (5) general commercial printing. The printing of labels and rolls for packaging which is about a $2.6 billion industry, is a 40/40/20 split between lithography, letterpress and gravure respectively. Actually, much of the letterpress is flexography which has been improved with the introduction and use of photopolymer plates, central impression cylinder presses and improved ink distribution systems.

Commercial printing is also benefitting from the new technology used by newspapers, magazines and books. After several years of experimentation, both Time and Newsweek have been successful in the satellite transmission of 4-color news photo. The type of equipment used for this work, the III AIDS system, is also being used for in-house organs, parts catalogs, telephone directories and multi-list real estate photo books. Image assembly, page composition and imposition, the printing industry's most time, labor and cost intensive operations, are being automated by new equipment like the Misomex 605 Image Assembler, computerized step and repeat machines, the Opticopy Page Imposer and the Rachwal Super 70 Paginator.

Most impressive have been the new electronic pre-press systems which will eventually fulfill the dream of going directly from the original copy to the plate without any intermediate photography, film or manual image assembly. These are the Sci-Tex Response 300, Hell Chromacon, Crosfield Magnascan 570, Dai Nippon, and KC-Digital Litho Systems. These systems are very sophisticated, complex and expensive, and it will be at least three to five years before they will be completely debugged and properly implemented.

Other improvements in operation and increases in productivity will result from the proper use of the new press information systems and the war on waste programs. Productivity will also be increased by the introduction of the new 32-page presses like the Harris 850 and new automated bindery systems.

Other Printing and Publishing

This segment of the industry includes engraving and plate printing, greeting card publishing, blankbooks, bookbinding, typesetting, electrotyping and stereotyping, lithographic plate services and miscellaneous printing. Despite increased costs and prices for products like greeting cards, the growth of lithography and the availability of low cost phototypesetting systems which are encouraging trends to in-house typesetting and page composition, this category is holding its growth pattern. The new sophisticated, electronic color page makeup systems are not only expensive but have capabilities beyond the needs of all but the very large printers and publishers, so they will shift many of the composition and layout functions from the other printers and publishers back to the tradeshops which can afford the systems and use them cost effectively.

Lithography vs. Gravure

Lithography continues to command the dominant share of printing, publishing and packaging production. It is about two and a half times gravure's share, and the ratio is expected to continue. Table 3 shows estimates of the proportion of the total industry, including printing, publishing, and packaging, but excluding quick and in-plant printing, shared by each of the printing processes and how the distribution between processes is expected to change over the next ten years.

TABLE 3 DISTRIBUTION BY PROCESS
Printing, Publishing and Packaging
(not including quick and in-plant printing)

	1981	1985	1990
Letterpress	20%	12%	5%
Flexography	14	15	18
Lithography	44	47	42 - 47
Gravure	17	20	20 - 25
Screen Printing and other processes	5	6	10

As is shown in Table 3, letterpress will continue to decrease in use as the other processes erode its market until by 1990 about the only uses for letterpress will be a few large daily newspapers, some label and packaging printing and business forms. The large daily newspaper market, which was expected to be dominated by letterpress, is questionable if the trends to offset and satellite printing plants continue. Flexography is rapidly replacing letterpress in book and label printing and could become a serious contender to letterpress and even lithography in newspaper printing. Flexography will continue to penetrate packaging, heat transfer printing and other publishing areas like newspaper inserts and low budget magazines so that by 1990 it will be the dominant relief process representing about 18% of all production.

Screen printing and other processes which include steel die engraving, die stamping, collotype, electrostatic, electronic and jet printing will grow to about 10% of total production by 1990. This category does not include copying, duplicating and other reproduction processes identified as reprography. These are the basis of the quick and in-plant markets which are growing at a more rapid rate than the printing industry. An estimated 15,000 quick printers have annual sales of about $2 billion. Estimates of in-plant printers range from 60,000 to 100,000 doing an estimated annual production of $12 billion.

Electronic and jet printing have been cited as new printing technologies which pose serious potential threats to the traditional printing processes. This is unquestionably true of products using variable type information such as sequential coding, marking, addressing, billing forms, computer letters, sweepstakes mailers, etc., from which the additional growth in this category is expected in the next 10 years. These processes will still be far from being able to print newspapers, magazines or other products requiring a large quantity of reproductions that should all look alike, because they are plateless processes in which the image must be produced each time a reproduction is made. When the reliability of electronics is improved, which could happen in the decade from 1990-2000, these can become very competitive processes in certain printing areas like short run books, business forms and commercial printing.

The main competitive area is still between lithography and gravure, and it will continue. New cylinder making developments in gravure like electromechanical engraving, powderless etching and laser engraving of plastic coated cylinders are making gravure attractive and cost effective in run categories of 300,000-500,000. At the same time run lengths are increasing in web offset with oven baked photopolymer offset plates consistently producing runs of a million or more on web offset presses. The main deterrent to more universal use of lithography for long run publications is paper waste which is still about twice as much for lithography as for letterpress and gravure but the war on waste programs are bringing this under control. This waste is the main factor that will limit lithography's peak at about 47% around 1985 where it will level off there or decrease to 42% by 1990, depending on what happens in gravure.

As discussed in the last two papers[2], the main problem facing gravure is the need for a practical, cost effective system of ink drying or curing that eliminates the fire and explosion hazards of the present solvent inks, which is causing the even more serious problem of high investment costs for new equipment and plants. So far no satisfactory solution to these related problems has been found. Unless this can be done, gravure will continue to be a process for large printers requiring not only expensive solvent recovery or incineration equipment, but also expensive presses and installation in special buildings or behind fire walls. This will continue to restrict capacity for gravure printing and its growth even though 30 new gravure presses will be installed in seven new gravure plants in the U.S.A.

in 1981. The range of figures in Table 3 for 1990 of 42%-47% for lithography and 20%-25% for gravure are based on whether gravure can solve its dilemma of solvent base inks which is the main cause of its high investment costs. If it does, the figures will be 42% for lithography and 25% for gravure. If not, lithography will continue to dominate the printing industry and will account for 47% of total production while gravure levels off at about 20%.

NEW DEVELOPMENTS

The changes in trends described in the printing and publishing sections will result from new developments and new applications of existing technology. The big changes will be in the development of software to broaden the utilization of computers to simplify operations, reduce their costs and increase productivity by eliminating hand work, consolidating steps or by automating them. The developments are in the areas of electronics and computer memories, electronic pre-press systems, electronic scanners, typesetting and page composition, pre-press proofing, electronic cameras, non-silver films, automated composition and imposition systems, platemaking, press control systems, ink drying and curing, automated bindery systems, computerized industrial robots, electronic printing systems, small business computers and environmental considerations.

Electronics, Computers, Memories

A brief look at the trends in the printing and publishing industries reveals the important role electronics and computers are playing in directing their future. Accounting for most of the application of electronics is the computer which is increasing in power as it reduces in size and cost. The computer has been responsible for the development and application of electronic business information systems, electronic text and page composition, electronic cameras, electronic color separation, electronic image assembly, electronic proofing, electronic platemaking, electronic press and bindery controls and electronic control of quality. The ability of computers to handle all these assignments is due to the preparation of the software that is used to instruct them on the requirements of each function. One of the deterrents to faster progress is the requirement in a number of these functions to handle and store huge quantities of data, especially where four-color screened or continuous tone separations are involved.

Changes are happening so fast in the development of computer power and the ability to program, handle and store data that people and companies working on the application of computer technology to graphic arts can't keep up with them. The floppy disk has been the standard storage medium for most computer systems used in graphic arts. It is gradually being obsoleted by the introduction of the 8-inch rigid disk which has 10 to 30 times the storage capacity and 10 times the speed at lower cost. Most equipment is still using the 8K bit chip and the 16K bit is just starting to be used while the 64K bit and 256K bit chips are approaching commercial availability. Bubble memories are being developed that will have the capability of storing 4 million bits on a fingernail size chip and 2-sided 12-inch optical memory disks with Direct-Read-After-Write (DRAW) capability have been announced with storage capacity of at least 2500 megabytes on both sides[7].

Not only will the new high density storage media speed up the development of computerized components, equipment and processes for the graphic arts, but they will help to reduce their costs as some of the new optical disks will be available at much lower cost than the other storage media. The manufacturer of a DRAW optical memory disk claims that by 1982 the disks will be available at costs of about two cents per megabyte. The manufacturer's comparisons of costs per megabyte of storing data on the optical disk vs. other computer memory media are given in Table 4.

TABLE 4 COMPUTER MEMORY COST COMPARISONS

Computer Memory Medium	Cost per Megabyte
Drexon D1201 - 12" optical disk	$0.02
Magnetic tape	$0.40
Magnetic disk	$3.00
Magnetic bubble	$200.00

Costs will certainly decrease as magnetic bubble memories are manufactured and used in quantity but how soon they will approach the cost of the other media is not known. In any case, the availability of low cost, high density storage media will encourage and stimulate software groups to accelerate development of computerized electronic functions for the graphic arts.

An emerging area of computer input which will be enhanced by the new storage media is voice recognition. In addition to higher storage densities, voice input also requires high data rate transmission lines such as are provided by fiber optics and laser transmitter diodes. This technology too is advancing[8]. A strand of glass fiber, five thousandths of an inch in diameter can transmit up to 672 conversations simultaneously. The cost of fiber optics has discouraged its use but it is coming down from $4 per fiber meter when it was introduced to $1 per fiber meter now, and expected to be $0.25 per fiber meter in a short time. Other technologies which can affect the efficiency and improved utility of computers are new output devices like ink jet printers and plasma display tubes which are expected to have some impact by 1985.

Typesetting and Page Composition

Typesetting which was the first target in the electronic invasion of graphic arts, is becoming the first beneficiary of the marvels of software as it merges typesetting with page composition. The market for all typesetting and related devices in 1980 (not including consumables and intercompany transfers) was predicted to be about $1 billion[9]. Of this about $470 million were for just typesetting machines, $445 million for input, page makeup and front end systems which are mainly the products of software, and the remaining $200 odd million was divided between type masters and service.

The evolution of typesetting is gradually approaching completion. The first true phototypesetter was the Photon 200B which was a direct input device priced at over $50,000. This was a high price for a one-person device so off-line typesetters were developed which could use input from four or more keyboards. In 1972 Compugraphic introduced its low cost (under $10,000) direct input Compu Writer which became the most successful typesetting machine since the Linotype with sales of over 17,000 worldwide. The direct input typesetter, however, still had the disadvantage of limited output, until peripherals were added to it. A video display terminal (VDT) was added in 1974; then record and playback capability first as paper tape and then as floppy disk; followed by automatic hyphenation and justification; and then came higher speed after which it could accept input from off-line devices. In 1979 Mergenthaler introduced its CRTronic direct input digitized typesetter at IMPRINTA 79 in Europe and in 1980 at PRINT 80 in the U.S.A. In less than two years since it has been introduced, over 2,500 units have been sold. It is expected that all direct input typesetters will be digitized in the near future, and with digitization will come an upgraded front end making the direct input typesetter a mini Document Processing System (DPS) with computer assisted page makeup (CAM) capability. This step will essentially complete the evolution of typesetting.

The other major developments in typesetting are the advancements in 'front end' or Document Processing Systems (DPS) and in page makeup. All are related. Front ends developed from cluster systems of video editing terminals which needed mass storage on magnetic disks that led to electronic file or data management. Eventually typesetters were connected directly with the clustered editing systems, programs were written for automatic hyphenation and justification, and the computer connecting all elements was programmed to format the typesetting. This system of hardware (video editors, computer, typesetter, line printer, disk units, etc) with software (data management, computerized typesetting and electronic editing) which was used to prepare material for typesetting, was called a 'Front End'. But with the typesetter connected, it is actually a total system dedicated to the production of typeset documents so the name 'Document Processing System' (DPS) is more appropriate. Almost 340 front-end systems were installed in the U.S.A. in 1979 and 700 are predicted for 1981.

Present page makeup systems use a stand-alone terminal which is like a video editor that allows you to interact with the makeup - i.e., it has feedback so you can see the result of a change immediately on the screen but the screen is passive. To make the actual changes, you must go through another editing terminal. This is a transitional configuration[10].

Soon page makeup terminals will be connected interactively into Document Processing Systems. They will also be connected with digitized Direct Input typesetters to produce a new category of Integrated Stand-Alone devices which will provide makeup interaction directly with typesetting.

Some of the software for these new systems will be provided by small companies staffed with proficient computer programmers. An example of such a company is Ed Berg and his Bedford Computer Corp. (Bedford, NH), which designed the Real Time Composition System displayed at PRINT 80. It is an interactive real time multi-terminal composition and data management system that saves time in keyboarding, editing, typesetting and page makeup; eliminates pasteup; simplifies complex math, tabular and display advertising composition; permits lower skill levels in operators; reduces material costs, and increases productivity. It uses existing hardware and can interface with most popular typesetters. Software is being developed to add graphics input into the system. There are other similar groups of programmers working in these types of applications.

Eventually word processing and graphics, both black and white and color will be integrated into a system that will accomplish the goal of the electronic pre-press systems of going directly from copy to press plate without engaging in any intermediate photographic or manual image assembly operations. An example of this type of system is the Information International Inc. (III) Automatic Illustrated Documentation System (AIDS). This system, which is being used by the weekly news magazines Newsweek, Time and U.S. News and World Report, and commercial users like Boeing, John Deere, McGraw-Hill Multi-List, Baird-Ward for telephone directories, etc., consists of a Text Editing and Composition System (TECS) which performs many of the front end functions (the magazines use ATEX front end systems); an Illustration Scanner which digitizes line art and continuous tone photographs that are later converted to halftones when received on the Video Comp 570; a File Manager which manages the data in the system; a Page Makeup Station which scales and positions text and illustrations; and a Video Comp 570 Pagesetter which is a computerized CRT typesetter capable of recording typefaces, halftones and line art on film or paper. The system can be used in-house or the data can be transmitted by phone lines, facsimile or satellite transmission. Color has been handled as continuous tone separations made on a scanner and transmitted as black and white images. Work is being done on direct input of color originals into the system from color separation scanners.

The bottom line for the developments electronics have brought to typesetting is the staggering decreases in price/performance or price as a factor of speed ratios in typesetting. Price of typesetting equipment per line per minute (lpm) of typesetting speed has gone from $4000 in 1960 to $50 today. Putting this in real terms, price for setting an average page has decreased from $50 in 1960 to $10 now, and it is expected to drop to $5 per page by 1985. Will a similar 'electronic revolution' happen in the other pre-press, press and bindery areas?[10]

Electronic Color Composition Systems

These are also known as Electronic Pre-Press Systems as they attempt to accomplish all the steps from the original copy to the press plate in one system of integrated units. These systems consist of an electronic color separation scanner or interface with a scanner, a digitizing tablet with cursor or tracer pen, computer storage, a high resolution video display station for local and area color correction and page composition manipulation, and an output device to record the image of the completed pages or plate imposition on film or plate. These systems cost about a million dollars which is expensive for the hardware, but when it is realized that the equivalent of about 400 man years of software had to be developed, the price seems justified. A number of such systems, varying in degree of sophistication are in use and in development. These include the Crosfield Magnascan 570 and Hell Chromacom announced at DRUPA-77, the Sci-Tex Response 300 introduced at PRINT 80, the KC-Digital Litho System, also shown at PRINT 80, and the Dainippon Screen modular system announced at the Lasers in Printing Conference in San Diego, California in September 1980.

Crosfield Magnascan 570 This system has made considerable progress during the past year. When shown at PRINT 80, the system did not have any facilities for local or area color correction during the makeup or composition function. All color correction had to be done on the scanner during the scanning operation. At IPEX, a color monitor was added to the system along with the ability to perform the equivalent of electronic dot etching, insert tint backgrounds and adjust color values along with page makeup, thus making the system interactive and comparable with the Hell Chromacom and Sci-Tex Response 300 in these capabilities. In the next stage of development Crosfield plans to separate the color retouching station from the page layout function, as they believe that these are separate functions, performed by different people with different kinds of skills and training[11]. Crosfield has 70 Magnascan 570 systems in the field with seven in the U.S.A.

Hell Chromacom System This is a completely interactive and integrated system with color correction and page makeup capability which is interfaced with the Hell DC300 or CP340 scanner and can output durectly on a Helio-Klischograph to make gravure cylinders directly from original copy. Hell claimed to have seven Chromacom systems in operation by March 1981. So far no system is in operation in the U.S.A. Reports of a system installed in West Germany are encouraging and optimistic.

Sci-Tex Response 300 This was the first of the pre-press systems to have completely interactive color correction and page makeup capabilities. There are 22 worldwide installations of Sci-Tex Response 300 systems, with 6 in the U.S.A. It is the only one of the active systems that can interface with any electronic scanner. A deficiency of all three active prepress systems, Crosfield, Hell, and Sci-Tex, has been the inability to input text into the system. At present type has to be set, scanned, digitized and then input to the systems. Sci-Tex is actively working on typesetting capabilities for the Response 300 system[11]. It appears to be ahead of Crosfield and Hell in this respect, having already output digitized type on its laser plotter. Sci-Tex is seeking a type font library to incorporate in the system. It has also modified the Response 300 to interface with Helio-Klischograph models K-200, K-201 and K-202 so the system can be used for direct color engraving or gravure cylinders.

KC Digital Litho System A prototype of this system was shown at PRINT 80. It is an integrated copy-to-plate system using the KC-Crystalplate as an output. The system can input text, graphics, and layout and production instructions and the outputs can be a soft page proof, a color proof, or KC-Crystalplates for printing on the press. It can interface with any modern text processing system. Graphics, color correction and page makeup are handled similar to the other systems. The major differences in the systems are the abilities (1) to produce hard proofs with toners or by printing the KC-Crystalplates, made on a digital platemaker, on a press with the appropriate inks, and (2) to make the actual printing plates without the need for the intermediate film(s) required in the other systems. The system is still in development but is claimed to be ready for market soon.

Dainippon Screen Process Not much is known about this process which was announced at the Lasers in Printing Conference in September 1980, except that it will be a modular system using the Dainippon SG701 scanner and that it is planned to introduce it at DRUPA 82. Dainippon Screen and 3M which owns COMTAL, manufacturer of color image processing systems, are cooperating in the development of the new pre-press system.

Electronic Color Scanners

While much of the effort in color reproduction is in the design of sophisticated electronic pre-press systems which use color scanners as input, there have been some significant changes in the color scanner market. PDI is back in business with agressive marketing of its new Compudot/Compuscale scanner first introduced to the U.S. market in 1978. Formerly a division of TIME, Inc., PDI was sold to a group of Japanese businessmen in 1980 who have generated new interest in this new scanner. Seven have been sold in the U.S.A. and 24 worldwide.

Crosfield has taken over distribution of its products in the U.S.A. from its former distributor, Ruthographics, and announced its long-awaited Magnascan 520 with electronic dot generation using lasers. Hell introduced its CP340 large format color scanner which will color-separate any flexible color original up to 25.6 inches square (66 cm) and output the color separation on film up to 44 x 50 inches (113 x 128 cm). The scanner can produce continuous tone or electronically (laser) generated screen images in reproduction scale from 25 to over 2000%. The CP340 linked with the Hell Chromacom will produce film flats for press plates directly from data stored on disk packs. Hell is predicted to sell about 700 scanners (DC-300, 299 and CP340), or a third of its worldwide market, in North America by the end of 1981. A new DC-350 laser scanner with conventional screen angles has been announced. It has been estimated that 48% of scanners in use have electronic (laser) dot generation. 40% use contact screens, and 12% have continuous tone output[10].

The use of small scanners is growing. Siemens has built a new plant in Hauppage, Long Island, New York, to build Hell DC299 scanners. Linotype Paul has sold 36 - 3040 scanners in the U.S.A. and is expected to sell 65 more in 1981. Dainippon Screen has sold 50 scanners in the U.S.A. and 1000 worldwide since they were introduced in 1979[5].

Color Previewers After being around for almost ten years, electronic color previewers are finally beginning to sell. Hazeltine has two color previewers: Color Separation Model 1620 which systhesizes color reproduction from continuous tone or halftone separations, and Scanner-Previewer Model PS300 which analyzes transparencies or reflection copy and generates settings for scanner operation. About 33 of these systems are installed in the U.S.A. and Hazeltine predicts a significant increase in sales because of the success of the units already in use. TOPPAN also offers two color previewers: Model CP525 to analyze continuous tone and halftone separations and Model ST525 to analyze the original for proper scanner settings. No data is available on sales of these units, but tests on the use of the CP525 to preview scanner halftone separations prior to making color proofs are reasonably successful.

Flatbed Scanners Other companies are designing components like flatbed scanners that can be used in custom electronic scanning systems. Optronics (Chelmsford, MA) which designed the first commercial electronic page composition system, Pagitron, that was used successfully by a newspaper in Finland for many years, is now devoting some of its resources to building components like flatbed scanners for contractors.

Another flatbed scanner has been announced by EIKONIX Corp. (Bedford, MA 01730). It is a new high-speed, high resolution image digitizing system using self-scan array technology. The company specializes in digital image processing and makes the flatbed scanner/digitizer and an output film plotter. The scanner, identified as EIKONIXSCAN 7800 resembles a heavy duty enlarger with the flatbed scanner in the top of a desk which contains the control electronics. It features computer controlled magnification, true and pseudo color capability, 12 or 8 bit software, selectable gray level resolution and real time recording of tones on input. The original to be scanned is placed on the flat easel and a lens mounted on a bellows focuses the image to size on the scanning area where scanning is performed by an array of detectors (CCD). The system can scan a subject up to 10 x 12 inches (25 x 30 cm) in 37 seconds with a 2048 element array capability. A 4096 element array is in development. The unit is being considered for use in graphic arts. EIKONIX is designing a system to scan and output writing systems for both black and white and color at 150 line screen resolution[11].

Electronic Cameras

The AM/ECRM Autokon 8400 laser camera was introduced to the newspaper industry at the ANPA-RI Production Management Conference and Exhibit in Houston, Texas, in 1975, and there are now over 500 units in use, mostly newspapers. It is an electronic optical device using lasers to reproduce text, line and single color halftone illustrations in screen rulings up to 100 lines per inch (40 lines per cm) to convert images to positive, negative, correct or lateral reversal modes, and to perform controlled distortion and anamorphic functions. It has the same capability of other electronic imaging systems of unsharp masking, edge enhancement, variable tone control and special screen element configurations like double dot, chain dot or random element screens. As manufacturers of text processing systems are scrambling for means to input graphics, a few have suddenly realized the capability of the AM/ECRM Autokon 8400 for this use. Autologic and Mergenthaler both displayed systems at the ANPA-RI show in 1980. The Autokon 8400 laser camera is also an integral part of the Mitel Corp. electronic publishing system (mentioned earlier in the section on Book Publishing and Printing and described in the section on Electronic Printing Systems). When the laser camera is expanded to a larger format size with 150-line screen capability, it will find many more uses in integrated text/graphics processing systems.

High Speed CCD Camera At the 1980 ANPA-RI Production Conference a new high speed TV camera using charge couple device (CCD) sensors was demonstrated that had a photographic speed equivalent to ASA 300,000 (ASA 400 is the fastest amateur film presently available). Bob Madden, award-winning photographer for National Geographic magazine, amazed everyone with real time images of the audience in the darkened auditorium. A similar camera was used by NASA for the amazing pictures from Mars, and National Geographic used it for a story using undersea photography. Its main deterrent is cost. A CCD chip needed to store the information from one photograph would cost $15,000. National Geographic Society expected to have an economy model CCD camera with a price tag of $25,000 completed by early 1981.

William D. Rhinehart, ANPA vice president/technical, stated that a reasonably priced black and white electronic CCD camera suitable for newspaper use would be available in 3-5 years[12]. He noted that instant photographs could be displayed on distant video screens using ordinary telephone connections. A photo editor could call up a picture on his VDT, electronically crop it and if working in color, make color corrections in it. With the full text and graphics pagination system all newspapers will have by the time the camera is available, the editor will send the picture directly to the page generator or typesetter. The electronic CCD camera will be the final link in the all-electronic pre-press system which will eliminate the use of all conventional materials, even the photographer, and make it possible to go from the original scene to the printing plate completely electronically. It may, however, find use in news magazines well before newspapers can afford it.

Non-Silver Films

Camera films will continue to use silver halides, although new films are being developed with reduced amounts of silver. The increasing cost of the petrochemical resin used for the film base is offsetting what possible savings there might be from the reduction of silver in emulsions. Non-silver films for contact use will cost as much as 40% less than silver halide films while processing machines and chemistry will be simpler and cheaper and less likely to violate water pollution regulations.

A number of non-silver films were introduced at PRINT 80. Many of them were of the familiar diazo type. There were five new films, however, that all seemed to be based on similar technology[13]. All had a polyester base with a vapor deposited thin metallic layer covered with a photopolymer coating. The new films were 3M Silverlook Contact Film (SLC) and Silverlook Duplicating Film (SLD), Agfa-Gevaert Rapilux NS Negative working film and Rapilux NSD duplicating film, and Kimoto K.D.P. negative working film. All were slow enough to be handled in plateroom illumination (yellow) and in white fluorescent light for short periods. All were claimed to have good dimensional stability, dot-for-dot reproduction, dot etching capability and high opacity or D-max. The films differed in processing characteristics indicating differences in photopolymer sensitizers. Of interest is the fact that the Agfa-Gevaert films are based on Energy Conversion Devices (S.R. Ovshinsky) technology the first known application of this controversial technology to graphic arts.

Pre-Press Proofing

DuPont, with its family of Cromalin materials and 3M with Color Key, Transfer Key and Matchprint, continue to dominate the pre-press color proofing market. Both DuPont and 3M have improved their materials in the past year. DuPont introduced its automatic toning system which has improved consistency of proofs and reduced the time of making them almost in half. They have also developed means of modifying the appearance of the background so it simulates the substrate on which the final job will be printed.

3M introduced an improved positive Color Key II process and a new system called Matchprint/Transfer Key at PRINT 80. Color Key II is a positive overlay system which has pigment transparency without blush to improve the hues of overprinting colors, very high resolution with the capability of reproducing 3% and 97% dots on 150 and 200 line screen images and it is processed in a low polluting, aqueous developer. Matchprint/Transfer Key is a single sheet system on the printer's paper with reproducible dot gain which is produced by the use of a spacer placed between the base sheet and the sandwich of the four-color Transfer-Key layers. Variations in dot gain can be simulated by using spacers of different thicknesses.

Another new pre-press color proofing system shown at PRINT 80 was the KC Color Proofer which uses electrostatic principles and toners similar to the old Remak process. To make a proof, a KC-Crystalplate[2] is charged, exposed to a color separation positive, and toned with the appropriate colored toner, after which the toner is transferred in register to the base sheet or actual printing paper. The system is equipped with four separate color toner trays so that 4-color proofs can be made in about 10 minutes. This system is scheduled for market about the end of 1981. A more sophisticated digital color proofing system using the same principles with the hardware of the KC-Digital Litho System is in development. This system is particularly suited to proofing for gravure as the KC process has continuous tone imaging capabilities.

Automated Imposition

Manual page composition, stripping or image assembly and page imposition are the most time, labor and cost intensive operations and most serious bottleneck, not only to pre-press but the whole production system. The electronic page makeup systems described are sophisticated and expensive ways to eliminate these bottlenecks. They not only eliminate the film assembly operations, they eliminate the film and the costs associated with its handling and transport. These new systems will find their greatest use among very large printers, in publishing operations where shorter deadlines are desired, where printing is done in satellite plants, or in trade shops that can handle large volumes of work. For the printer doing average commercial printing and packaging the conventional processes of composing jobs from multiple copies that are reproduced as films on a scanner or camera and assembling them into page layouts for step and repeat or flats for platemaking will be his way of working for a long time, and what he needs are ways of speeding up the processes by eliminating much of the handwork.

To accomplish this a number of new devices for automating these operations have been introduced during the past few years. These have been in the areas of automated step and repeat machines, optical imposition, and microstripping. First, the X - Y movements of step and repeat (or step and advance) machines were automated by the use of punched tapes, punched cards, and most recently computer keyboards. Next computer programs were prepared for exposing the subjects on the plate in the proper order as well as position. The latest development has been the design of an automated page assembly system like the Misomex Multinex 605 which is tied in with the Misomex Autostripper Step and Repeat Machine.

The Opticopy optical imposition unit which can compose pages and expose them in position on a large film ready for exposure to the plate has found a market in plants printing books, publications, labels and other specialties, where it is often used as a step and repeat machine.

The latest development in the area of microstripping in which the pages are photographed on microfilm, processed, stored and reprojected in the proper position on a printing plate is the Rachwal Super 70 Paginator introduced at PRINT 80. It is a second generation Paginator which uses 70 mm film instead of 35 mm. Since the 70 mm film has four times the area of 35 mm film, it requires less reduction so halftones can be reproduced and film savings are still appreciable. Plate sizes up to 55 x 77 inches (140 x 196 cm) can be used; individual exposures are about 5 seconds; an 8-page plate can be exposed in under 3 minutes. This system will find some application in book printing.

Platemaking

Lithography There are three major areas of development in lithographic platemaking: waterless plates; water developed plates; high speed plates. Waterless plates offer other advantages in addition to the elimination of dampening systems which are mainly less color variation and reduced waste. Water developed plates eliminate problems with water pollution. High speed plates are needed to complete the cycle in the sophisticated copy-to-plate electronic pre-press systems, for laser platemaking and for the computer-to-plate systems electronics and computers make possible.

TORAY is meeting with some success in the tests being run in the U.S.A. using its waterless plates which consist of inked image areas and silicone rubber in the non-image areas. Runs as high as 250,000 have been claimed in these tests and runs of 50,000 to 75,000 are common. The main problem affecting acceptance of these plates in the U.S.A. is the fact that they are made from positives. TORAY claims to be working on a negative process but so far no negative plates have been offered.

3M introduced a waterless plate, press and ink system for short run duplication at PRINT 80. Called the 3M Dry Duplicating System, it consists of a paper base plate covered with a coating of silicone rubber on which a toner image is applied in an electrostatic copier. After fusing, the plate is printed on a duplicating press with special ink for runs up to 1000.

NAPP was one of the first plate manufacturers to introduce a water developed plate. The plate has not been very popular but the technology has been used to develop other products like a proofing paper and contact film. 3M introduced its Hydrolith water developed plate in 1979 and an improved longer run version (150,000) on anodized aluminium at PRINT 80. Polychrome introduced its Aqualith water developed plate also at PRINT 80. Kalle introduced its Azosol N35 plate at Ipex 80.

It is an environmentally compatible (i.e. water developable) negative working, electrochemically grained and anodized plate suitable for medium and long runs. The plate will be available in the U.S.A. in 1981 and eventually will be manufactured by Kalle's U.S. affiliate, Azoplate Corp.

Attempts to produce high speed diazo and photopolymer plates have not been very successful. Diazos, especially, show no amplification and sensitivity is limited to the UV region of the spectrum. Photopolymer plates are similar except that it has been possible to extend the sensitivity into the visible range to allow laser exposure using the powerful visible lines of the argon laser (488 mm)[14]. Kodak and 3M have both developed plates of this type but while the plate speeds are increased about 3 orders of magnitude, they are still 2 to 3 orders too slow for economical use in laser platemakers.

At PRINT 80, Fuji Photo Film Co introduced a new high speed plate, HSP-N, consisting of an anodized aluminum photopolymer plate overcoated with a silver halide emulsion. The plate can be exposed to a negative or Argon laser and processed in an alkaline developer and fix. Overall exposure produces a negative image in the photopolymer layer after which the silver halide layer is removed and the photopolymer coating is processed to produce a positive printing image on the plate. Sci-Tex demonstrated some of these plates at PRINT 80 which had been produced directly on the Laser-Plotter of the Response 300 system.

Another means of obtaining higher exposure speed in plates is with the use of electrostatic principles. Kalle has developed the Elfasol plate which uses an organic photoconductor and liquid toner. It has an exposure speed still 2-3 orders of magnitude slower than lith film but it is fast enough for use with argon ion lasers in EOCOM laser platemakers. It is also used in EOCOM's computer-to-plate system.

The KC-Crystalplate also uses a photoconductive surface, cadmium sulfide, which has an exposure speed on the order of lith film and can be exposed with lower energy lasers like Helium-Cadmium (He/Cd). The cadmium sulfide can also be doped in deposition so it is panchromatic and can be exposed by Helium-Neon (He/Ne) lasers. The plates, however, are not available for use in other than KC systems and these are not scheduled for market entry until 1982 or later.

There are other types of lithographic plates which have been introduced and are in various stages of testing or use such as baked photopolymer long-run plates; dry no-processing, peel-off, non-polluting plates; screenless plates; presensitized multi-metal plates; but there are no new developments in any of these areas since the last report[2].

Gravure The new developments in gravure cylinder making are laser engraving, wraparound segments, and litho-to-gravure conversions. Crosfield has been experimenting with several types of laser engraved cylinders. The latest one is a plastic coated cylinder on which the image is engraved in a variable depth helical groove by a laser driven by a scanner or other front end system. The cylinders are currently undergoing field testing in England. The system, if successful, could be instrumental in promoting gravure for shorter run publication and commercial printing.

An improved gravure wraparound plate which was originally introduced by Daetwyler at DRUPA 77 was shown at PRINT 80. It is actually a cylinder segment as it must have a precision fit on the cylinder to avoid doctor blade problems. If successful, these plate segments will be a boon to publication printers as they will allow page changes for regional editions without having to remake cylinders. A prominent publication printer in the U.S.A. is currently field testing the plates. What needs to be determined is whether the wraparound segments can run at current gravure press speeds. If not, what advantage of versatility and cost is gained will soon be lost.

A system of conversion of lithography to gravure was introduced by Frank Sportelli in 1962[15]. Adaptations of this system are now being promoted in Japan, Germany, France and Italy. The system is based on using the simpler lithographic pre-press techniques for color separation, correction and proofing to produce a film that can be converted for gravure by a light diffusion process or on carbon tissue. The merits of the process are being evaluated[16].

Relief Plates No new developments have been reported in letterpress and flexographic platemaking. A number of large users of newspaper photopolymer plates, like the New York Times, Los Angeles Times, Chicago Tribune, are converting to lithography. While photopolymer plates for flexography have been available for some time, their acceptance has been slow. There have been problems with learning how to use them and with ink receptivity, transfer and solvent resistance on the press. Laser engraved flexographic plates appear to be limited to book and decorative printing with seamless patterns but the technology inherently produces image elements with straight sidewalls which restrict the image resolution or detail which can be reproduced. Flexography, however, is a growing process and a promising trend is the development of larger plate sizes and pin register systems throughout the pre-press steps to facilitate plate mounting and printing in register.

Press Control Systems

The use of electronics and computers to reduce costs and increase the productivity of pre-press operations has not gone unnoticed in the pressroom and other production areas like the bindery. The developments in the press area have been in pre-makeready and ink-presetting systems, color measurement and control, and press information systems.

Pre-Makeready Systems The MGD PCS/PAR and Harris Telecolor electronic pre-makeready systems for pre-setting ink controls and shortening makereadies have been in use for several years[2]. At least four other manufacturers have pre-makeready systems or are developing them.

The newest addition to these systems is the introduction by Harris of its Densicontrol to complement the Telecolor system. The Harris Densicontrol, which consists of a film table scanner that accepts any size positive or negative film flats and drop-in panel for the existing Telecolor, scans the flat and generates the proper initial settings for the inker keys. The data for each ink key path is stored in memory and, along with the ink color and web identification is transmitted to the Telecolor which automatically sets the keys and fountain roll. After the final settings are made on the inkers, they are transmitted back to the Densicontrol memory where they are stored for reruns or lifted jobs. One film table scanner can serve 10 presses with Telecolor and an optional data storage extension using a dual floppy disk drive can increase the number of jobs for presetting from 10 to 100 which involves initial and final ink settings for up to 1600 plates.

Measurement and Color Control Very little new has been added to the color measurement and control systems mentioned in the last paper[2]. At a conference on Process Control in Chicago in September 1980 sponsored by the Research and Engineering Council of the Graphic Arts[17] it was reported that about 450 Roland CCI systems were installed worldwide and an RCI system used on a press in a trade shop to print color proofs helped to complete makereadies in about 30 minutes and 6 to 7 28x40 inch (71 x 102 cm) forms could be run in an 8 hour shift. The system could also be used to measure and control dot gain, trapping and gray balance. Similar experience was reported with the use of the Heidelberg CPCI and CPCII systems. A report of the Tobias SCR Scanning Densitometer indicated substantial savings in color measurement time (54 readings across a 40 inch (102 cm) sheet in 8 secs) and 50% savings in paper for each makeready (from 2000 to 1000 sheets).

A system to increase consistency of ink flow and printing quality was introduced at IPEX 80. It is an ink unit temperature control system developed by Grapho-Metronics of Munich, West Germany, use of which is claimed to result in shorter start-up time, longer intervals between blanket washes, reduced waste after press stops, and improvements in quality and productivity. Constant ink temperature insures constant ink tack and viscosity and improves control of dampening. The system is available in the U.S.A. on Roland and MAN sheet-fed presses sold by the Miehle Products Group of Rockwell International.

At PRINT 80, Kodak announced its new Color Analysis Service in which a specially designed test target printed alongside production work on a press is sent to Kodak for measurement on a spectrophotometer to derive information on tone reproduction, color hues, saturation, ink trapping, dot gain, gray balance, color correction, aim points and other color factors. A computer is programmed to analyze the data and generate curve shapes and aim points to produce color separations for optimum tone reproduction, color and gray balance. Use of this system will help to improve color reproduction by standardizing printing on the press.

Other marvels of computer software displayed at PRINT 80 were new 'smart' densitometers which not only read reflection densities but produce instant conversion of the density readings to values for ink hue error, grayness, and trapping using Preucil's equations for these functions[18].

Press Information Systems MGD, a division of Rockwell International, has announced a new newspaper press and distribution management information system to improve the efficiency of newspaper plants. Harris announced its PresFax management information system in PRINT 80 that uses microcomputer controlled data recording systems to accurately record and analyze waste and production providing management with the information necessary to measure and control waste and productivity. Hantscho will introduce its Color Set 80 Management Information System in 1981. Hurletron-Altair has two press management information systems, the Hurletron Print Management System 700 for gravure and the Q.C. Control System 745 for web offset which use distributed intelligence to report complete press events and materials information of importance and use to reduce waste and increase productivity.

At the R&EC Process Control Conference in September 1980[17] was a fitting review of the accomplishments and deficiencies of process control systems in a report of progress on a nine-year study at Case-Hoyt Printing Company (Rochester, NY) of nine target areas for process control. The areas were: (1) control of substrate on web press; (2) remote ink control; (3) remote unit register; (4) data acquisition system; (5) closed loop register control; (6) closed loop water control; (7) color control; (8) closed loop color control; (9) system to control input and output.

About half of these objectives have been accomplished. There has been some improvement in productivity but there has been more downtime from electronic failures due partly to inadequate skills in the operators. Despite failures, makeready time has been reduced 5% with an objective of 40% for the 80's; makeready waste is 6% more than restart waste - objective is for them to be equal; non-chargeable time is 100% more - objective is to reduce it to 50%; productivity is up to the objective, 25%; running waste has been reduced 10% - objective is 40%. Overall objective for the 80's - annual savings of $500,000. With such insistent and systematic attacks on inefficiency and wastefulness, these savings will be realized, along with corresponding increases in productivity.

Ink Drying and Curing

The super quick-set inks[2] have been very effective in speeding up the setting of inks (up to 10 times faster than conventional quick-set inks) and in promoting the use of infra-red drying systems. But they have resurrected an old problem - gloss ghosting which is a phenomenon caused by ink printed on one side of a sheet affecting the setting of ink printed on the opposite side so a variation in gloss appears corresponding to the printing on the back side. It generally happens during the first few hours of the drying cycle and is dependent on the type and volatility of ink solvents and their reaction with paper surfaces or coatings. Super quick-set inks have such volatile solvents which set so rapidly that they have aggravated the gloss ghosting phenomenon by the practice of backing jobs up too quickly after printing. A number of research institutes have attacked this problem. GATF believes it has a solution in the application of selective varnishes after printing.

Two presses have been equipped with electron beam (EB) curing units but results of tests are not available. While on the subject of EB, there is some work being done in research institutes in the U.S.A. and Europe on the use of EB for exposing printing plates but no information is available on results or sponsors. There are obvious advantages to the use of EB exposure over lasers. Any coating could be used as EB is independent of spectral sensitivity or exposure speed effects.

Automated Bindery Systems

Binding has been a highly labor intensive operation and has been considered a necessary evil by many printers. It is not unusual on many printing jobs to find 35% of the labor hours spent in the pressroom and 65% in the bindery. And conversely, 65% of new equipment dollars is going into the pressroom and only 10% in the bindery. This situation is changing. With many printers having similar press equipment and more typesetting and preparatory work being done by the publisher or buyer, the bindery is becoming a main area of product differentiation between printers. Two American printers have computerized warehouses.

Binderies have more automatic units for counting, bundling and materials handling; bundling machines; bundle loaders or automatic hopper loaders; automatic bundle distribution systems; automatic palletizers and depalletizers. There are microprocessor controlled cutting machines with automatic jogging, loading and unloading, and folders are increasing in speed and accuracy. There is an increase in in-line processing and on-press addressing by electronic printing methods like ink jet printing. Adhesive soft binding will continue to increase at the expense of hand bound books. Even newspapers are improving bindery operations. The growth of inserts plus the advantages of zoned advertising has created a massive sorting and collating operation which has forced newspapers to use new sophisticated computer controlled processing equipment.

Computerized Industrial Robots

Computer controlled industrial robots have been used very successfully in many industries and are just being introduced for the printing industry. At PRINT 80 and Ipex 80 Wamac Machinery AB displayed robots programmed for such functions as moving completed signatures from stacker outputs to pallets and skids. They are being operated by companies like Greater Buffalo Press, Rand McNally, and World Color Press to support perfect binders, book lines and deliveries from web offset presses. Costs of the robots range from $75,000 to $100,000 and some users have reported paybacks as short as five months[19].

Electronic Printing Systems

Ink Jet Printing is an established technology which is used extensively for computer printout, document generation, personalized promotional pieces, numbering, addressing, coding and imprinting of forms, invoices and the like. These are all applications involving variable information and it is expected to find greater utilization in this type of uses. The process has prospects for further improvements in speed and quality. The market is presently dominated by two systems, A. B. Dick Videojet and Mead Dijit, but there are now 17 manufacturers of ink jet printing systems so some advancements and competition can be expected in this area in the next few years.

Electronic Printing Systems are evolving from electrostatic duplicators like the Xerox 9400 and Kodak Ektaprint 150 which have completely replaced offset duplicators in many in-plant and quick printing operations. The next step in these is equipment to print in sizes 11 x 17 inches, two sides, at speeds of about 2 per second or 7200 per hour. These systems operate from reflection copy. New advanced systems have been developed which are known as intelligent copiers and operate directly from computer stored digital information. Examples of these are the Xerox 9700, IBM 6670 and the Wang Image Printer. These systems have been around for several years and there were about 500 Xerox 9700's in use in the U.S.A. at the end of 1980.

The Xerox 9700 consists of a system controller with system disk for input buffering, digital processor memory for forms storage and electronic circuits for data input and system control. An image is produced by a laser beam which creates characters and forms from a font memory where images are stored digitally in dot patterns. Characters are built up in a dot matrix of 300 x 300 dots/inch and pages are printed xerographically with toners. Xerox has introduced a computer software package for the Xerox 9700 called XICS (Integrated Composition System) which accepts text from any terminal or word processing system and has page makeup and editing capabilities.

What has been called the world's first completely electronic publishing system has been assembled by Mitel Corporation in Canada to meet its need in the publication of technical documents[6]. The system consists of a Xerox 9700; software from Krontek Corp. to digitize graphics for input to the 9700 from an AM/ECRM Autokon 8400 laser camera equipped with an interface directly to a PDP11/34 computer; and a Kurzweil omnifont OCR to convert existing documentation produced on a Compugraphic 7700 typesetter to a computer compatible magnetic tape format. The printed page is produced directly from the keyboard entry of data without any typesetting paper, photography, platemaking, printing press or ink. The system is claimed to be simple enough that one or two people with little training or knowledge of traditional printing or publishing could operate it, and yet the software is powerful enough to handle complex indexes, typography and page formats. Systems like this will be used for short run business forms, customized promotional literature and short run utility book publishing.

Small Business Computers

Five new small business computer systems with software specifically developed for the printer have been introduced recently at regional graphic arts equipment exhibits in San Francisco, California, and Orlando, Florida, to compete with systems already in use such as those marketed by Heidelberg, Printers Software and Stewy Systems[19].

Printers Computer Systems (Longmont, Colorado) introduced its software which runs on a Radio Shack computer with separate packages for estimating production control and financial control at a cost of under $4,000 for each package.

California Micro Products (Brea, California) offered an estimating/inventory program for use on an Apple II system for $995 which was designed specifically for the quick printer but could have some application for in-plant printers.

B & B Computing Services (Reseda, California) offered a system called 'Estimator' utilizing Vector Graphics equipment which includes estimating/costing, general ledger, accounts receivable, payroll, inventory management, and word processing capability.

Alpha Tract I, marketed by Graphitech Computer Systems (West Palm Beach, Florida) uses a Texas Instruments PC-100 printer and TI-59 magnetic card programmable calculator and software for estimating, copyfitting, camera rescreening analysis, and payroll. The system sells for $1200. An Alpha Track II more sophisticated software package is under development for use on a Radio Shack TRS-80 System and will sell for under $2000.

Applied Micro Systems (Salt Lake City, Utah) demonstrated the PRINTERS MCS (Printers Management Control System) which consists of software for Digital Equipment (DEC) hardware to include estimating, costing, job tracking, production analysis, sales analysis, accounts payable, accounts receivable, inventory control, payroll, general ledger and utility programs.

The proliferation of these small business computer systems is an indication of the scope of the software systems that are being developed for printers and a warning to them to be selective in their choice of systems to use.

Environmental Considerations

Printers and manufacturers of printing equipment and supplies will continue to be plagued by governmental regulations regarding health and safety and air, water and solid waste pollution. The industry is fortunate to have the facilities and staff of the Graphic Arts Technical Foundation (GATF) to filter, monitor and publish the information on environmental and safety and health regulations and the Environmental Conservation Board (ECP) to help pay for these activities.

Gravure and flexography have been the two major targets of the Environmental Protection Agency (EPA) in 1979 and 1980. Permissible organic solvent emissions from plants will be substantially reduced by 1982. New gravure installations will be subject to more severe restrictions than existing plants, further increasing the investment costs of new plants and equipment and discouraging growth of gravure. EPA will complete a study in 1981 that may result in new regulations for heatset letterpress and web offset operations[5].

Except for deep-etch and bimetal platemaking, and gravure cylinder preparation, where heavy metals like chromium and copper are involved, no federal pretreatment standards for water are scheduled for 1981. Local municipal codes, however, may still require pretreatment of effluents of specific pollutants. Printers with captive ink-making facilities may be required to comply with the same 1980 federal regulations as ink manufacturers do which calls for contract hauling of clean-up wastes.

With regard to solid and hazardous wastes, federal regulations allow exemptions if not more than 2200 lbs (1000 kilograms) of hazardous wastes are disposed of monthly which should exempt most printers. All states, however, will not abide by the Federal restrictions. Some state legislatures will eliminate or reduce these examptions, making printers responsible for the disposal of their solid waste. To make matters worse, some haulers will refuse loads that contain hazardous wastes.

The Occupational Safety and Health Administration (OSHA) has promised a resolution of the amendment to the noise standard sometime in 1981. The acceptable noise level for an eight hour exposure is expected to be reduced from 90 dBA to 85 dBA. Already a number of web offset presses have been enclosed so that noise levels are below these limits. One of the most important events that can affect the severity of OSHA's future regulations is a recent decision by the U.S. Supreme Court in the case Benzene Standard vs. OSHA. According to a report by GATF[5], the Court ruled "that OSHA must be able to show that a more restrictive standard will eliminate or reduce the health risk caused by a given material controlled at the level of the existing standard. This decision which shifts the burden of proof from the regulated to the regulator will slow the process of issuing new standards." It is expected that the new administration will be more sympathetic to industry and the need for increased productivity and may delay the enforcement of many restrictive regulations.

SUMMARY

The printing industry and its major users, publishing and packaging, will continue to grow at about the same rate as the Gross National Product for the near term. During this period the industry will be busy adapting the new computer and imaging technologies to the improvement of its processes and productivity. The big need is for software and as this is developed many new automated systems will evolve to reproduce the original directly on the plate or cylinder without conventional typesetting, photography, image assembly, page imposition and the materials and labor these entail. Plates will be printed on a press which will be madeready and run almost automatically, and the first printed product will be bound and finished in-line also almost automatically. In addition, software is developing new processes of electronic printing which can supplement and enhance the present printing processes. The next five years will be a period of dynamic growth when existing technology is exploited to the fullest and the seeds of new technology will be sown.

REFERENCES

(1) 1981 U.S. Industrial Outlook, U.S. Government Printing Office, Washington, DC 20402

(2) M. H. Bruno, 'Status of Printing in the U.S.A. - 1979', Advances in Printing Science and Technology, Vol. 15, edited by W. H. Banks, Pentech Press, London (1980)

(3) M. H. Bruno, 'ANPA-RI Forums 80', What's New(s) in Graphic Communications, No. 27, p. 2, Nashua, NH, July-August (1980)

(4) Graph Comm 80, 'Interaction of Markets and Technology on Graphic Communications 1980-1985', Batelle Report, PIA, Arlington, VA (1980)

(5) 1981 Technology Forecast, GATF, Pittsburgh, PA, December (1980)

(6) 'Forecast 1981: Making Way for the Coming Electronic Revolution', American Printer/Lithographer, Maclean-Hunter, Chicago, IL, December (1980)

(7) M. H. Bruno, 'New Optical Memory Disk for Laser Recording', What's New(s) in Graphic Communications, No. 30, p. 4, Nashua, NH, January-February (1981)

(8) J. E. Rowe, 'Newspapers and Technology', paper presented at ANPA-RI Production Management Conference, June (1979)

(9) F. Romano, 'Typesetting Moves to the Systems Approach', American Printer/Lithographer, Maclean-Hunter, Chicago, IL, August (1980)

(10) F. Romano, 'Printing Industry Trends Almanac', PIA, Arlington, VA, November (1980)

(11) The Seybold Report, Seybold Publications, Media, PA., Vol. 10, No. 13, p. 13-19, (1981)

(12) M. H. Bruno, 'New Electronic Still Camera', What's New(s) in

Graphic Communications, No. 27, p. 2-3, Nashua, NH, July-August (1980)

(13) M. H. Bruno, 'The New Technology: PRINT 80 - The Beginning of the Software Age', American Printer/Lithographer, Maclean-Hunter Chicago, IL, p. 42-43, June (1980)

(14) M. S. Kaplan, 'Laser Imaging of Graphic Arts Materials', 1977 TAGA Proceedings, TAGA, Rochester, NY, p. 90-111

(15) F. Sportelli, U.S. Patent 3,024,111 (1962)

(16) J. Chevalier, 'Revival of Gravure Conversions using Fast Offset Techniques to Cut Costs', GRI Report M97, GRI Annual Meeting, November (1977)

(17) M. H. Bruno, 'R&EC Process Control Conference', What's New(s) in Graphic Communications, No. 28, p. 2, Nashua, NH, September-October (1980)

(18) GATF Research Progress No. 38, GATF, Pittsburgh, PA, April (1957

(19) W. C. Lamparter, 'Print Manufacturing Goes Electronic', The Printing Manager, NAPL, Teaneck, NJ, December (1980)

DISCUSSION

B. Corsham   Mr Bruno, in his review of the different manufacturers developing and supplying electronic page make-up systems, mentioned that there were some strange relationships being established in this field. He stated that Dainippon Screen, who have announced they will introduce a system in Japan in Autumn 1981, have a software purchasing agreement with 3M who now own Comtal. This is particularly interesting in the light of the fact that P.D.I. uses Comtal Twin Display equipment in a new page make-up system which has just been introduced for sale; a system which, incidentally, does not appear in Mr Bruno's list.

Can Mr Bruno tell us any more about these commercial relationships than he states in his report?

M. H. Bruno   As far as I can learn, the arrangement between Dainippon Screen and 3M involves the supply of software from Comtal and marketing help from 3M for Dainippon Screen's colour composition system. Also my understanding is that Comtal is still free to supply hardware and some software to manufacturers of image processing systems like P.D.I. and Coulter System Corp. which also uses this equipment in its Digital Litho System.

J. A. Bristow   Flexography has traditionally been used on materials such as corrugated board, toilet tissues, etc., which are not primarily intended as printing materials and where no extreme printability demands are made. There appears to be a growing tendency for flexography to be used for quality printing on carton board, coated boards, etc. Complaints are beginning to be heard sometimes that 'something is wrong' from a printability point of view. Now demands are being made on the paper or board with regard to flexographic printability. It appears that we must soon initiate a new field of research into 'paper and board for flexography'. Do you confirm this tendency?

M. H. Bruno It has only been since the development of photopolymer plates and new ink metering systems like reverse angle doctor blades for flexography that it has become capable of the improved quality which has made it eligible for use in publishing and packaging. As it is used more for quality products using process color, differences in print quality will become more noticeable. Like any other printing process, paper and board surface and compressibility characteristics can affect print quality and as flexo is used more, these effects will need to be studied and controlled.

W. D. Schaeffer In your description of the graphic arts industry and division among the processes, you indicated separate records are used for packaging. In view of the number of organisations represented by IARIGAI, would you comment on the adequacy of statistics on packaging comparable to the Bureau of Census data on printing and publishing.

M. H. Bruno Government statistics on packaging are very inadequate, incomplete and confusing. Printing is not broken up and some packaging items like labels are included in commercial printing.

W. Hansen Can you report on water based news inks replacing inks based on mineral oil?

M. H. Bruno Water emulsion inks have been developed and used for letterpress and offset newspaper and commercial printing using up to 30% water. As far as I know no water based inks have been used for newspaper printing except a flexographic test run at ANPA Research Institute.

J. L. Poujade Regarding newspaper conversion to offset, can circulation figures be considered as a good approximation for newsprint tonnages used?

M. H. Bruno No! Figures used to be published on tonnage, but since light weight newsprint was introduced ANPA reports are all based on circulation.

A. F. Nielsen Are you aware of any suppliers of equipment working on a total system, combining digitizing of text, picture and pagination via graphic display terminals and a computer-to-plate system?

M. H. Bruno Information International Inc (III) introduced a new newspaper system at the ANPA-RI exhibit in Atlantic City, but it does not go directly to plate. Autologic has an agreement with 3M to produce high speed plates that can be used in a complete copy to plate system. EOCOM, Muirhead and LogEScan all have the capability for computer to plate systems and with the proper interfaces with existing front end systems it should be possible to go directly from copy to plate.

LOADS ON PAPER IN A PRINTING PROCESS

L. O. Larsson, TFL, Sweden

ABSTRACT

The loads on paper in a printing process are of various kinds. Conventionally mechanical loads are thought of when loads on paper are discussed. Here, it is proposed that there is actually a number of requirements which can be regarded as loads on paper.

As illustrative examples (1) the increased demands (loads) that a non-optimal choice of ink puts on the paper and (2) the increased loads that web-tension disturbances induce on paper webs in a rotary printing press are discussed.

The optimisation of the paper-ink-press system as well as the proper adjustment of the press machinery and the proper mode of running it are suggested as vital to the economy of the printing operation.

INTRODUCTION

Attention is drawn to some factors which unnecessarily make the printer's work more complicated and difficult, forcing him to make greater demands on the properties of paper in the printing process, which often brings with it increased costs.

The material and experience employed in this survey are drawn from the field of newspaper printing but the underlying ideas and conclusions should have an analogous bearing on other branches of printing.

LOADS ON PAPER IN A GENERAL SENSE

Usually loads on paper, particularly loads on a paper web, are thought of in terms of mechanical loads, i.e. stresses and tensions. Furthermore, what is termed a load is frequently associated with some excessiveness which increases the demand on the capacity of the paper to withstand the load, or to function satisfactorily under the load. By the same token factors other than mechanical forces can then be thought of as loads on the paper in the press. A merit of this way of looking at these factors is that it helps to reveal the nature of the corresponding requirements on paper and to discover ways to deal with them.

'REQUIREMENT' LOADS ON PAPER IN A PRINTING MACHINE

When the requirements and loads on paper in a printing machine are spoken of, we usually refer to those requirements or loads which are influenced and largely determined by the printing process, its phenomena and materials and not by the demands for the satisfactory performance of the printed end product.

Such requirements are seldom absolute as they depend upon a specific situation when the paper will interact with other materials or certain equipment; be printed with a certain ink on a certain press, etc. The assumed paper properties have often been given names ending in '-ability', which seems often to compensate for the lack of definition of the real material properties.

As the requirements are influenced by the individual presses and materials used on them and by the fact that the production conditions vary from printer to printer, it is impossible for the paper manufacturer to meet all requirements explicitly. It is also, in many cases, impossible or very difficult to find a paper test that guarantees that the paper has the required property in each particular printing situation. As an example, it has not been found possible to devise a universally reliable lint test for offset paper.

This situation may easily lead to too high general requirements on the paper - to be on 'the safe side' in all instances. If the requirements are then codified in specifications, they will enforce maintenance of the property values so specified, and thereby very likely lead also to increased costs.

It is obvious that different printing procedures impose their own requirements on the paper. Lists of requirements that are deemed to be the most pertinent are often compiled, e.g. Table 1. They are never complete, often cryptic and sometimes partly questionable. The table is included here not because it is thought to provide some new ideas but rather because it provides a common and fairly recent (1980) list of important properties of paper in different printing procedures for the participants of this conference to discuss and to improve.

From Table 1 we can infer that, in practice, common to letterpress, offset and gravure printing, the important properties of paper fall into two groups. One group refers to effects of the interaction between paper and ink. The other group refers to effects related to the mechanical properties of the paper. The loads on the paper corresponding to these two groups of properties will now be discussed, using as illustrative examples studies of, in the first case, print quality characteristics descriptive of ink quality, and in the second case, web tensions in rotary presses.

TABLE 1 IMPORTANT PROPERTIES OF PAPER IN DIFFERENT PRINTING PROCEDURES

Common properties for all printing procedures:		
	Uniformity	
	Moisture content	
	Opacity	
	Ink absorbency	
	Chemical inertia	
	Same properties on both sides	
	Mechanical properties	
Sheets:	Webs:	
Flatness	Uniformity	
Orthogonality	Burst resistance (alt. tensile strength)	
	Elasticity	
Specific properties:		
Offset	Letterpress	Gravure
Surface strength, dry and wet	Smoothness	Smoothness
Compressibility	Compressibility	Surface compressibility
Z-strength		
Linting resistance	Surface strength	Wettability
Dimensional stability		
pH		

LOADS DUE TO PAPER - INK INTERACTION

It is obvious that the different printing procedures pose different demands or loads on the paper. In all procedures, however, there is a prerequisite (1) that the print on one side of the paper is not visible in a disturbing way on the opposite side, and (2) that the ink in the printed areas of the paper surface resists all forms of smearing so that the print quality is retained. These phenomena consequently deserve special attention to R and D by the ink industry.

In so-called wood containing papers where there is a strong tendency towards the use of lower grammage due to the soaring costs of energy and raw fibre, the load posed on the opacity of the paper is particularly striking.

In a paper[1] given at the IARIGAI conference 1977, it was shown that commercial letterpress newsinks, used in actual newspaper printing,

Symbols used in Figures 1 and 2:

- ink especially designed for direct lithography
- ink adjusted to direct lithography
- letterpress ink deemed usable in direct lithography
- unspecified ink used in direct lithography

Figures are codes for identification of the individual inks which are represented by the boxes.

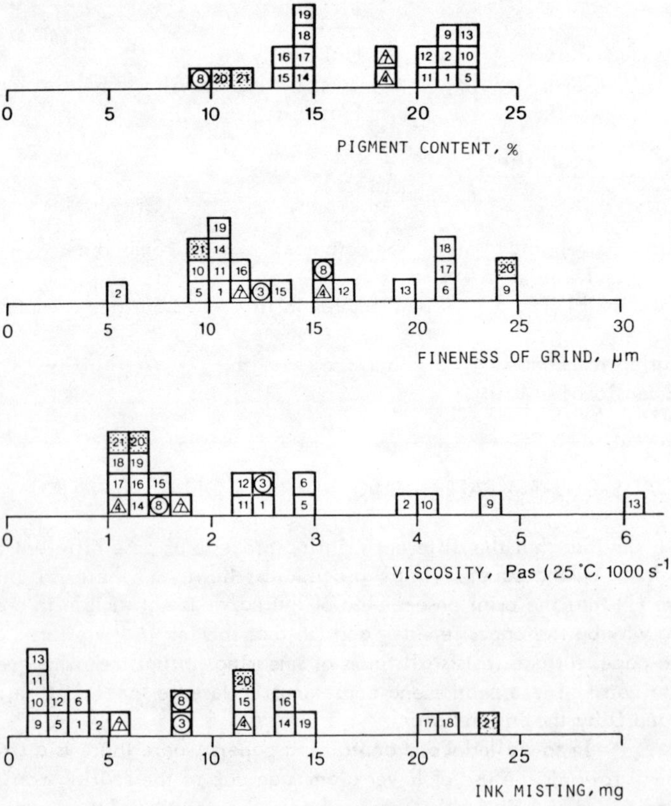

Fig. 1. Histograms showing (a) the pigment content, (b) fineness of grind, (c) the viscosity, and (d) the misting of the inks (2).

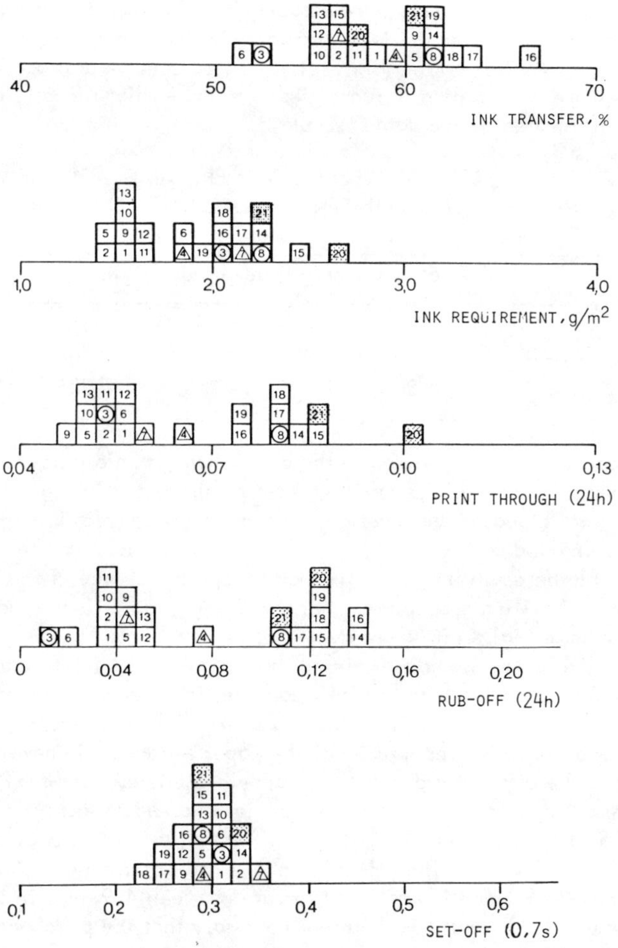

Fig. 2. Histograms showing (a) ink transfer, (b) ink requirements, (c) print through (24 h), (d) rub-off (24 h), and (e) set-off (0.7 s) for the inks at print density 0.90 (2).

display a wide range of values in a number of important quality respects and consequently exert a severe load on the paper. It was demonstrated that it is technically feasible and easy to choose the inks optimally with respect to selected parameters. It was also pointed out that the most important quality characteristics in this context can readily be compared with each other on the basis of simple standard test.

Letterpress printing of newspapers has diminished in importance in several countries over the last decades, whereas lithography has found an increasingly wider application. The inks used in these procedures have the same wide ranges of properties as found earlier for letterpress inks, as illustrated by the data in Table 2.

TABLE 2 DIFFERENCES BETWEEN COMMERCIAL BLACK NEWSINKS IN DIFFERENT PRINTING PROCEDURES IN APPROXIMATE FIGURES (Percentages of mean values)

	Ink requirement	Print through	Rub-off	Set-off
Letterpress	60	70	170	60
Direct lithography	60	70	170	60
Offset	50	40	150	60

The properties of newsinks used in the direct lithographic procedure for newspaper printing are illustrated by a set of histograms in Figures 1 and 2. They are based on an investigation[2] of 21 ink samples from Sweden, Norway, England and the USA. A reduction of the load on the paper is also possible here in many cases through an optimal choice of the ink on the basis of a testing procedure. A complicating factor is the fact that inks for commercial printing are available as ready only for large standard jobs; in other cases the printer himself has to adjust the ink to the paper.

The interplay between ink and water in lithographic procedures is another area in which difficulties may arise, rightly or wrongly, to claims for a better ink receptivity of the paper surface. In newspaper printing in the offset and direct lithography procedures (used for 89% of all newspapers, 78% of all newspaper copies and 75% of all printed newsprint in Sweden at the end of 1980), tap water is used in most cases for the fountain solution. The pH-value of the tap water may vary within wide limits (in Sweden it was found to vary between 4.9 and, in an extreme case, 10.1), and it is therefore necessary that the pH of the fountain solution is carefully observed.

Finally, it should be mentioned that inks also play an important role in phenomena such as linting, napping, plucking, etc. in which the load on the paper is mechanical. The rheological properties come into focus here and the tackiness is often discussed. The method of reasoning used in the above-mentioned examples should, in principle, apply in the discussion of any crucial ink property. There still seems to be opportunities for optimizing ink properties in most situations.

For the sake of completeness, it should be mentioned that there are a number of other factors that, if overlooked or neglected, may cause considerable loads on the paper in the form of increased demands. Examples of such factors are underpacking on the printing or impression cylinder, printing pressure, evenness of the printing medium, correct handling and conditioning of the paper so that, for example, it is not overdried, thereby causing it to print less well.

LOADS DUE TO WEB TENSION

The effects of improper web tension in a rotary press are web breaks and impaired production efficiency. Both these effects have grave economic implications.

TABLE 3 WEB BREAK FREQUENCIES (number of breaks per 100 reels) in different printing procedures divided according to causes

	Total web break frequency	Paper faults	Press and press handling faults	Unknown causes	Reel change faults
Newspaper printing and alike on wood-containing papers in letterpress, offset and direct lithography	5.6	1.6	1.4	1.4	1.2
Printing of magazines and weeklies in					
Offset					
Printing house 1	3.7	3.3	0.3	0.1	-
Printing house 2	9	3	2	2	2
Printing house 3	28	8	4	10	6
Gravure					
Printing house 4	7	3	2	2	< 1
Printing house 5	17	3	1	11	2
Printing house 6	17	3	2	6	6

The extent to which web breaks occur varies appreciably among printing plants and the level of web break frequency is not the same for different types of printing. Typical mean values based on regular statistics in Swedish printing houses are given in Table 3. The figures indicate considerable scope for improvements and savings. Worth noting are the high figures for gravure. Very few figures for web break frequencies in gravure have been published but they have been stated to be high. The high figures under the heading 'Unknown causes' for printing houses 3, 5 and 6 demonstrate the difficulty in deciding the cause of a break. It is believed, on good grounds, that a great number of the breaks in the categories 'Press and press handling faults' and 'Unknown causes' are due to the web tension conditions in the press. There is, therefore, good reason to study web tension in a press and how it is influenced by the press conditions and the way the press is run. Here, some findings of investigations of types of web tension loads will be given as examples.

Nominal web tension value

To pull the paper web through the printing press, a certain web tension must be maintained. In newspaper presses this tension usually is around 0.1 kN/m web width in letterpress, offset and direct lithography presses; in some cases, however, around 0.2 kN/m. In gravure presses the web tension is often around 0.3 kN/m. These figures are nominal: in all presses the tension varies in time and from point to point along the web path through the press.

In most presses the normal web tension is not measured particularly accurately, if measured at all. Measuring systems are often of a slowly reacting type, insensitive to rapid changes in the web tension.

It has been found experimentally that web break frequency increases exponentially with web tension or, equivalently, that the mean period between consecutive breaks diminishes with the web tension. A reduction of web tension by 0.07 kN/m would on average reduce the web break frequency to half of its previous value. The practical possibilities of lowering the web tension are, however, limited by the fact that a too low web tension would lead to web instability, eg, the web might flutter in a manner impairing the production speed, or even cause a web break. There is thus an optimum web tension giving a number of web breaks and the maximum production.

Rapid disturbances of web tension and web movement The remarks in the previous section concern the appropriate mean web tension. The variation in web tension must, however, lie within limits. There are a number of types of variation ranging in duration from seconds, caused for example by a change of the press speed, to those of some ten milliseconds corresponding to the frequency of disturbance of a rotating machine element, such as an eccentric roller. Parallel to the web tension variations are variations in the movement of the web, such as oscillations in speed and flutter. Such instabilities and variations in web tension are specific to individual presses and may account for the fact that web break frequencies differ considerably among press installations even when the same make of paper is run on them.

Disturbances of web tension during normal runs

In an apparently normal press run, the web tension in different positions in the press may contain frequency components of comparatively high amplitudes. This is examplified by the diagrams in Figure 3 based on an investigation[3] of the web tension in a gravure printing press with two reel stands and twelve printing units. The web tension was measured during ordinary production runs at a number of positions, namely at the reel stand, before and after the printing units, before, between and after

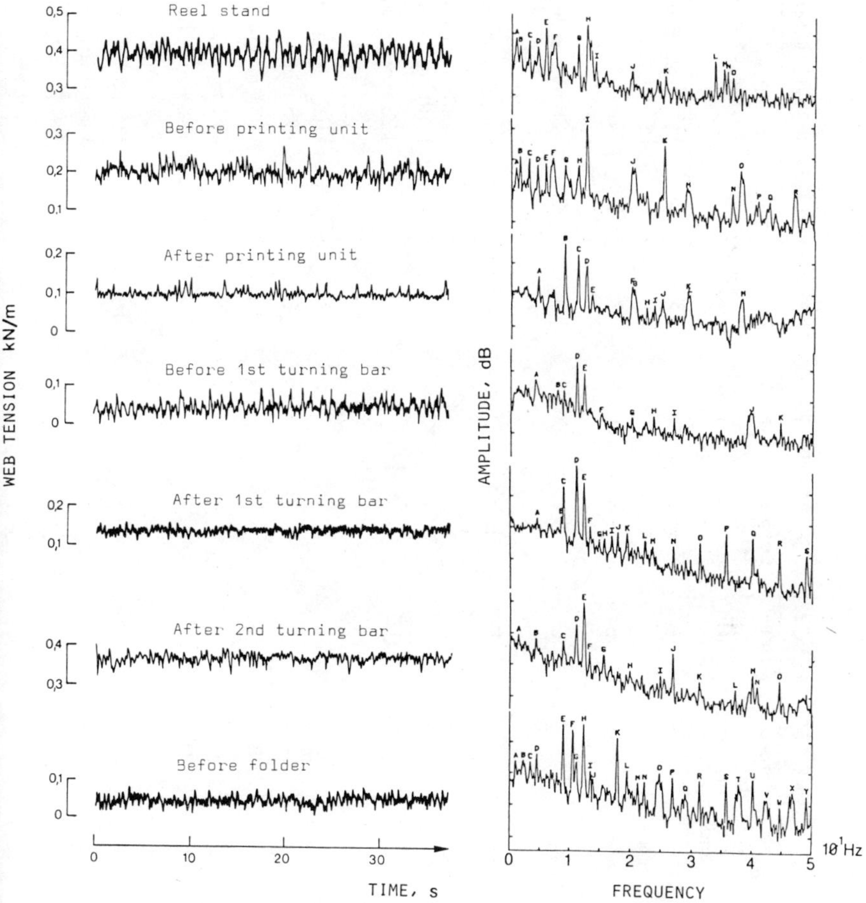

Fig. 3. Time histories of web tensions at different positions in a gravure press running at a constant speed and the corresponding frequency spectra (3).

TABLE 4 IDENTIFIED SOURCES OF SOME COMPONENTS OF THE WEB-TENSION DISTURBANCES (3)

Diameter of rotating element (mm)	Frequency (Hz)	At reel stand	Before 1st printing unit	After 1st printing unit	Before 1st turning bar	After 1st turning bar	After 2nd turning bar	At folder	Source of disturbance
317	4.50			A	A	A	B	D	Printing cylinder Folder
161-158	8.875-9.000		G	B	C	C	C	E	Cooler
128-127	11.125-11.250	G	H	C	D	D	D	G	128 mm roller
115-114	12.375-12.500			D	E	E	E	H	Drag roller Printing cylinder 3rd harmonic
80	17.875					J		K	Cooler 1st harmonic Printing cylinder, 4th harmonic
64	22.250			H		L		N	128 mm roller 1st harmonic
	26.750				I	N	J	P	Cooler 2nd harmonic Pipe roller 1st harmonic
	44.625				K	R	O	W	128 mm roller 3rd harmonic
1037	1.375	B	B						1037 mm roller
496	2.875	C	C						496 mm roller
336	4.250	D	D						336 mm roller
254	5.625	E	E						254 mm roller
113	12.625	H	I						Infeed roller
	25.125	K	K						Infeed roller 1st harmonic
	37.750		O						Infeed roller 2nd harmonic

the turning bars and at the folder. The web tension signals displayed in the left column of Figure 3 were frequency-analysed. The frequency spectra are to the right in the figure. The analysis revealed peaks at a number of frequencies that coincided with rotational or resonance frequencies of certain machine elements, whereby it was possible to locate the sources of the disturbances (see Table 4). The sequence of frequency spectra from the different measuring poisitions along the paper web gives an impression of how the web tension is influenced by the different parts of the press. It is interesting to note that many of the disturbing web tension components originating at the reel stand disappear or are attenuated during the passage of the web through the printing unit, but that new disturbances are introduced further on along the path of the web.

All disturbances create loads on the paper and are potential sources of web breaks. They must therefore be minimised in every press.

Disturbance of web tension due to a defective machine part

In a press with a great number of parts, it may easily happen that one of these parts becomes defective or out of adjustment. It is not always readily observed or detected that such an incident has occurred. In the study of the gravure press already referred to in the last section, one peak in a frequency spectrum was due to the combined effect of a disturbance from the printing cylinder and one from a pipe roller. The amplitude of the corresponding web-tension component was, as a consequence, too high and dangerous. Inspection of the pipe roller showed that the bearing was defective. After changing the bearing the amplitude was reduced to about one third, with the consequence that the previous production-impairing frequency of web breaks fell. The techniques employed for examining web tension in the press proved to be a powerful tool that should be worth using in troublesome web-break situations. The above example shows the need for and usefulness of a careful checking of the machinery.

Disturbance of web tension due to press manoeuvre

Press maneouvres are potential causes of web breaks because they are often accompanied by disturbances of the web tension.

Reel changes often cause severe disturbances. Figure 4 displays the web tension and the web flutter in an offset press [4] and Figure 5 in a web change situation. The changes of web tension which follow a reel change usually influence the web behaviour in the press. Already, minor changes in web tension may alter the conditions for the stability of the web in a decisive way. This is clearly seen from the graphs in Figure 4. The small increase of the web tension at position 4 just after the reel change, diminished the flutter to approximately one third of the previous value. The graphs also show that the disturbance originating from the pasting of the webs was much less than that caused by the band brake coming into action on the paper reel after the pasting procedure.

In the same press, the circumstances at a web break were record-

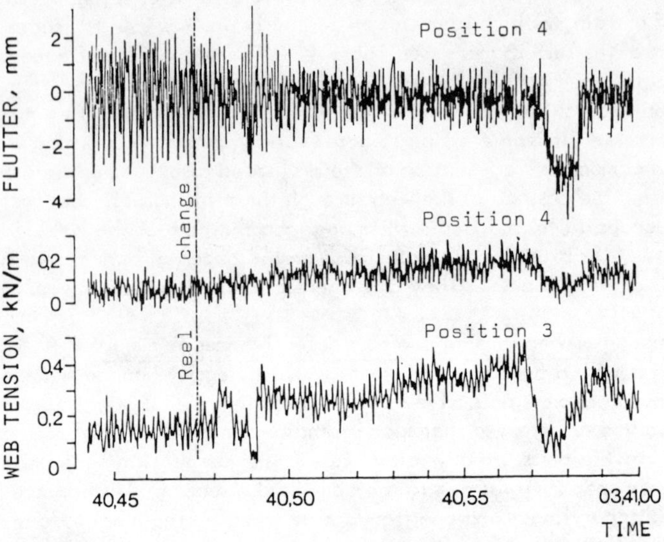

Fig. 4. Time histories of web flutter and web tensions at a reel change in an offset press (4).

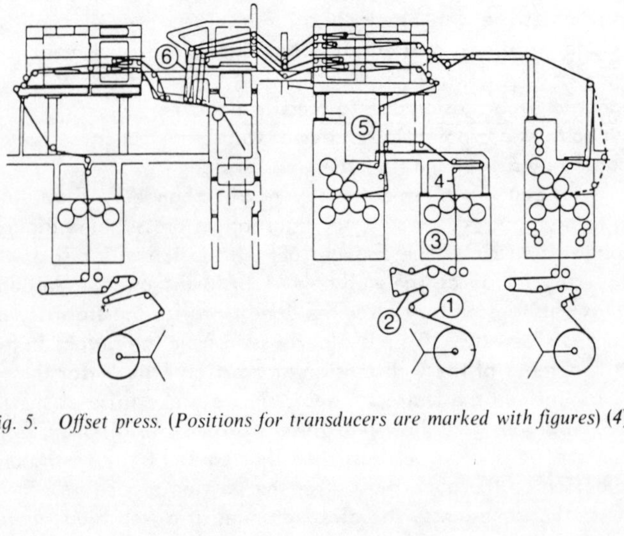

Fig. 5. Offset press. (Positions for transducers are marked with figures) (4).

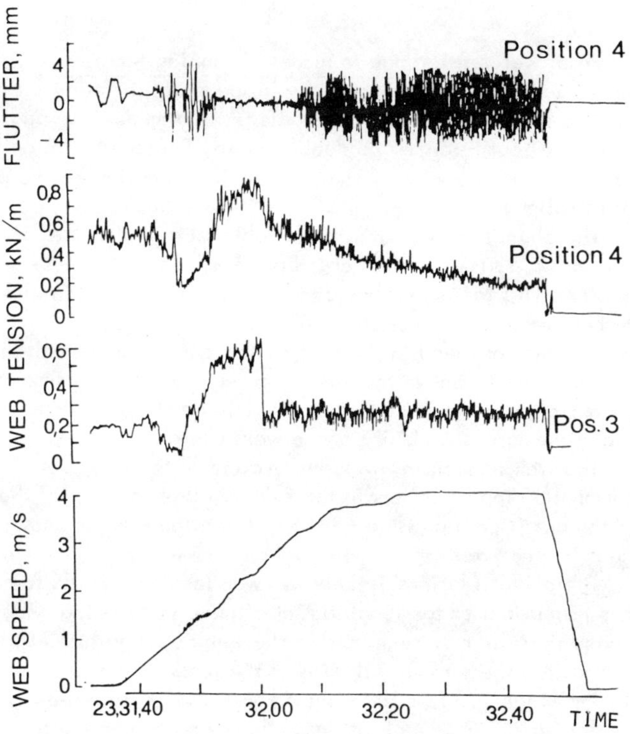

FLUTTER, mm
4
0
-4
Position 4
WEB TENSION, kN/m
0,8
0,6
0,4
0,2
0
Position 4
0,6
0,4
0,2
0
Pos. 3
WEB SPEED, m/s
4
3
2
1
0
23,31,40
32,00
32,20
32,40
TIME

Fig. 6.

ed, see Figure 6. During the acceleration period of a start up of the press, the web tension was greatly increased after the impression cylinder came into action (at time 23.31.48). Ten seconds later the fountain roller was activated and the web tension between the printing units (position 4) started to decrease. The excess of fountain solution given to the paper induced an elongation of the web resulting in a strong flutter and finally a web break.

These examples demonstrate the importance of proper press handling.

Disturbance of web tension due to modes of running the press

By means of a series of experiments the influence of different press parameters on the web tension has been studied[5] in a modern newspaper offset press, double circumference and double width, equipped with perfector and satellite printing units. Some results with a bearing on the paper load are as follows:

The changes of web tension which occurred after the printing units were not the same for all webs, but depended on the number of webs in the press. This fact shows the complex interplay between different parts of the press in the creation of the load on the paper.

Changes of speed in the drag rollers caused only small changes in the web tension. However, in some cases this was very important since it created wrinkles in the paper, which are starting points for paper breaks, and the lateral stability of the webs was affected.

The web tension in the superstructure of the press could be varied by means of the nipping rollers in the folder, either by way of their speed or their contact (line-) pressure. The web tension was very susceptible to changes in line pressure.

There is a relationship between web tension and the formation of web wrinkles; a fact of far-reaching importance in avoiding press stops. It was found that the maximum tension the web could withstand without breaking was not the same in all parts of the press. This was approximately three times higher before the printing units than before the RTF ('roller on top of former'). This fact displays another paradoxical situation. The explanation lies in different distributions of the tensile load in the cross direction of the paper.

To understand the full importance of the web tension as a load on the paper web, it is necessary to continue the present investigation of the tension distribution over the width of the paper web.

Note on measuring techniques In order to investigate web tension, web flutter, web velocity, etc, in a press, as reported above, rather advanced measuring equipment was needed. TFL has developed a system comprising transducers, measurement systems and techniques for analysis. A prerequisite for successful studies of this type is also an experienced staff. To meet the requests from those who wish to make such investigations TFL has therefore set up a group to take on work for printers, press manufacturers, etc.

THE ECONOMIC IMPORTANCE OF THE LOADS ON PAPER IN PRINTING

Every printing procedure must work satisfactorily in order to be economically feasible. There are two main requirements, quality of the printed product and production efficiency.

The first category of loads on paper discussed in this presentation is primarily related to ink quality. It has been shown that a proper choice of ink reduces the load on the paper. This load is in fact both physical and economical. A proper choice is understood here as that which gives the lowest possible total cost for paper, ink and printing in achieving the specified quality of the printed product. It has been advocated earlier that the system paper, ink and press is a system to be optimised[6].

The second category discussed, mechanical loads, has a prime bearing on production efficiency. A printing run with many stops is time consuming. Web breaks are synonymous with waste of paper. Both factors greatly worsen the cost situation.

The general increase in costs of materials and wages strengthens the need for the printing industry to exercise a thorough scrutiny of the function and interaction of all materials and machinery used in printing. They have always to be judged against the balance between cost and function. The research institutes are responsible for providing some of the information for this.

REFERENCES

1. L. O. Larsson, 'Ink quality - importance and problems', Advances in Printing Science and Technology, Vol. 14, Pentech Press (1979)

2. H. Busk, L. O. Larsson and P. O. Trollsås, 'Survey of differences in properties of black newsinks used in the direct lithography procedure' (in Swedish), TFL Report 5:15, Swedish Newsprint Research Centre (TFL), Djursholm (1980)

3. L. Eriksson, 'Studies of web tension in a gravure printing press - a case study' (in Swedish), TFL Report 3:11, Swedish Newsprint Research Centre (TFL), Djursholm (1979)

4. L. Eriksson, 'Studies of web tensions in an offset newspaper printing press - a case study' (in Swedish), TFL Report 3:15, Swedish Newsprint Research Centre (TFL), Djursholm (1980)

5. L. Eriksson, L. O. Larsson and J. Å. Viglund, 'Web tension measurements in a modern offset newspaper printing press - a contribution to an improved background of controlling the web tension' (in Swedish), TFL Report 3:17, Swedish Newsprint Research Centre (TFL), Djursholm (1981)

6. L. O. Larsson, 'Paper performance - a matter of balance', Svensk Papperstidning, No. 4, 83, 101-105 (1980)

DISCUSSION

B. Corsham The occurrence of web breaks in a rotary printing press can perhaps be likened to an unplanned event or accident. Such an event happens when a number of unconnected circumstances are mutually incompatible and occur simultaneously. A break occurs when the level of stress exerted at a point in the paper momentarily exceeds the stress that that point can sustain. When you consider how often a break occurs along the length of the web it is a pretty rare event. However, to the pressroom manager it is all too frequent. When a break occurs on a gravure press printing at 40,000 revolutions per hour a lot of production copies are lost and it is difficult to recover a sample of the paper at the break point. Insofar as the instrument which Dr Larsson has used is an effective way of sensing the web tension and cyclic changes of the tension, it must be a useful tool for establishing the level of tension at different points across the web and perhaps identifying the causes of fluctuations in tension. Effective use of such equipment could lead to improved press design. Could Dr Larsson tell us to what extent his equipment will be available to press manufacturers to study tension loads on paper in their presses?

L. O. Larsson The answer to your question is, in fact, already in the paper: TFL has set up a special group to take on work for or together with printers, press manufacturers, etc. who want to have their presses studied in this way. Such work, which will take the form of special research work on a contract basis and comprise, for instance, case-studies, will benefit from being coupled to our current research activities in this field. The systems for measuring, recording and analysis will be developed further.

R. Ginman IFRA did some work on newsprint runnability about ten years ago. They also constructed a tension gauge. I do not remember the details of the construction of this, but can you tell me how your gauge differs from the IFRA gauge?

L. O. Larsson The two gauges work on different principles. For patent reasons I cannot go into details just now.

W. D. Schaeffer Your elegant analysis of web tension results in a noise frequency spectrum for correlation with mechanical press sources. Have you attempted to use noise frequency analysis as such to correlate with your results from web tension analysis?

L. O. Larsson The frequency components which we deemed most important were those with amplitudes well above the noise level. Therefore, up to now, we have confined the analysis to identifying such components.

J. L. Poujade In the section of your paper 'Disturbances of web tension during normal runs', you have written 'It is interesting to note .... along the path of the web.' Could we infer from this statement that disturbance and subsequent web breaks occurring after the printing unit have very few chances to be due to paper or reel defects which generally create disturbances just after the reelstand?

L. O. Larsson What we can conclude is that the web-tension disturbances induced before the printing unit (e.g. those originating from the reelstand) in this particular case were 'filtered away' to a great extent by the printing nip and thus prevented from propagating further into the press at full strength.

There are reasons, I think, to assume that printing units in general have this effect on web tension disturbances. As we make more investigations of printing presses of different types, data to support this assumption or to disprove it will be available.

W. Hansen Have you noted any differences in web tension disturbances when printing by letterpress and when printing by offset? The question refers to eventual influence of the dampening solution on web tensions.

L. O. Larsson A very definite effect due to the dampening solution is to be seen in Figure 6. The increase of the amplitude of the flutter is an effect of excessive dampening solution during the start-up of the press. The decrease of the tension as displayed by the two tension curves of the same figure has also been ascribed to the dampening solution.

There are, however, no curves showing the effect of more regular amounts of dampening solution. We have not made any systematic comparative investigation between letterpress and offset in the respect which you asked about. It is very likely, however, due to indications, that there are some effects on web tension specific to the offset process.

M. B. Lyne Under uniform tension, newsprint has a tensile strength that on a weight basis is within an order of magnitude of sheet steel. The tensions you have measured at one point are possibly quite different from the tensions in adjacent positions across the web and these skew tensions may exceed the tensile strength of the paper locally. Have you measured tensions in adjacent positions simultaneously to gain some information about skew tensions?

L. O. Larsson The tension distribution across the web has been seen to be very important in the way you suggested in your question. Our results not only show that, but also that the web behaviour, e.g. the creation of wrinkles, is strongly dependent upon the tension distribution.

Your mention of the well-known comparison between the strength of paper and that of steel underlines the idea that paper in general has strength enough to take the load of the nominal web tension (or the tension set by the press control system) and that it might be at least as much worth while to devote efforts to reducing the loads of a more rapid, transient nature, such as those mentioned in the examples given in my paper.

NEW MODELS FOR OFFSET LITHOGRAPHY

U. Lindqvist, S. Karttunen and J. Virtanen, Technical Research Centre of Finland Graphic Arts Laboratory, Espoo, Finland

ABSTRACT

Studies have been made of ink and water compatibility and on-press behaviour with a selection of European news inks. The GTL model for water flows in a offset unit has been verified quantitatively, and full scale water transfer has been compared in respect of printing conditions, ink/water interactions at laboratory level, and ink properties.

The aim of the current studies is to characterize fountain solutions by their dynamic surface tension, and the development of offset unit simulators with both ink and water flows being taken into account. The theoretical postulates, possibilities and limitations of these approaches are outlined.

INTRODUCTION

The theory of lith-offset printing is incomplete. Many press manufacturers design inkers and dampeners by rule of thumb, rather than upon theoretical models. Litho-offset differs from other printing processes, with its very thin ink films - down to one micrometer on coated papers. Moreover there are two liquids employed, namely, ink and fountain solution.

Ink is a relatively thick pigment-in-varnish suspension which, in addition, contains some emulsified and surface water as a result of dampening. To some extent, water and ink are mixed, although the mixtures (water-in-ink and ink or its components in water) are very unstable. GTL has done a great deal of work in the creation of models and partial theories for water transfer in litho presses. Some theories have also been tested and applied in development of inks, fountain solutions, dampener and inker constructions and control systems design [1].

After some approximate simulations had been made of the water transfer [3], based on the water flow model prepared by the same authors [2], it was evident that a combined inking-dampening simulator should be developed for more effective studies. The GTL model is based upon a number of several documented observations on both water pick-up for the inking train, and water transfer by means of non-image and image areas (or combinations of these two) onto printed paper. It was a semi-quantitative model, which has some quantitative features. Essentially, it shows how water is transferred through the inker:

(1) by being picked-up from the plate non-image areas by the inker forme rollers, which happens as many times as there are forme rollers - while water is left to protect the non-image area from tinting, which is the original purpose of the water feed;

(2) by being mixed with the ink on the image areas in the inker forme roller nips, where some water is entering as free surface water from the inker (together with the ink fed); most of it comes onto the ink remaining on the plate image areas in the blanket/plate nip, and through contacts with the dampening forme rollers (before entering the inker forme rollers).

The 'ink-water balance', as it is termed, is necessary, as the press has to run with a quantity of water in equilibrium accumulated in the inker. This equilibrium cannot be constant, if changes occur in the water and ink temperatures and rheology, as happens in long runs. The printing tolerance can be defined as 'the range between minimum water feed where tinting begins, and the maximum water feed where water marking begins. Tinting results from too little water in the non-image areas, whereas water marking is a disturbance in ink transfer due to too much water being in and on the ink layers in the image areas. Under good conditions of litho offset production, the tolerance is usually quite adequate. In practice, some tinting or water marking is allowable, although with a visible loss in print quality.

This paper presents some new basic models for employment as modules for ink/water transfer simulators, and other research and development results regarding water/ink interaction in offset-litho processes.

FULL SCALE TRIALS OF INK/WATER INTERACTIONS

As a part of a wider study (Lindqvist and Oittinen[4]), full scale test runs of eight European offset newsinks have been made on a Goss Community press. In these runs, the amount of water transferred to the paper in image and non-image areas was determined for each ink at low and high water feed, corresponding to the printing tolerance, i.e. the lowest and highest water feed, which does not give any tinting or water marks. The tolerance may be regarded as a measure of the insensitivity of the ink to disturbances in the ink/water balance. In the same way, it could be postulated that a good ink should allow a high water transfer to the paper in the image areas at the upper limit, and a low water feed and transfer

TABLE 1 Results obtained in a full scale test run on a Goss Community

Ink No.	Ink amount on paper	TEST PRINTING IN FULL SCALE: Water amount transferred, Upper level, image areas	Upper level, non-image areas	Under level, image areas	Under level, non-image areas	Water feed on press, Maximum	Minimum	Tolerance	Optimum	X: Alkaline fountain solution	LABORATORY TEST RESULTS: Relative polarity	Ink viscosity ($500 s^{-1}$)	Ink/water interaction, Sun Chemicals: 1 min	10 min	E_∞	k	Emulsification with fountain solution	Emulsification with distilled water	Tack maximum	Interfacial tension
		I_{max}	NI_{max}	I_{min}	NI_{min}	$i_{o_{max}}$	$i_{o_{min}}$	$T = \Delta i_o$	$i_{o_{opt}}$											
	(g/m^2)	(g/m^2)	(g/m^2)	(g/m^2)	(g/m^2)						(%)	(pas)	(%)	(%)	(%)	min^{-1}	(%)	(%)		(mN/m)
2	1.8	0.637	0.705	0.154	0.024	30	12	18	24		0.6	11.9	31.6	149.8	221	0.12	22.2	21.7	7	14.0
4	-	0.777	0.697	0.127	0.016	30	9	21	21		11.4	13.6						35.9	6	16.2
11	1.0	0.568	0.236	0.160	0.029	33	13	20	20	X	18.8	6.5	46.2	160.0	170	0.33	34.2	35.7	22	9.0
16	1.3	1.019	0.540	0.308	0.169	30	15	15	22		1.7	4.3	30.8	76.4	71	0.53	23.1	11.8	23	18.0
18	1.6	0.855	0.627	0.217	0.087	30	12	18	21		10.4	4.3	40.0	138.6	134	0.34	37.4	39.0	13	12.6
22	1.1	0.376	0.295	0.145	0.052	31	14	17	22	X	7.8	5.4	32.2	93.2	92	0.42	33.2	33.0	4	17.8
28	1.6	0.794	0.415	0.277	0.040	32	12	20	19		2.7	5.8						10.5	10	15.1
29	-	0.718	0.616	0.168	0.038	31	11	20	21	X	3.1	5.4	21.4	61.4	55	0.45	21.6	18.5	3	16.4
$\overline{X}$	1.4	0.718	0.516	0.195	0.057	31	12	19	21		7.1	7.2	33.7	113.2	124	0.37	28.6	25.8	11	14.9
$S/\overline{X}$, %	22.6	271.0	35.1	33.9	87.7	3.2	16.7	10.5	4.8		88.7	50.0	25.2	36.7	51.6	37.8	24.8	44.6	27.7	20.1

at the lower, i.e. tinting limit. The results obtained are presented in Table 1. Figure 1 indicates the relationships between the variations in water transfer parameters on the press, and the following three categories of interaction indicator:

(1) Ink properties

(2) Laboratory test of ink/water interactions

(3) Press command variables.

Studies of water flows

Quantitatively, the amounts of water transferred show close agreement with the model of Karttunen and Lindqvist [2] and with Pyliotis' findings [5]. About 0.2 g/m^2 is transferred through non-image areas and about 0.5 g/m^2 through the image areas. Water transfer is governed by both the ink and the water feed. Here, the variations in ink feed are attributable to differences in the NCI-level between the inks. At a high water feed (close to water marks), ink feed and ink film thickness become the dominating factors, while the water feed is more significant at the tinting limit. This may be explained by the ability of ink to hold water being critical only when an excess of water is present. At low water feed, the water content of the ink does not reach the level of saturation; consequently, the amount of water becomes the critical factor.

It is worthwhile to note the marked effect of ink feed upon water transfer in the non-image areas. However, this may be explained by the interaction between ink and water feed. High inking requires high water feed, as there is a major risk of ink tinting non-image areas. It is evident that both the maximum and the minimum levels for water feed are in direct proportion to the ink feed, whereas the printing tolerance is less dependent.

Predicting ink/water interactions

It appears from Figure 1 that ink viscosity and relative polarity account for the variations in water transfer and water feed to almost the same degree as do the laboratory tests for the interactions. The laboratory tests applied can be divided into two main categories: mechanical emulsification tests - such as determination of the water content of ink as a function of mixing time [7], or the Tack-o-Scope emulsification and surface chemical interaction tests - such as estimation of the interfacial tension between ink and water [8]. Furthermore, it is observable that these categories of test account almost for the scatter between the inks. These tests describe surface chemical properties and interactions under static conditions, and do not take into account any mechanical interactions arising from excess work in nips.

In essence, emulsification tests and ink viscosity measurements are made for estimating the mechanical mixing of water in ink. It also appears that these methods describe almost the same parts of the full-scale

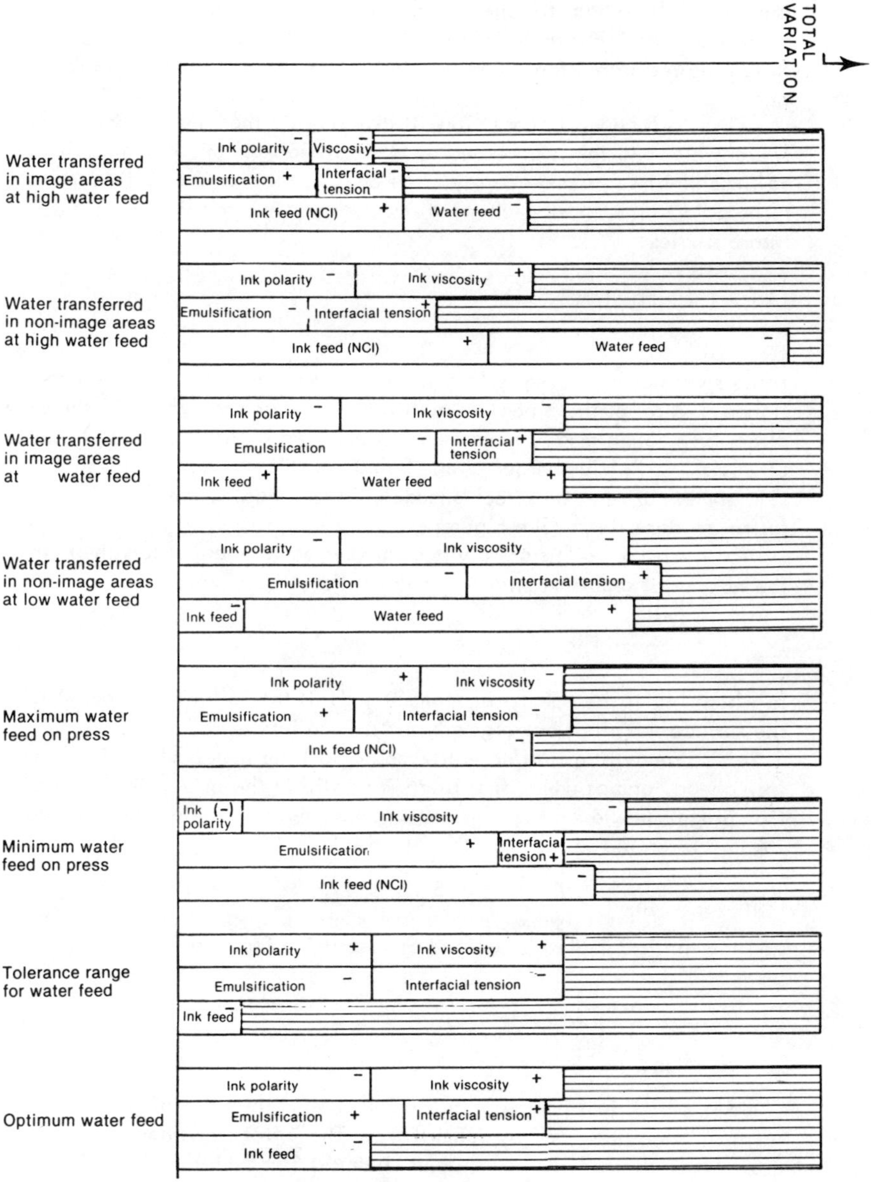

Fig. 1. Relative significance of some variables (ink properties, interactive properties, command variables) for ink/water interaction in full scale. Signs express the direction of the effects in the study in question.

behaviour, which may indicate that they also describe the same mechanisms. It follows that to some extent these methods are alternatives to each other, and that their real value is dependent upon the degree to which the test conditions and the printing conditions in the press are identical.

As a general conclusion, it may be said that the interaction between water and ink is governed by both physico-chemical and mechanical mechanisms. Consequently, it is obvious that more attention needs to be paid to the physico-chemical mechanisms in dynamic conditions in future studies.

OFFSET UNIT SIMULATOR

General

Models simulating inking systems have to some extent already been designed. Most of these neglect both water flow and movement in the cross direction. Such models have been documented by Mill[9], Hull[10], Rech[11], Lindqvist[12], Reichenberger[13] and Solna[14]. The model employed by GTL somewhat resembles Rech's model. Our model ranges from ink transfer to ink densities[2] (See Figure 2).

All the rollers are divided into segments of equal length, with the number of segments on a particular roller A being

$$N_A = n\left[\frac{\pi d_A}{\lambda} + 0.5\right] \qquad (1)$$

in which N_A is the number of segments, d_A is the roller diameter, λ is the surface length of one segment and n is an integer.

Simulation is effected by moving all the rollers of the inker one step ahead, and assuming that there is no slip in the nips. In the simulation programme, each segment corresponds to a matrix element. As no ink can be stored in the nip, the film thicknesses i are related by

$$i_{A_{in}} + i_{B_{in}} = i_{A_{out}} + i_{B_{out}} \qquad (2)$$

and the nip is completely defined by a splitting factor a:

$$a_A = \frac{i_{A_{out}}}{i_{A_{in}} + i_{B_{in}}} \qquad (3)$$

A model which takes account of the ink flow in the cross direction has recently been presented by Schmitt[15]. The water flows have been discussed by Pyliotis[5], Karttunen and Lindqvist[2, 3], Liebing[16] and McPhee[17], although no simulation programme has been presented to date.

Mechanisms to be taken into account

The fountain solution transferred to the inking unit will in part appear as free surface water, and in part in emulsified form. The distribution of

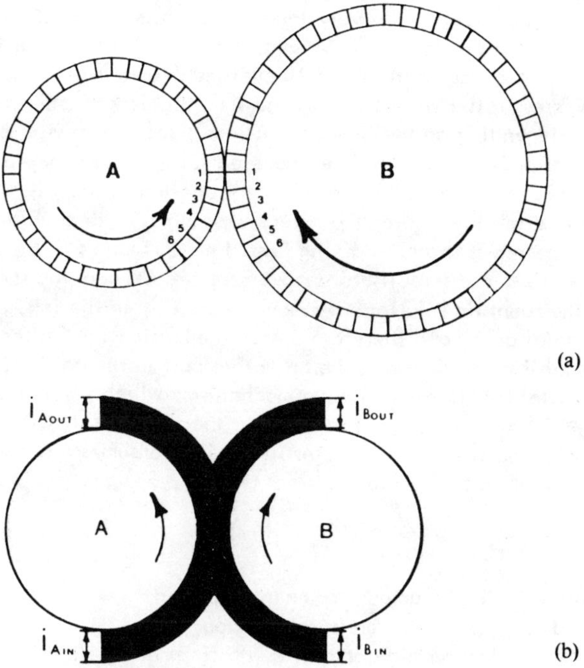

$$i_{A_{IN}} + i_{B_{IN}} = i_{A_{OUT}} + i_{B_{OUT}} \quad (1)$$

$$i_{A_{OUT}} = a_{A_{OUT}} \left(i_{A_{IN}} + i_{B_{IN}} \right) \quad (2)$$

Fig. 2. Simulation model for a roller inking unit: (a) principle, (b) ink split model.

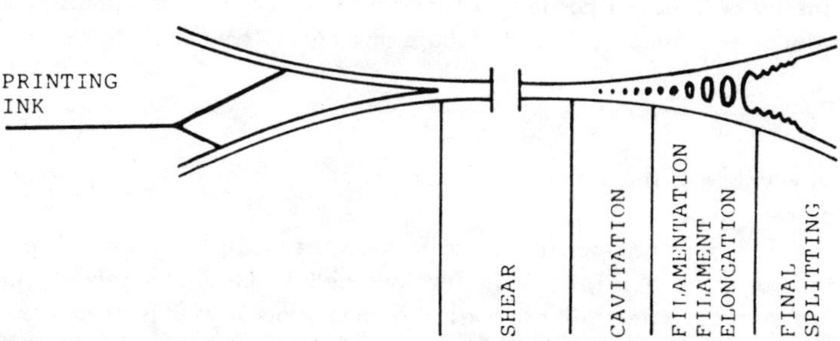

Fig. 3. Two-dimensional graph of roller nip.

water in the ink film may vary markedly between different points in the nip, as the nip contact is very long in comparison with the ink layer thicknesses, which are about 10 - 20 microns, and with the quantities of water. Inker nips are very thin (about 10 microns), tunnels are of considerable width (press roller widths are around 1 m) and length (something around 5 mm, depending on the pressure settings), see Figure 3. As a rule, the nip is formed between a rubber-covered roller and copper-plated steel drum; copper is one of the most water-repellent of metals. In the tunnel, water and ink layers are mixed very efficiently, although the time of mixing in the nip is short. Mixing is still more rigorous in the roller/drum nip, where a cross-oscillatory movement occurs. In the shear zone of the nip, the fountain solution may move vigorously in the ink, creating a new interfacial area, and therefore unstable emulsions. On the other hand, the emulsified water may take an active part in the cavitation, and act like nuclei for filamentation, a mechanism, which would explain the relatively high amount of surface water at the outlet of the nip.

Emulsification of water in the ink consumes a share k of the energy applied to the nip

$$kE = k \iiint_{BLH} p(x,y,z)\,dx\,dy\,dz = \Delta A \gamma_{iw} \qquad (4)$$

where E is the amount of energy applied
- p is the nip pressure function
- L is the nip contact length (in printing direction)
- B is the roller width
- H is the impression in the nip
- γ_{iw} is the interfacial tension between ink and water under the dynamic conditions of the nip
- ΔA is the area created in the nip.

On substituting normal values for the nip parameters the energy used in a nip is of magnitude 0.5 mJ per mm roller length. As water droplet sizes up to 50 microns have been reported[17], it can be concluded that the emulsification of a homogeneous water film to droplets ($d = 10\,\mu m$) requires energy of 0.06 mJ per mm roll length, i.e. 12% of the energy would be enough to form a new, but perhaps momentary emulsion.

$$E = \left(\frac{3w}{r} - 2\right) LB\gamma_{iw} \qquad (5)$$

in which w is the water film thickness at the inlet of the nip and r is the droplet radius.

It is obvious that some of the water must be in emulsified form at the outlet of the nip. A distinction needs to be drawn between emulsified and surface water in the model. Some surface water is always present.

An estimate should be made of the evaporation of the surface water between the nips, see Figure 4. Under normal conditions (23°C, 50% rH), the evaporation rate of water was found to be of the order of

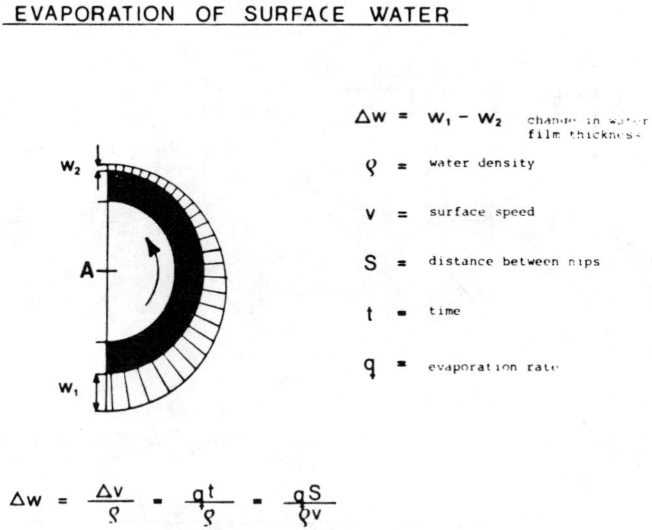

Fig. 4. Evaporation of surface water between two nips.

(a) Change in film thickness.

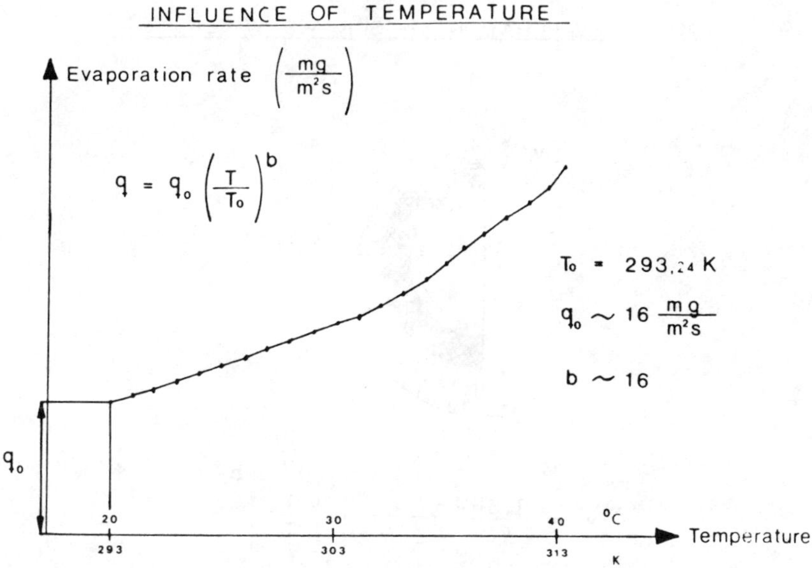

(b) Effect of temperature on evaporation rate.

INFLUENCE OF SPEED

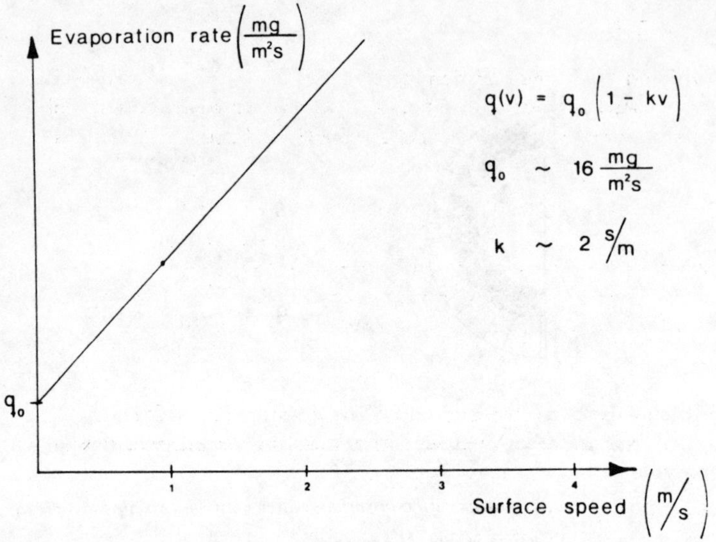

(c) *Effect of speed on evaporation rate.*

TOTAL EVAPORATION OF SURFACE WATER

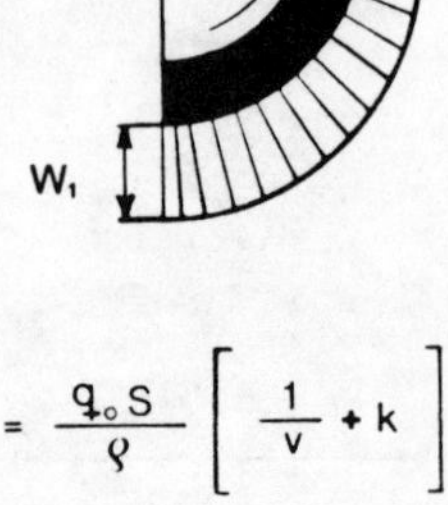

$$\Delta W = \frac{q_o S}{\rho}\left[\frac{1}{v} + k\right]\left[\frac{T}{T_o}\right]^b$$

(d) *Combined effect of speed and temperature on evaporation.*

$15.7\ mg/m^2s$. The effect of temperature and damping pressure can be expressed as

$$\frac{q_2}{q_1} = \frac{p_2 T_1}{p_1 T_2} \tag{6}$$

in which q is the evaporation rate.

As the relationship between damping pressure and the absolute temperature almost follows a power function, Equation 6 assumes the form

$$q = q_o \left(\frac{T}{T_o}\right)^b \tag{7}$$

in which b is a constant; for normal dampening solutions $b \sim 16$.

The evaporation rate increases almost linearly with increase in the surface speed of the roller. This may be expressed by

$$q = q_o (1 + k\ v) \tag{8}$$

in which v is surface speed, and k is a constant ($k \sim 2\ ^s/m$).

In a distance S between two nips, this corresponds to an evaporation volume V:

$$V = q_o S \left(\frac{1}{v} + k\right) \tag{9}$$

The reduction in water film thickness Δw between two nips with both surface speed and temperature being taken into account, can be expressed as follows:

$$\Delta w = \frac{q_o S}{\rho} \left(\frac{1}{v} + k\right) \left(\frac{T}{T_o}\right)^b \tag{10}$$

Emulsion break-down An analysis has also been made of the possibility of emulsified droplets reverting to surface water between nips. The movements of the droplets in the ink can be estimated from Stoke's equation, which by reason of the alternating gravitation field in the ink must be written (see Figure 5) as:

$$u = \frac{D^2 (\rho_w - \rho_i)}{18\eta} \left[g \sin \left(\frac{v}{R} t\right) + \frac{v^2}{R} \right] \tag{11}$$

where u is the separation speed of the droplets
- ρ_i is the density of ink
- ρ_w is the density of dampening solution
- η is the ink viscosity
- g is the gravitational acceleration
- R is the roller radius
- D is the particle diameter.

Calculations using realistic values for the parameters ($\Delta\rho$ =

0.1 kg/dm^3, η = 5 Pas, D = 5 μm, R = 5 cm, L= 15 cm, v = 4 m/s) have demonstrated that the distance which a normal water droplet can move within the ink during the time between two nips is of the order of 1 nm - that is to say, negligible. For this reason emulsion break-down can be expected in the nips alone.

Specification of the model

In principle, the new model is based upon the same ink-splitting postulates as those expressed in Figure 2. However, for water-film split and water distribution some additional assumptions must be made. At the inlet side of the nip, the fountain solution is partly emulsified, and in part forms a surface water film. The total amount of water in the nip is thus:

$$W = w_{A_{in}} + w_{B_{in}} + i_{A_{in}} e_{A_{in}} + i_{B_{in}} e_{B_{in}} \qquad (12)$$

where W is the total water amount

w is the film thickness of surface water, and

e is the degree of emulsification.

On the outlet side of the nip, it is assumed that the water distribution function obeys the following general rules:

(1) A particular share p (characteristic of each individual nip) of the total water amount is in emulsified form; the share 1-p is free surface water.

(2) The surface water film thickness is of the same magnitude on both rollers.

(3) The degree of emulsification is identical on both rollers.

These postulates can be expressed in the form (see Figure 6):

$$e_{A_{out}} = e_{B_{out}} = \frac{(1-p)W}{i_{A_{in}} + i_{B_{in}}} \qquad (13)$$

$$w_{A_{out}} = w_{B_{out}} = \frac{1}{2} pW$$

It is now feasible to build a simulation model with these assumptions and the evaporation being observed (Equation 10). The output values of the model are print density with corresponding variations, and water film thickness transferred to the paper in the image and non-image areas.

The results obtained in simulating operations will be presented later. Ink film transfer to the paper is calculated in accordance with the Walker-Fetsko model, and print density is estimated from Oittinen's model [32], which is a modification of the Tollenaar equation.

EMULSION BREAK-DOWN BETWEEN THE NIPS

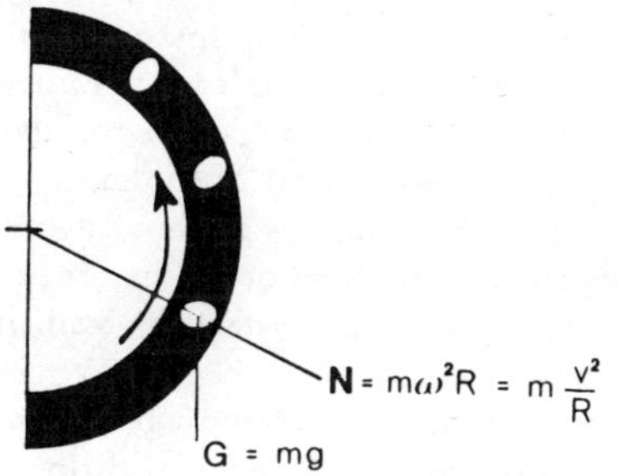

STOKE'S LAW

$$U = \frac{D^2 (\rho_w - \rho_i)}{18\eta} \left[g \sin\left(\frac{v}{R} t \right) + \frac{v^2}{R} \right]$$

Fig. 5. The possibility for emulsified water to return into surface water between the nips.

WATER SPLIT

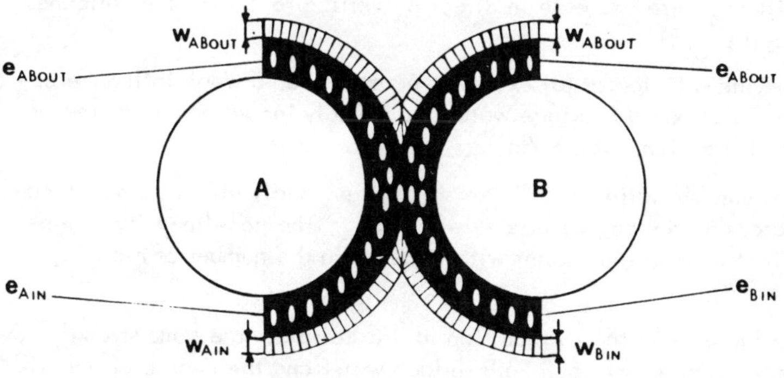

$$w_{AIN} + w_{BIN} + e_{AIN} i_{AIN} + e_{BIN} i_{BIN} = 2 w_{ABOUT} + e_{ABOUT} \left(i_{AOUT} + i_{BOUT} \right) \quad (3)$$

$$p \left(w_{AIN} + w_{BIN} + e_{AIN} i_{AIN} + e_{BIN} i_{BIN} \right) = e_{ABOUT} \left(i_{AOUT} + i_{BOUT} \right) \quad (4)$$

Fig. 6. Specification of a simulation model for an offset unit.

Discussion

The principles of the model are illustrated in Figure 6. Of course, the three main principles assumed to be a characteristic of the water distribution function at the outlet of a nip involve an over simplification of the situation, which limits its application, at least to nip figures, which are strongly dynamic. In addition, ink and water flow across the web is completely ignored. Nevertheless, the opportunity to choose the ink split parameter, emulsification parameter and evaporation parameters separately for each nip must permit of the use of the model in any existing - or planned - printing unit. A further advantage of the model is the possibility of leading the ink feed and water feed to any of the rollers continuously or periodically, along with the possibility of obtaining information on print density direct.

A final evaluation of the model is obtainable after an extensive test of the programme in different units alone. At this stage, the hypotheses have been tested for a simple roller system (Figure 7(a)). In this system, the main flows of ink and water pass from inlet A, through nips B and C, to paper nip D. It is assumed that the ink split parameter (a) is equal in each nip; the same applies for the emulsification ratio (p), and the emulsification speed (q). When both ink and water feed are continuous, the situation closely resembles the printing of solid areas.

From this, an interpretation of the parameters can be made as follows:

(1) Increasing water feed linearly increases both the amount of surface water and the proportion of water emulsified in the ink. Ink feed does not affect the surface water, and the degree of emulsification will of course decrease in direct proportion to the ink film thickness (Figure 7(b)).

(2) The ink split factor (a) exercises no more than a weak influence upon the amount of surface water, but visibly increases the degree of emulsification (Figure 7(c)).

(3) The emulsification coefficient (p) will markedly affect both surface water and the degree of emulsification. The non-linearity of the effect is more significant with increase in the number of nips (Figure 7(d)).

(4) An increase in the evaporation rate (q) displays the same strongly decreasing effect upon both surface water and the degree of emulsification, as a result of the reduced amount of water present (Figure 7(e)).

(5) Finally, if no ink is present, the situation somewhat resembles that of the non-image areas on the plate. In this case, the critical factor becomes the relative evaporation of 'surface' water between the nips (Figure 7(f)).

From this, we can draw the general conclusion that the effect of the para-

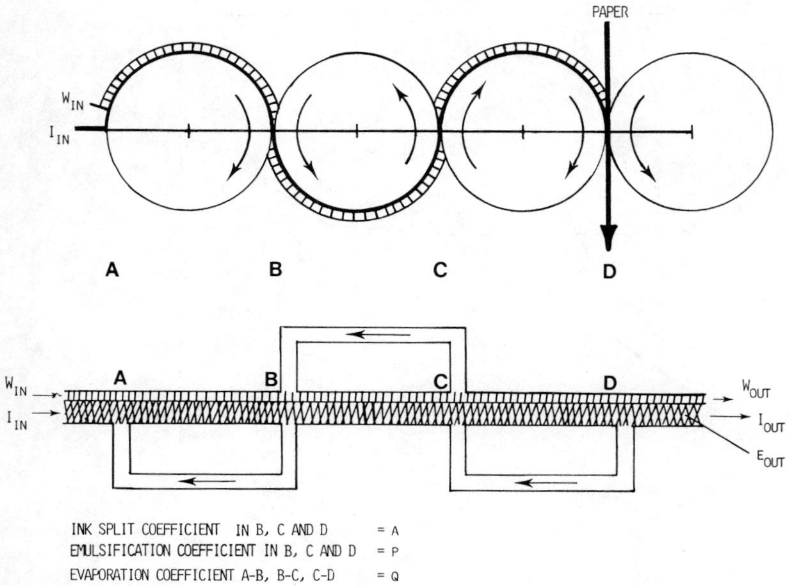

Fig. 7. Testing of the model in a simple roller system.

Fig. 7(a). Roller system and flow diagram.

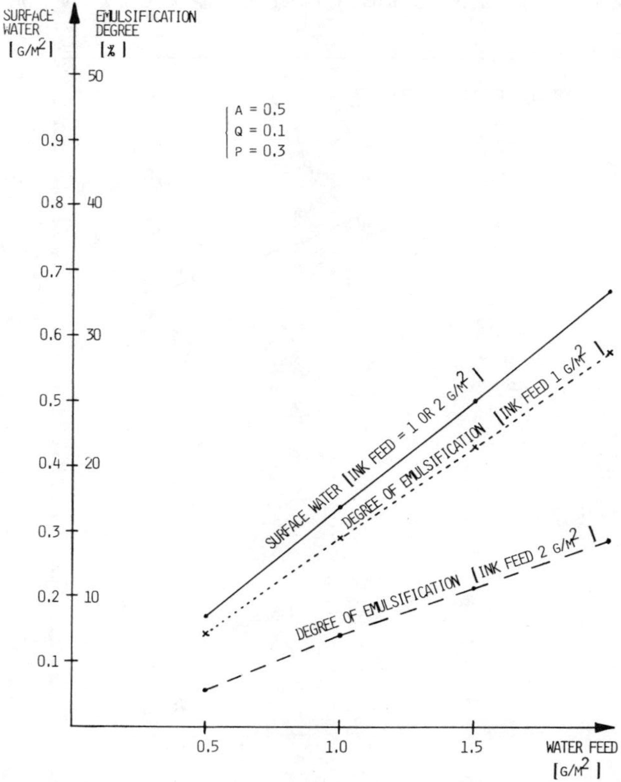

Fig. 7(b). Effect of ink and water feed on surface water and degree of emulsification.

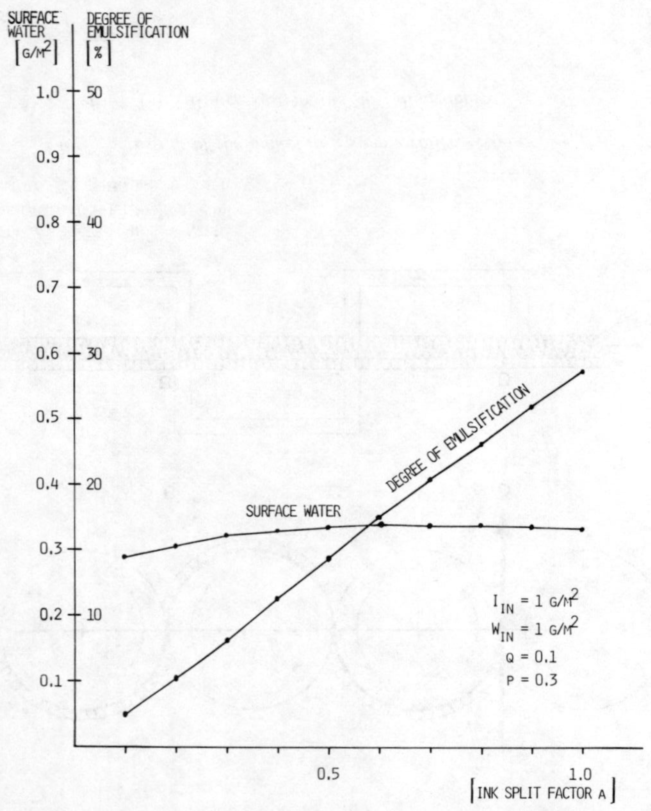

Fig. 7(c). Effect of ink split factor on surface water and degree of emulsification.

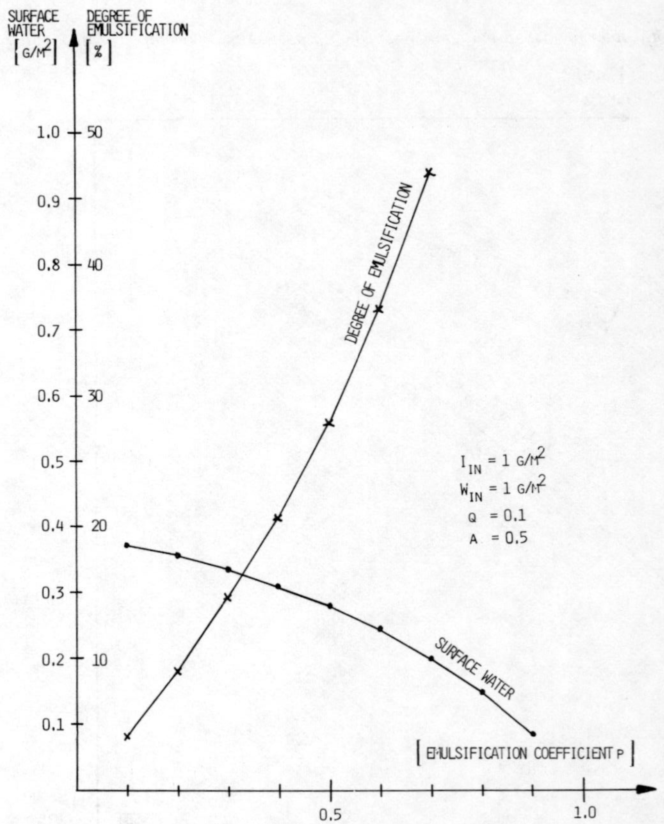

Fig. 7(d). Effect of emulsification coefficient on surface water and the degree of emulsification.

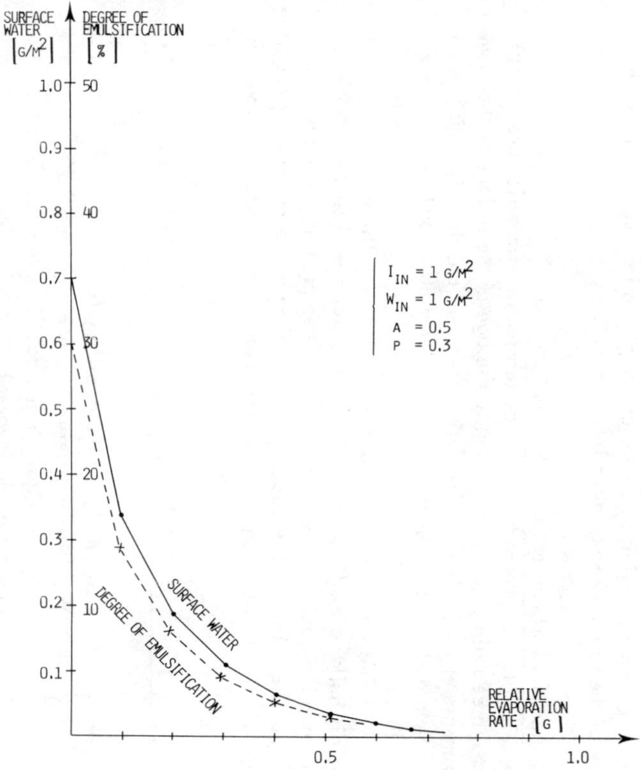

Fig. 7(e). Effect of relative evaporation rate on surface water and the degree of emulsification

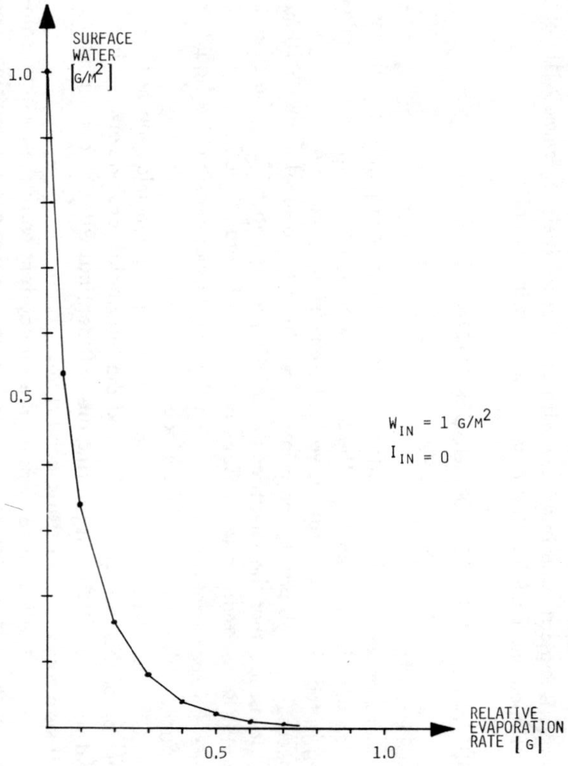

Fig. 7(f). Case (e) for no ink film present.

meters is logical, and in accord with practice. The programme will be continued and model-tested in a number of printing units.

INK/WATER INTERACTION UNDER DYNAMIC CONDITIONS

General

It seems necessary to make surface energy measurements under conditions which are highly dynamic. The spreading of water on ink, frequent mixing in the nips, and break-down of the emulsion all occur at very high speeds. In web offset, the printing speed reaches a level of about 10 m/s. If we assume that the length of the printing nip is about 1 cm, the time available to create a new surface area in the fountain solution is about 1 ms. Experimentally, it is known that the surface tension of a liquid changes with the age of the surface.

As the conditions in the printing nip are dynamic, the surface tension of the fountain solution and the interfacial tension between ink and fountain solution in printing are not identical with the corresponding values measured under static conditions.

It is difficult to measure the surface tension of the fountain solution direct during printing as the rollers are running and the printing zone is very narrow. In principle, it could be measured in the press, for example by the application of laser beam techniques. It is thus preferable to measure surface tension by means of a separate instrument, and with the age of the surface being taken into account.

Methods for surface tension determination

In principle, the methods for surface tension determination are classifiable into three main groups, according to the age of the surface at the moment of measurement. These three main categories are then dynamic, semi-dynamic and static, with time limits of 1 and 30 seconds respectively (Table 2).

The limits of these groups are not very definite, and scientists use dissimilar classifications. Among the static methods, the Du Noüy ring and the capillary rise methods have been used for the measurement of fountain solutions. The capillary rise method is one of the most accurate methods for air/water phases. In this method, the surface tension γ is derived from

$$\gamma = \frac{r_i \; \Delta\rho \, g \, h \, f_o}{2 \cos \theta} \qquad (14)$$

where r_i = inner radius of the capillary (m)
$\Delta\rho$ = the density difference between phases (kg/m^3)
h = height of liquid in the capillary (m)
θ = contact angle (degree)
g = acceleration of gravity (m/s^2)
f_o = correction factor.

TABLE 2

DYNAMIC	SEMIDYNAMIC	STATIC
Oscillating jet Thomas (17)	Drop weight Pierson (20)	Du Noüy ring Harkins (25)
Collision of jets Bikerman (18)	Contact angle Scan-method (21)	Wilhelmy plate Adamson (23)
Max bubble pressure Kloubek (19)	Capillary fall Murakami (22)	Capillary rise Shaw (26)
	Capillary waves Adamson (23)	Pendant drop Adamson (23)
	Inclined plate Van den Bogaert (24)	Sessile drop Adamson (23)
		Rotating drop Cayias (27)
		Laser beam reflection Rödel (28)

0 — 1 — 30 — Surface age (s)

The correction factor f_o is often approximated for reasons of simplification, Shaw[26].

Du Noüy's ring method has been used for measurement of the interfacial tension between oil and water phases, and the surface tension between air and water phases. Mathematically, the relationship between surface tension and measured parameters has the form:

$$\Delta G = \frac{4\pi R_1 \gamma}{f_1} \tag{15}$$

where ΔG = weight difference of ring (kgm/s^2)
R_1 = the ring radius (m)
f_1 = correction factor.

As a rule, the surface tension is obtained direct from a scale on the ring-tensiometer.

As regards semi-dynamic methods, the drop-weight method must be mentioned. It is useful for foaming fountain solutions. In this case the surface tension is expressed by:

$$\gamma = mg/2\pi R_o f_2 \tag{16}$$

where m = mass of drop (kg)
g = gravitational acceleration (m/s^2)
R_o = outer radius of capillary (m)
f_2 = correction factor.

Correction factors f_2 have been tabulated, Lando[29].

One of the most important of the dynamic methods is based upon interpretation of the oscillating jet formed when a liquid jet passes through an elliptic orifice. The surface age is thereby defined as the time needed by the jet to move from the orifice to the centre line of a wave. This oscillating jet method enables measurement of the surface tension of surfaces aged in the range 1 to 30 ms.

If the velocity of flow f, the density of liquid ρ, and the parameters of the jet are known, the surface tension can be expressed in the form, Thomas[17]:

$$\gamma = \frac{2 \cdot 10^3 \rho f^2 (1 + 1.542 A^2) \psi}{3 r_i \lambda^2 + 5\pi^2 r_i^3} \qquad (17)$$

where f = velocity of flow in jet (m^3/s)
ρ = density of liquid (kg/m^3)
$r_i = \frac{1}{2}(r_{max} + r_{min}) \left[1 - \frac{1}{6}(A^2)\right]$ medium radius of jet (m)
$A = (r_{max} - r_{min}) / (r_{max} + r_{min})$ amplitude of jet
λ = wavelength of jet (m)
$\psi = 1 + 2(\eta\lambda/\rho v)^{3/2} + 3(\eta\lambda/\rho v)^2$ viscosity factor
η = viscosity of liquid (Pas)

If $\eta < 0.001$ Pas then $\psi \approx 1$. Parameters of the oscillating jet are expressed in Figure 8.

By way of example, the parameters of the jet can be measured by photography of the jet, or direct by reflection of the shadow of the jet on to a matrix made of photodiodes. In our Institute we are currently designing an instrument for measuring the dynamic surface tension of fountair solutions. This instrument is based upon the oscillating-jet principle, and is controlled by microprocessors. The shadow of the jet is focused on the photodiode matrix by lenses. A photomultiplier and lenses have also been employed earlier, Netzel[30].

Preliminary results

In preliminary tests, we measured the surface tensions of isopropanol solutions and solutions containing non-ionic surfactants of the Tergitol-15 S series (manufactured by Union Carbide) and which have also been tested in fountain solutions. For comparative purposes, anionic Teepol HB 6 (manufactured by Shell) has also been tested. The methods applied were those of the Du Noüy ring (surface age 60 s), capillary rise (60 s), drop weight (1, 15 and 30 s) and oscillating jet (preliminary test 3 ms).

It is discernible from the results, Virtanen[31], that the surface tension values of isopropanol solutions were almost identical for all the methods used, whereas the surface tension of surfactant solutions depended greatly upon the method in use. The surface tension of surfactant solutions was greatly affected by surface age.

The pH-values of a surfactant solution did not seem to influence surface tension; moreover, the dependence of surface tension on molecular weight in the Tergitol series is not linear. A short summary of the results obtained is given in Figures 9, 10 and 11.

We intend to continue our measurements with the oscillating jet instrument mentioned above. Further to this, pilot printing tests will be done to discover correlations between the dynamic surface tension of fountain solutions and their behaviour on a press. In this respect, special attention will be paid to the differences between solutions of IPA and surfactants. This will enable a better understanding of the ink/water interactions in printing. It is well known that surfactants with varying dynamic surface tensions may influence wetting speeds, emulsion and surface water film formation, and emulsion break-down.

CONCLUSIONS

Experimental data derived from full-scale trials on an offset unit have verified quantitatively the model presented earlier for water flows, giving maximum water transfer in the image areas. The variation in amounts of water transferred between different offset news inks could be predicted reasonably well from two laboratory tests - one depicting the surface-energetic properties of the water and ink, and the other one describing the mechanical mixing of the two phases.

With these experimental findings as basis, a new model for an offset unit was designed, in which a distinction is drawn between emulsified and free surface water at each roller nip. The model accounts directly for the evaporation of surface water, while the effect of surface-energetic properties on the formation of emulsions and emulsion breakdown can be inferred only indirectly by a suitable choice of the emulsification parameter (p). A high degree of emulsification at the outlet side of a nip obviously corresponds to thorough mixing of ink and water, and consequently to a low interfacial tension.

Preliminary calculations made with the model on a simple three-roller system have shown explicable effects for all the variables involved. The model has now been transferred to a general data programme and will be tested in different offset units. At this point, it seems that application of the model is primarily limited by its static nature. It is evident that dynamic wetting phenomena play a significant part in the formation of surface water films, water-in-ink emulsions and emulsion break-down.

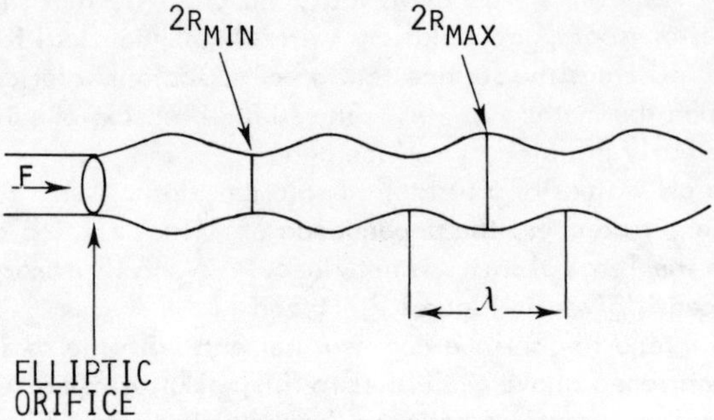

Fig. 8. Parameters of the oscillating jet.

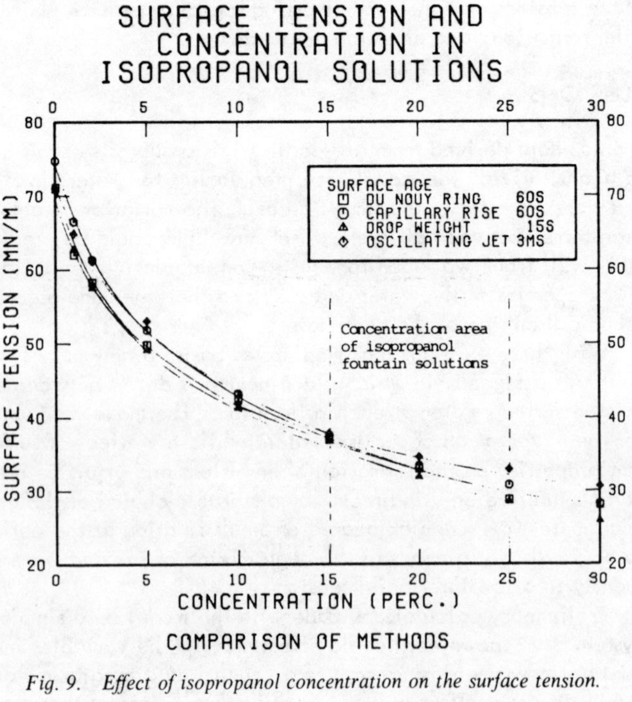

Fig. 9. Effect of isopropanol concentration on the surface tension.

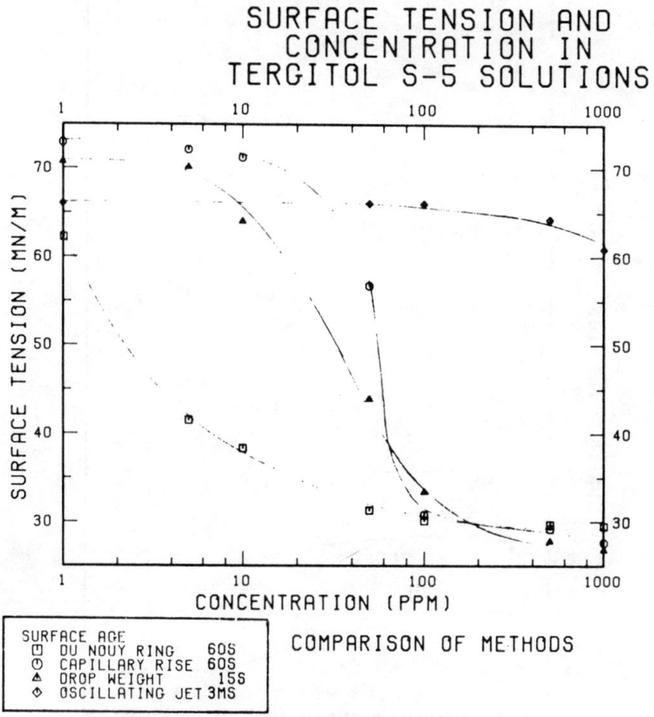

Fig. 10. Effect of surfactant concentration on the surface tension.

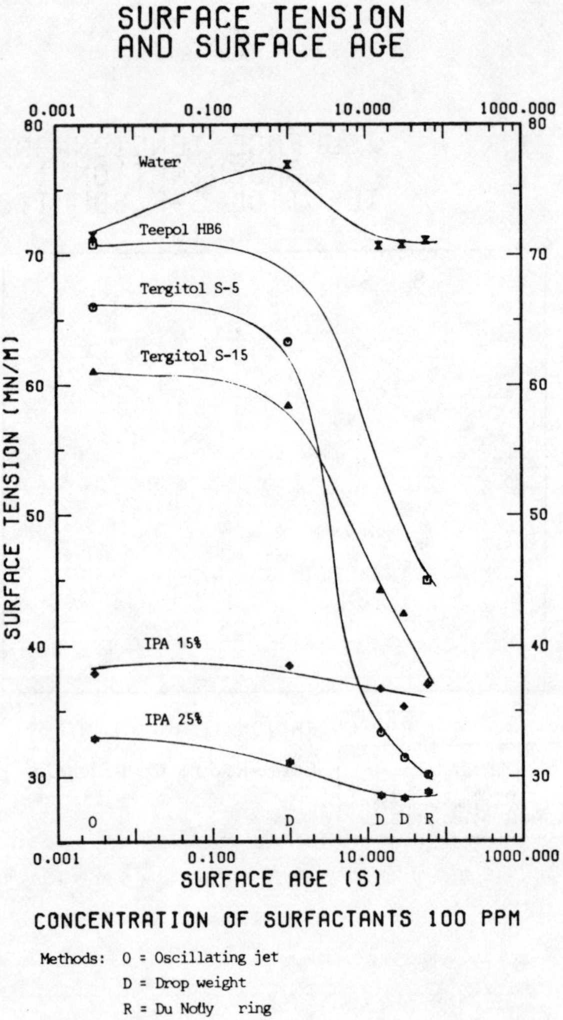

Fig. 11. Effect of surface age on the surface tension.

For this reason, the next step is further development of the model to account for complex dynamic phenomena in and between the nips. This calls for more adequate characterization of the materials involved in terms of dynamic parameters. Our first approach to this problem concerns the dynamic surface tension of the fountain solution. With a view to matching the dwell-time in a roller nip, the surface tensions of surfaces not exceeding 3 ms in age need to be measured. Preliminary tests at our Institute have also indicated large differences in this respect between water solutions of isopropanol (IPA) and surfactants. To date, the surfactant solutions tested behave in almost the way as distilled water within this time range, whereas IPA-solutions usually reach (static) equilibrium.

The method of an oscillating jet was selected from among the techniques to estimate dynamic surface tension. A device employing photodiodes for interpretation of the jet has been designed. In further studies, special attention will be paid to the difference between surfactants. This may enable us to study the behaviour of different materials in the press with the model described.

NOMENCLATURE

i_A = ink film thickness (μm) on roller A

N_A = number of segments on roller A

d_A = diameter of roller A (m)

a_A = ink split factor on roller A

k = proportionality factor

E = energy applied to the nip (J)

$p(x, y, z)$ = pressure distribution function in a nip (Pa)

L = nip width (μm)

B = roller length (m)

H = impression in nip (μm)

γ = interfacial tension (mN/m)

ΔA = area created in the nip

r = radius of emulsion droplet of water in ink (μm)

q = evaporation rate (mg/m^2s)

T = absolute temperature (K)

b = constant

V = volume (cm^3)

S = distance between two nips (m)

ρ = density (kg/m^3)

u = speed of droplets in a liquid (m/s)

η = viscosity (Pa s)

R = roller radius (m)

D = particle diameter (μ m)

W = total amount of water in a nip (mm^3)

w = film thickness of surface water (μm)

e = degree of emulsification (%) in a given ink film

p = share of total water amount emulsifying in a given nip

r_i = inner radius of capillary (m)

$\Delta\rho$ = density difference between air and liquid phases (kg/m^3)

h = height of liquid in capillary (m)

θ = contact angle (degree)

g = acceleration of gravity (m/s^2)

f_o = correction factor in capillary rise method

ΔG = weight difference of ring (kg m/s^2)

R_1 = ring radius (m)

f_1 = correction factor in Du Noüy's ring method

m = mass of droplet (kg)

R_o = outer radius of capillary (m)

f_2 = correction factor in drop weight method

f = velocity of flow in jet (m^3/s)

r_j = medium radius of jet (m)

A = amplitude of jet

λ = wavelength of jet (m)

ψ = viscosity factor of liquid in jet

v = velocity (m/s)

REFERENCES

1. Lindqvist, H., Lindqvist, U., Karttunen, S., Lehtonen, T. and Simomaa, K., TAGA Proceedings (1980).

2. Karttunen, S., Lindqvist, U., 'Interfacial phenomena in Litho-offset printing', Part 1. Graphic Arts in Finland, 7(1978)2.

3. Karttunen, S., Lindqvist, U., 'Water flow and surfactant effects in offset litho printing', Advances in Printing Science and Technology, Vol. 15, Ed. W. H. Banks, Pentech Press, London (1980).

4. Lindqvist, U., Oittinen, P., 'Undersökning av svarta off-settidn-

ingsfärger'. An STC-IFRA-NATS commission at GTL and FPPRI, Espoo (1980) (In Swedish).

5. Pyliotis, D., 'Wechselwirkung zwischen Druckfarbe und Feuchtmittel beim Offsetverfahren', FOGRA Forschungsbericht 5.205, Munich (1978) (in German).

6. Surland, Aa., A laboratory test method for prediction of lithographic ink performance. TAGA proceedings (1980).

7. Laraignou, R., 'Interfacial tension measurements in offset inks related to halftone printing in offset', Advances in Printing Science and Technology, Vol. 6, Ed. W. H. Banks, Pergamon Press, Oxford, (1964).

8. Mill, C.C., 'An experimental test of a theory of ink distribution', Advances in Printing Science and Technology, Vol. 1, Ed. W. H. Banks, Pergamon Press, Oxford, (1961).

9. Hull, S., 'The theoretical analysis and practical evaluation of roller ink distribution system', TAGA Proceedings (1968).

10. Rech, H., Beiträge zur experimentellen und rechnischen Untersuchung des Farbtransportes in Walzenfarbwerken von Druckmaschinen, Technical University, Darmstadt 1971, Institutsbericht (1971) (in German).

11. Lindqvist, H., Telavärilaitteen simulointimallin verifiointi täysmittakaavaisella offsetpainokoneella. Graduate thesis (M.Sc.), Helsinki University of Technology, Espoo (1976) (in Finnish).

12. Reichenberger, R., 'Vorgänge im Walzenspalt unter Berücksichtigung der seitlichen Verreibung im Farbwerk einer Offsetdruckmaschine', Papier und Druck 29 (1980) a, p. 5. (In German).

13. Anon., 'Solna Offset simulerar färgverket med hjälp av data', Grafiskt Forum 78 (1973) 2.

14. Schmitt, H. J., Steuerung der Farbdosierung bei Offsetdruckwerken durch rechnergestützte Modellvorhersagen, Karlsruhe Kernforschungszentrum (1979) (in German).

15. Liebing, H., Riedl, R., 'Zur vorbestimmten Feuchtung im Offsetdruck', Papier un Druck 27 (1978) 7, (in German).

16. McPhee, J., 'An engineer's analysis of the lithographic printing process', TAGA Proceedings (1979).

17. Thomas, W. D. E. et al., 'An Oscillating Jet Relative Method for Determining Dynamic Surface Tensions', J. Colloid Interface Sci., 50 (1975): 3.

18. Bikerman, J. J., Surface Chemistry for Industrial Research, Academic Press Inc., New York (1948).

19. Kloubek, J., 'Measurement of the Dynamic Surface Tension by the Maximum Bubble Pressure Method IV', J. Colloid Interface Sci., 41 (1972): 1.

20. Pierson, F.W. et al., 'Studies of the Drop-Weight Method for Surfactant Solutions II. Experimental Results for Water and Surfactant Solutions', J. Colloid Interface Sci., 54 (1976): 2.

21. Scandinavian Pulp, Paper and Board Testing Committee, 'Method SCAN P 18:66, Contact Angle of Water on Paper and Paperboard', Paperi ja Puu - Papper och Trä 48 (1966):6.

22. Murakami, Y., 'Determination of Dynamic Surface Tension of Liquid in Gravitational Flow along an Open Capillary by Adachi's 'Capillary Fall' Method', Proc. of 5th Int. Congress on Rheology 1968, Ed. Onogi, S., 181-9, University of Tokyo Press, Japan (1969).

23. Adamson, A.W., 'Physical Chemistry of Surfaces', John Wiley & Sons (1976).

24. Van den Bogaert, R. et al., 'Dynamic Surface Tensions of Sodium Myristate Solutions', J. Phys. Chem. 83 (1979):17.

25. Harkins, W.D. et al., 'A Method for the Determination of Surface and Interfacial Tension from the Maximum Pull on a Ring', J. Amer. Chem. Soc. 52 (1930):2.

26. Shaw, D.J., 'Introduction to Colloid and Surface Chemistry', Butterworths (1970).

27. Cayias, J.L. et al., 'The Measurement of Low Interfacial Tension via the Spinning Drop Technique', ACS Symposium Series 8 (1975).

28. Rödel, K. et al., 'Zur Bestimmung der Oberflächenspannung von reinen Flüssigkeiten und Tensidlösungen nach der Reflexionsmethode', Tenside Detergents 15 (1978):3. (in German).

29. Lando, J.L. et al., 'Tabulated Correction Factors for the Drop-Weight-Volume Determination of Surface and Interfacial Tensions', J. Colloid Interface Sci., 25 (1967).

30. Netzel, D.A. et al., 'Adsorption Studies of Surfactants at the Liquid-Vapor Interface: Apparatus and Method for Rapidly Determining the Dynamic Surface Tension', J. Colloid Sci., 19 (1964).

31. Virtanen, J., 'Static and Dynamic Surface Tension in Water Solutions of Surfactants and Isopropanol', Graduate thesis of M.Sc. (Chem), (1980) (in Finnish).

32. Oittinen, P., 'A tentative study of back transfer in two colour printing', Graphic Arts in Finland 1 (1972) 2.

DISCUSSION

W. H. Banks Was the tinting level well defined experimentally?

U. Lindqvist It was defined by consulting the printer.

J. M. Fetsko Surland and Bassemir found that the rate of water pickup is important. Work at NPIRI suggested that the rate of emulsion break-down is important. How do these findings fit into your model?

U. Lindqvist In the simulation programme the rate of water pick-up (emulsification) and emulsion break-down in the nips can so far be taken into account only indirectly by choosing the emulsification parameters of each individual nip. At this stage the model is not suitable for testing materials.

W. Hansen In your model you calculate the amount of water evaporating on the assumption of a relative humidity in the air of 50%. A far higher humidity, maybe up to 100%, could be expected around the rollers. Is this important to the calculations of film transfer?

U. Lindqvist The effect you mentioned can be taken into account in the model only indirectly via the effect of temperature, surface speed or simply by choosing the evaporation rate level. For the purpose of the model I do not think the effect is critical.

L. O. Larsson New approaches to a problem are always interesting. find your paper stimulating. I would like your comments on the following:

(1) You mentioned in your Table 1 the amount of ink on the paper. How did you assess this?

(2) The break-down of an emulsion is in general, I think, a comparatively slow process, and one would not expect such a thing to happen in the short time intervals between the nips, where the passage time usually is less than 0.15. Do you have any comment?

(3) The agreement between the results of the full-scale experiments (Table 1) and the theoretically derived ones are not particularly good, e.g. the amount of water emulsified in the ink, and the corresponding amount in laboratory experiments. What in your opinion is the reason for this? Could it be insufficiencies in the experimental methods (e.g. insufficient accuracy) or neglected factors in the theoretical models?

U. Lindqvist

(1) The quantities were calculated from the ink milage values observed in the laboratory.

(2) I fully agree. This is exactly what I stated in my short mathematical analysis of emulsion break-down.

(3) The amount of water transferred to the paper on the press, e.g. in image areas at high water feed i_{max}, depends on a number of factors not included in laboratory tests such as ink feed, water feed, emulsion break-down in the nips, evaporation of surface water between the nips and the treatment of the emulsion in the many nips on the press. In addition, the formation of emulsions in any laboratory test markedly deviates from the situation in the nips. Since none of the emulsions moreover is stable, a good agreement between full scale and laboratory tests can hardly be expected. The amount of water transferred to the paper (in full scale) has been determined using tracers and should be rather accurate. The model itself can of course not have influenced the agreement between laboratory and full scale.

J. Aspler

(1) How is the 'relative polarity' mentioned in Figure 1 determined? Also, what ink component or property is 'printing tolerance' derived from?

(2) Do you have any data on the dynamic surface tension of an offset ink? Also, would you not expect any solution of a macromolecular surfactant to show a dynamic surface tension due to the time required for it to migrate to the newly-formed surface? (ct. Chang and Gray, J. Colloid and Interface Science, 67, 255 (1978).

U. Lindqvist

(1) Relative polarity means here the polarity component of the surface energy of the ink, and it is determined from contact angle measurements using model liquids. In the study referred to, the relative polarity and surface tension showed significant mutual correlation, so it is not possible to make a strict distinction between the effects. The 'Printing tolerance' depends on both physico-chemical and flow properties of the ink, but cannot be mathematically derived from any particular measurement. The relative effects of the different ink components have not been calculated.

(2) The dynamic surface tension of printing inks cannot be determined with the method in question due to their high viscosity. Solutions of macromolecular surfactants will, of course, always show a dynamic surface tension compared to IPA for the reasons you mentioned. The behaviour of a surfactant solution cannot, however, be depicted by molecular length alone, and we believe this method might serve as a tool for classifying the suitability of surfactant for fountain solutions.

OPTIMUM CONDITIONS FOR COLOUR SEPARATION IN GRAPHIC ARTS REPRODUCTION

K. Schläpfer, UGRA, St. Gallen, Switzerland.

ABSTRACT

In order to study the colour separation process, a computer program has been developed to simulate colour reproduction in four-colour printing.

In the first part, the effect of different sets of colour separation filters has been investigated. It has been found that filter curves which are derived from colour-matching functions are far better suited than the currently used broad-band and narrow-band filters.

The second part deals with mathematical models suited for the calculation of exact colourimetric reproduction. A new model is presented which allows the spectral reflectance curve of an exact colourimetric match for a specified illuminant to be calculated.

As an application of this model the degree of metamerism in the case of exact colourimetric reproduction is studied. Since the mathematical model is able to calculate the colourimetric match for all possible three-colour combinations with the four process inks, some advantages of the concept of total UCR (reproduction with only two colours plus black) are discussed.

IDEAL FILTERS AND PRIMARIES IN GRAPHIC ARTS REPRODUCTION

The obvious aim of the colour separation step is to record three images which, when printed with the three process inks, give a faithful reproduction of the original. This goal, however, can only be achieved in theory, i.e. if the process inks have unrealisable reflectance curves and if the separation filters have transmittance curves with negative portions.

Strictly speaking, the process inks must consist of dyes having rectangular reflectance curves. Moreover, the separation filters have to be identical with the colour-matching functions for the secondary colours of the chosen set of process inks. A possible combination of theoretical process inks and colour-matching functions is shown in Figures 1 and 2.

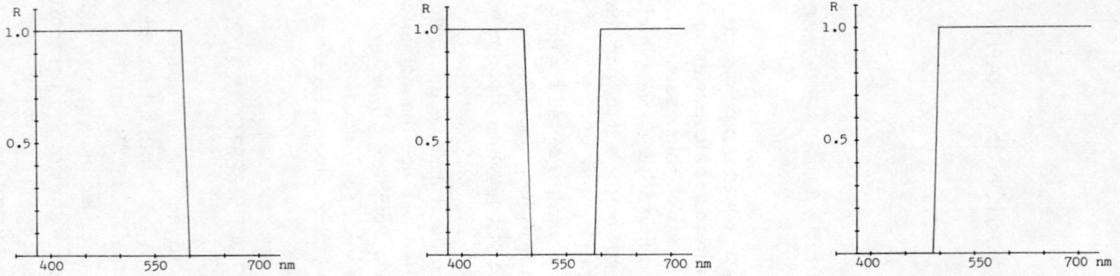

Fig. 1. *Theoretical primary colors for halftone printing.*

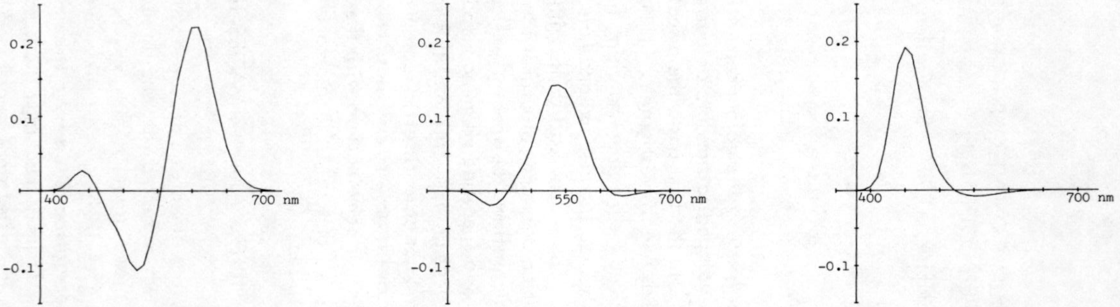

Fig. 2. *Color-matching functions for the theoretical secondary colors in halftone printing.*

These requirements for exact colour reproduction hold for graphic arts reproduction as well as for colour photography, whereas the requirements for colour television are less rigorous. In colour television, exact reproduction can be achieved with any set of three primary colours, if the separation sensitivities are colour-matching functions of the chosen set of primaries.

The limiting condition, however, is that the sensitivity curves must have negative portions in every trichromatic reproduction system.

To achieve exact colourimetric reproduction with positive separation sensitivities, iterative correction techniques are required (see section 4).

A COMPUTER PROGRAM TO STUDY THE GRAPHIC ARTS REPRODUCTION PROCESS

Even though a colourimetrically exact reproduction is not achievable with substractive processes, the question is of interest as to what extent this theoretical goal can be approached by optimizing practical colour separation conditions.

To investigate this, a computer program has been developed to simulate the reproduction process. The program uses the following variables:

(1) the original
(2) colour separation filters
(3) spectral sensitivity of the separation film
(4) taking illuminant
(5) colour correction
(6) tone reproduction curve
(7) screening characteristic curve
(8) undercolour removal (UCR)
(9) dot area dependence on density of the black printer
(10) primary colours of the process inks (including the substrate)
(11) characteristic curve in platemaking and printing
(12) viewing illuminant.

The original to be reproduced is introduced as a reflectance curve consisting of 35 values ranging from 380 nm to 720 nm. A series of test colours, i.e. the 17 colours specified in DIN 6169, Part 5 (1976), is permanently stored. These test colours are:

No. 1	Flesh tone	No. 10	Saturated yellow
No. 2	Light brown	No. 11	Saturated green
No. 3	Yellow-green	No. 12	Saturated blue
No. 4	Green	No. 13	Skin tone
No. 5	Blue-green	No. 14	Foliage green
No. 6	Light blue	No. 15	Dark grey
No. 7	Light purple	No. 16	Medium grey
No. 8	Red-purple	No. 17	Light grey
No. 9	Saturated red		

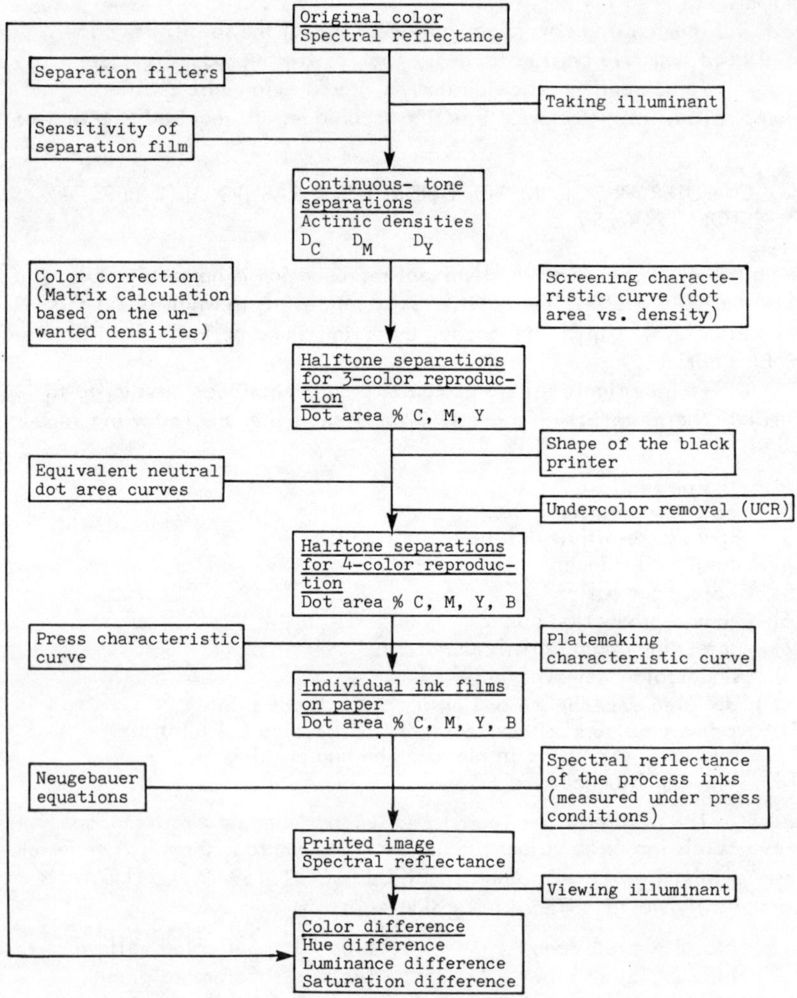

Fig. 3. Calculation of color reproduction in printing.

The colour separation process is simulated by multiplying the spectral values of the original with those of the separation filters, the taking illuminant and the film sensitivity. This calculation provides the three actinic density values.

In a second step, colour correction is carried out by means of matrix algebra starting from the actinic densities of the chosen set of inks. To eliminate the unwanted densities of the primary colours, first order masking equations (3 x 3 matrix) are applied. Furthermore, the 50% tone of the primary colours is corrected in the gradation by adjusting the wanted density.

In the next step the corrected actinic densities are converted to dot area values. To do this, a screening characteristic and a tone reproduction curve are introduced. The lowest of the three dot area values is then used to calculate the black printer. As parameters the lowest grey density of the original to respond to the black printer, the maximum dot area and the degree of UCR can be varied.

When the dot areas of the four halftone separations are determined, the next step is to calculate the dot area on the paper, based on a given platemaking and press characteristic curve. To calculate the reflectance curve of the four-colour print, the spectral reflectance values of the process inks (measured on a particular substrate) are used.

A flow chart illustrating the simulation program is shown in Figure 3.

OPTIMUM SEPARATION FILTERS FOR CONVENTIONAL COLOUR SEPARATION

The subject of optimum separation filters has been investigated by many authors, the best known among them being Brewer and Hanson[23] and Kowaliski[6]. Moreover, a comprehensive treatment of this subject can be found in the textbooks of Yule[9] and Hunt[4]. Most work, however, has been done in the area of colour photography rather than in graphic arts reproduction. Although both processes are based on the subtractive principle, they are different as to the primaries and the separation sensitivities. Hence, conclusions for optimum separation sensitivities in colour photography cannot be directly extended to graphic arts reproduction.

Another reason justifying this study is that earlier researchers described the result of colour reproduction in terms of density values rather than in colourimetric terms, probably because programmable calculators were not so accessible at that time.

Starting from the ideal separation filters (Figure 2), one could assume that the best filters for the practical process inks would have colour-matching curves. The feature of colour-matching curves is that they record the colours as the eye sees them. However, since filters

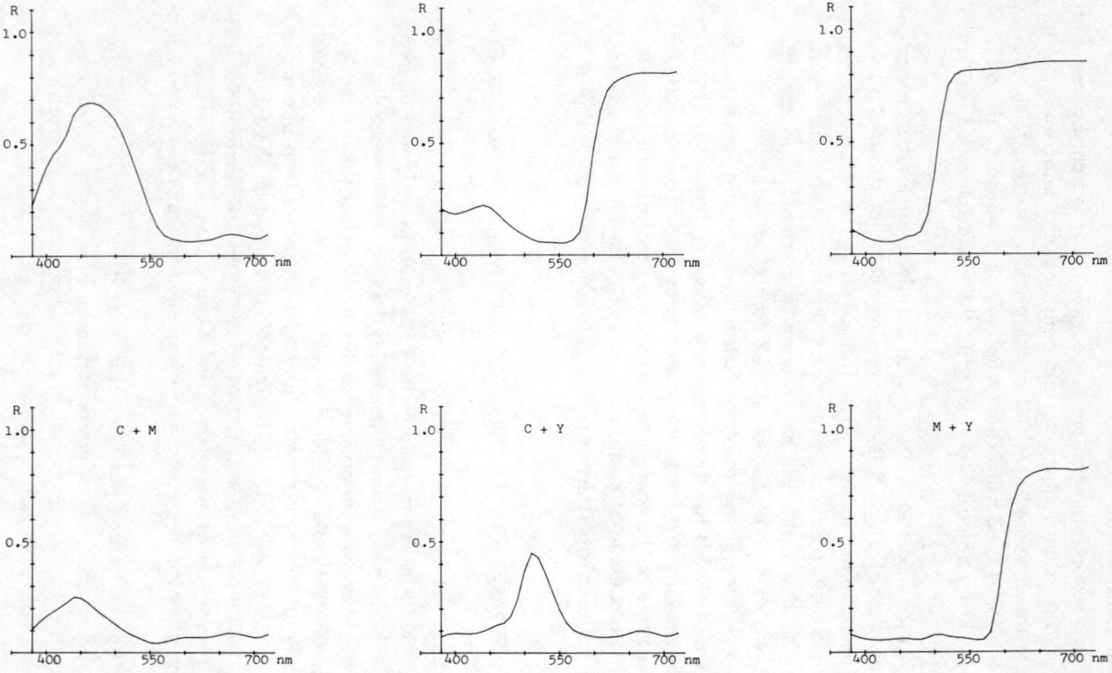

Fig. 4. Primary and secondary colors of the European process inks (as measured in Brunner offset Standard No. 1).

with negative portions cannot be realized, the best that can be done is to take the negative portions as zero. This, in turn, lowers their colourimetric quality.

To test the effect of separation filters, it is common practice to use the primary colours of the process inks as originals. As a measure of the quality of separation filters the magnitudes of unwanted densities is then assessed, arguing that a high level of unwanted densities requires undesirable stronger masking.

Based on this view, any set of colour-matching curves is not well suited as they produce higher unwanted densities than do currently used separation filters.

Yule[9] assessed colour separation filters according to three criteria, i.e.

(1) colour separation efficiency
(2) colour correction efficiency
(3) Colourimetric quality.

A good colour separation efficiency means that the three primaries, when recorded as originals, produce low values of unwanted densities. The colour correction efficiency describes the degree of masking required to eliminate the unwanted densities. Separation efficiency and correction efficiency are related properties, as low unwanted densities require less masking. The colourimetric quality finally describes how well the filters correspond to colour-matching curves.

Considering all these criteria, the commonly recognized view is that filters with a good separation efficiency for the primary inks (when recorded as originals) are more important than a good colourimetric factor, because the latter is difficult to achieve in any case. This view can be accepted as long as the process inks are tested as original colours. The topical role of the primary colours is justified because the masking requirements can be directly derived from them. From the viewpoint of importance as picture colours, however, their significance is only marginal.

In this study the effect of separation filters is tested with picture colours rather than with the process primaries. The reproduction is simulated with the European process inks standardised for letterpress and lithography. The reflectance curves of the chosen set of process inks are shown in Figure 4.

As a measure of the reproduction quality the colour difference between the original colour and the three-colour reproduction with C + M + Y is assessed. The black has been deliberately omitted, since when used as a fourth colour it is an arbitrary quantity. Similar to Yule, three criteria are tested:

(1) colour reproduction quality with full colour correction
(2) colour reproduction quality when no colour correction is applied
(3) masking requirements in terms of wanted and unwanted densities.

Six sets of separation filters have been tested which can be divided into two groups:

(1) Colourimetric filters:

First, the colour-matching curves for the European process inks have been calculated (Figure 5). Since this set of curves has distinct negative portions, a second set of colour-matching curves with less negative portions has been chosen (Figure 6). This set was derived according to the considerations known as the Ives-Abney-Yule compromise (Hunt [4], p. 89). The compromise consists of shifting the matching stimuli to the spectrum locus, which diminishes the negative portions of the colour-matching curves.

As the new matching stimuli have the same dominant wavelength as the European process inks (Figure 7), the hue errors resulting from these colour-matching curves should be minimal. For colour separation, both sets of colour-matching curves are used without negative portions. As a set of filters without negative portions the CIE colour-matching functions based on the primaries X, Y, Z were included (Figure 8).

(2) Spectral filters:

The term 'spectral' refers to the fact that these filters have transmittance curves adjusted to the reflectance curves of the process inks.

Monochromatic filters were selected as an ideal realisation of spectral filters (Figure 9).

Two sets of Wratten filters were chosen as practical filters, i.e. the narrow-band set No. 70, 99 and 50 (Figure 10) and the broad-band set No. 25, 58 and 47 (Figure 11).

In Table 1, the colourimetric filters (except the CIE filters) are compared with the spectral filters with respect to their band-width and dominant wavelength. As can be seen, the green and blue curves of the colourimetric filters match very closely the Wratten broad-band filters, while the red curve of the colour-matching function has a considerably shorter dominant wavelength.

The first criterion investigated was the magnitude of unwanted and wanted densities. As can be seen from Table 2, the colourimetric filters show higher values of unwanted densities than the spectral filters. However, the two sets of colourimetric filters derived from the process inks are not much worse than the Wratten filters. Only the CIE tristimulus filters may create problems in masking due to the high level of unwanted densities.

In the next step, the six sets of filters were tested by reproducing a series of test colours. Of the 17 test colours listed in Section 2, the following ten were used:

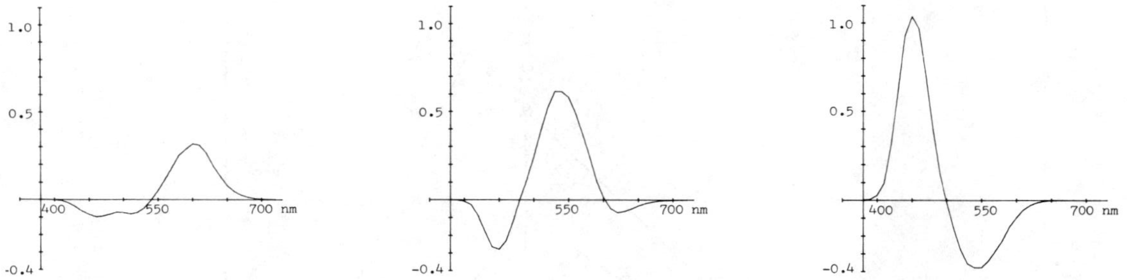

Fig. 5. *Color-matching functions for the secondary colors of the European process inks.*

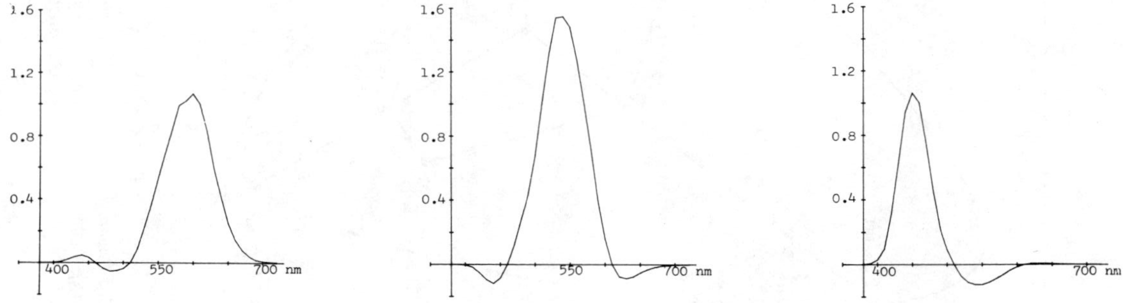

Fig. 6. *Color-matching functions for the spectrum colours having the same dominant wavelength as the secondary colors of the European process inks.*

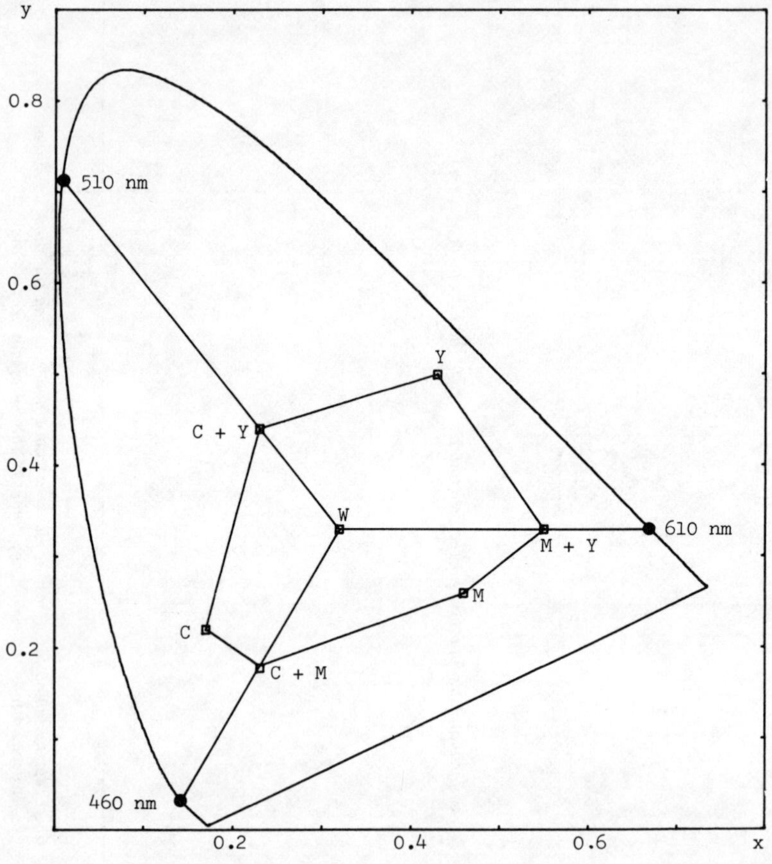

W = paper white

Fig. 7. Gamut of the European process inks (as measured in Brunner Offset Standard No. 1) and spectrum colors having the same dominant wavelength as the secondary colors of the European process inks.

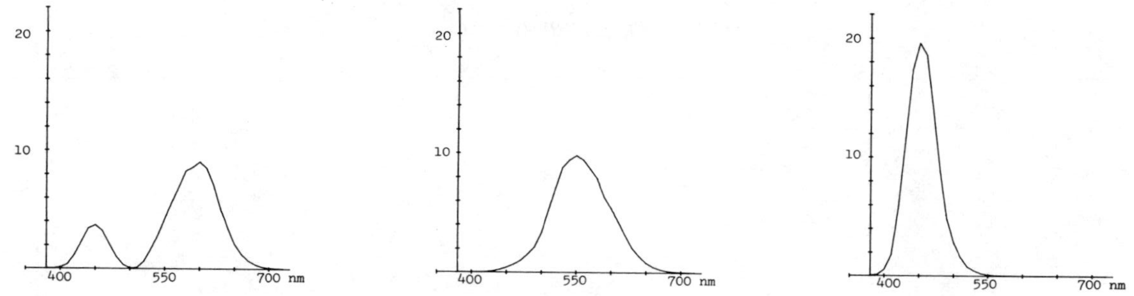

Fig. 8. *Color-matching functions for the CIE primaries X, Y, Z.*

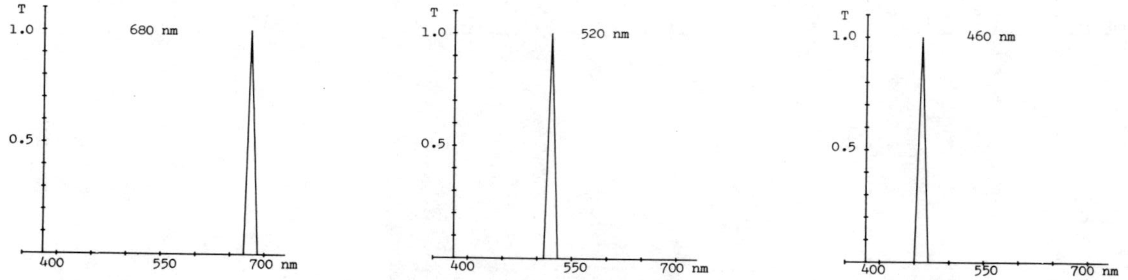

Fig. 9. *Theoretical monochromatic filters.*

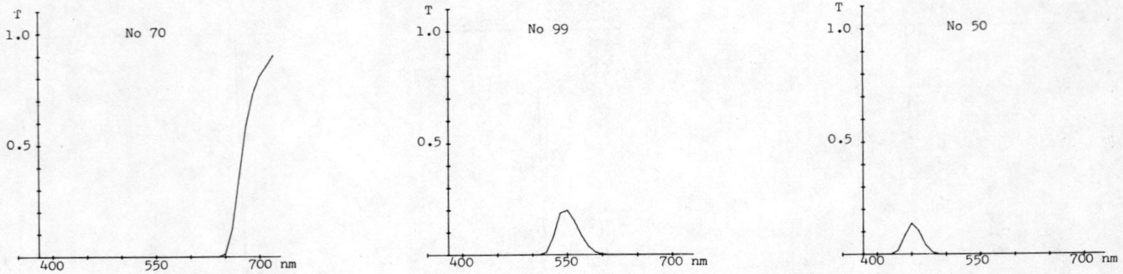

Fig. 10. Wratten narrow-band filters.

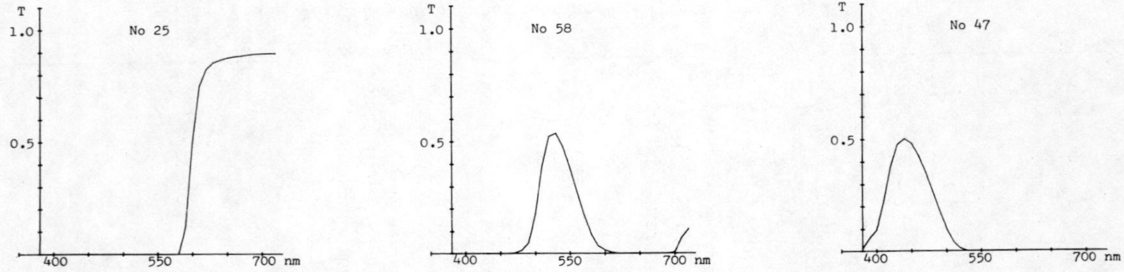

Fig. 11. Wratten broad-band filters.

No. 1	Flesh tone	No. 10	Saturated yellow
No. 2	Light brown	No. 11	Saturated green
No. 4	Green	No. 13	Skin tone
No. 5	Blue-green	No. 16	Medium grey
No. 6	Light blue	No. 17	Light grey

The remainder of the 17 test colours was not reproducible within the gamut of the process inks.

TABLE 1 Bandwidth and dominant wavelength of Wratten separation filters and of possible colourimetric filters

	Bandwidth (nm)	Dominant wavelength (nm)
Wratten broad-band:		
Red No. 25	130	710
Green No. 58	120	530
Blue No. 47	140	440
Wratten narrow-band:		
Red No. 70	70	720
Green No. 99	50	550
Blue No. 50	40	450
Colour-matching curves without negative portions for R.G.B. of the European process inks:		
Red	130	600
Green	110	530
Blue	90	450
Colour-matching curves without negative portions for the spectrum colours with the same dominant wavelength as R.G.B. of the European process inks:		
Red	180	600
Green	130	540
Blue	120	450

The average colour differences resulting from these ten colours when reproduced with C + M + Y are listed in Table 3 (without colour correction) and Table 4 (with full correction).

From Table 4 it can be seen that the two colourimetric filter sets referring to the process inks give far better results than all three sets of spectral filters. The CIE tristimulus filters, in turn, are worse than the spectral filters and, taken all together, they must be considered as unsuited for colour separation.

TABLE 2 Unwanted densities obtained with different sets of separation filters using the European process inks

Set of separation filters	Sum of wanted densities	Sum of unwanted densities	Unwanted densities in % of wanted densities	Maximum single value of unwanted density %
Spectral filters:				
Monochromatic 460 nm, 520 nm, 680 nm	3.16	1.06	34	62
Narrow-band Wratten No. 50, 90, 70	3.22	1.35	42	57
Broad-band Wratten No. 47, 58, 25	2.87	1.27	44	66
Colourimetric filters without negative portions based on the primaries:				
CIE X, Y, Z	2.18	1.80	83	80
Secondary colours of the European process inks	3.16	1.48	47	59
Spectrum colours of the same dominant wavelength as the secondary colours of the European process inks	2.89	1.57	54	63

Comparing Tables 3 and 4 it is interesting to see the influence of colour correction on the three colour parameters, hue, saturation and luminance. The essential effect of colour correction is to improve saturation and luminance, whereas the hue errors remain largely unaffected.

As to the magnitude of the colour differences resulting from the reproduction process, including masking, the result must be considered as disappointing. However, even though a close match of the original colour cannot be achieved, all means to improve the reproduction result should be taken into account. One contribution could be the use of filters based on colour-matching functions without negative portions, as shown in Figure 5.

MATHEMATICAL MODELS TO CALCULATE EXACT COLOURIMETRIC REPRODUCTION

As already mentioned, an exact colourimetric reproduction cannot be achieved if existing primary colours are used and if colour correction is made by means of photographic masking techniques. The best that can be done by masking is to achieve exact reproduction for four selected

TABLE 3 Average colour difference between original and reproduction for 10 DIN test colours obtained with different sets of separation filters without colour correction. (Computer-simulated reproduction with the European process inks without black).

Set of separation filters	Colour difference in CIELAB units			
	Total difference	Luminance difference	Saturation difference	Hue difference
Spectral filters:				
Monochromatic 460 nm, 520 nm, 680 nm	17.0	7.7	9.1	7.2
Narrow-band Wratten, No. 50, 99, 70	20.1	6.6	12.8	8.8
Broad-band Wratten, No. 47, 58, 25	20.8	8.1	14.6	5.3
Colourimetric filters without negative portions based on the primaries:				
CIE X, Y, Z	32.3	15.3	23.0	8.8
Secondary colours of the European process inks	15.1	9.3	9.8	4.1
Spectrum colours of the same dominant wavelength as the secondary colours of the European process inks	16.8	11.0	10.6	3.6

TABLE 4 Average colour difference between original and reproduction for 10 DIN test colours obtained with different sets of separation filters with colour correction. (Computer-simulated reproduction with the European process inks without black).

Set of separation filters	Colour difference in CIELAB units			
	Total difference	Luminance difference	Saturation difference	Hue difference
Spectral filters:				
Monochromatic 460 nm, 520 nm, 680 nm	12.8	4.8	6.2	7.9
Narrow-band Wratten, No. 50, 99, 70	12.8	4.0	7.8	5.8
Broad-band Wratten, No. 47, 58, 25	11.8	3.1	7.7	6.3
Colourimetric filters without negative portions based on the primaries:				
CIE X, Y, Z	13.5	2.7	9.4	6.4
Secondary colours of the European process inks	7.0	1.4	5.1	3.3
Spectrum colours of the same dominant wavelength as the secondary colours of the European process inks	7.4	1.7	5.3	3.5

colours. In this case 12 masks are required. Commonly, exact reproduction is only attempted for three colours, requiring 9 masks.

With electronic colour scanners, however, the dot areas required to achieve exact colour reproduction can be calculated for any colour within the gamut of the process inks. To do this, a numerical relationship between the original and the dot areas required to match the original must be available.

The best known relationship is given by the Neugebauer equations (Hunt[4], p. 534) from which the three tristimulus values of the reproduction from the dot areas and the tristimulus values of the primaries and their subtractive combinations can be calculated.

As an alternative, the spectral reflectance curve of the reproduction can be calculated. In this case, the primaries and their subtractive combinations must be specified in terms of spectral values (Figure 12).

Direct calculation of the tristimulus values is much simpler than the calculation of the reflectance values. However, the latter approach is often more useful, for instance to study problems of metamerism.

A quite different approach for determining the dot area of the reproduction has been described by Korman and Yule[5]. They start from a colour chart consisting of 512 colour patches and calculate the dot area for any colour by linear interpolation. To do this, the 512 colour patches must be stored in the computer in terms of tristimulus and dot areas. The method of Korman requires a colour chart for each combination of paper and ink which must be numerically specified and entered into the computer. In this respect the Neugebauer equations are more flexible as only 16 colours must be specified. It was recognized, however, that the original Neugebauer equations lead to results that deviate considerably from experimentally produced reproductions. As a consequence, attempts have been made to introduce correction terms.

A set of modified Neugebauer equations has been proposed by Pobbaravsky and Pearson[7]. They used three exponents which, however, were not able to eliminate fully the errors of the calculation. Tests in our Institute using this approach aimed at further improving the exponential correction terms. Based on the European process inks a new set of exponents was derived which still gave a maximum deviation of $\pm$3% in terms of dot area.

Since the correction terms do not improve the Neugebauer equations but make them more complicated and not more efficient, attempts have been made to establish purely empirical relationships between the dot area of the process inks and the colour to be reproduced. A very recent proposal of Warner[10] which seems to overcome some weaknesses of earlier models, however, needs quite a number of empirical coefficients which must be determined experimentally.

A mathematical model better suited, is the Neugebauer concept applied to the calculation of spectral values. Although this method requires correction terms as well, the departures from experimental values

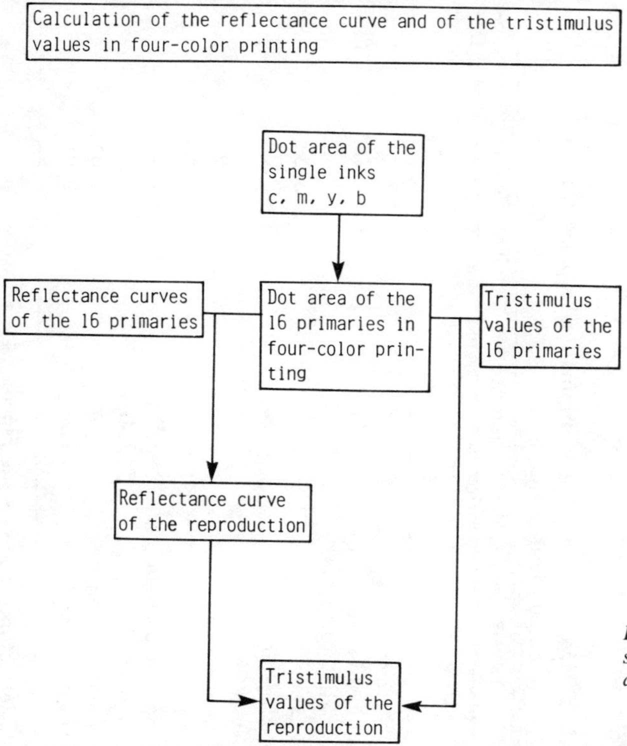

Fig. 12. *Calculation of the reflectance curve and of the tristimulus values in four-color printing.*

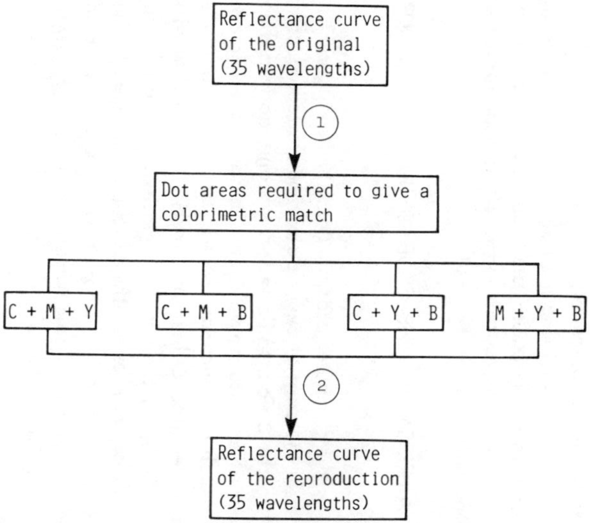

Fig. 13. *Calculation of the colorimetrically identical reproduction in four-color printing specified as reflectance curve. (1) Iterative solution of the modified Neugebauer equations as described by Pobboravsky in TAGA Proc. 1972, (2) iterative correction of the dot areas to compensate for the errors of the Neugebauer method.*

are much smaller and usually do not exceed ± 1% in terms of dot area. For the calculation of the dot areas required to match a specified original colour an approach is used combining the colourimetric and the spectral calculation according to Neugebauer (see Figure 13). For fundamental studies of the reproduction process this model offers wide facilities as it is very accurate and based on spectral values.

DEGREE OF METAMERISM IN THE CASE OF EXACT COLOUR REPRODUCTION

As an application of the computer program described in Figure 13 the degree of metamerism in the case of exact colourimetric reproduction has been investigated. The colour to be tested was entered into the computer in terms of spectral reflectance values. The program then calculated the colourimetric match for illuminant D_{65} in terms of reflectance values, and this for all possible three-colour combinations obtainable with the four inks. In the next step, the tristimulus values for the illuminant A were calculated. As a final result, the colour difference between illuminant A and D_{65} specified in CIELAB units was computed. Examples of reflectance curves calculated as colourimetric matches are shown in Figures 14 and 15 for the DIN test, colours No. 1 (flesh tone) and No. 16 (medium grey).

In Table 5, as a measure for the degree of metamerism, the colour difference between illuminant D_{65} and A is shown in terms of CIELAB units for 12 DIN test colours.

TABLE 5 Metamerism between original and reproduction in three-colour halftone printing

Test colour DIN 6169/5	Colour difference for illuminant A when the reproduction is a colourimetric match for illuminant D65	
	C + M + Y	Black + 2 colours
(1) Flesh tone*	4.06	2.33
(2) Light brown	4.69	1.40
(3) Yellow-green		4.43
(4) Green	3.91	3.38
(5) Blue-green	4.36	3.07
(6) Light blue	1.98	
(8) Red-purple		0.76
(9) Saturated yellow	3.09	2.74
(11) Saturated green	2.99	2.40
(13) Skin tone	2.47	
(14) Foliage green		3.65
(16) Medium grey*	5.30	0.91

*Reflectance curves - see Figures 14 - 17

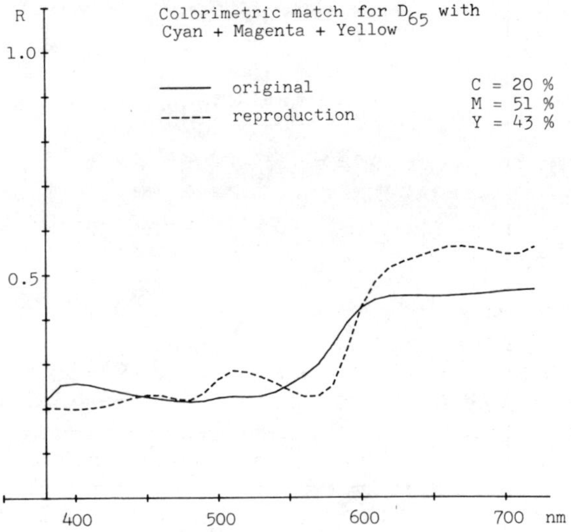

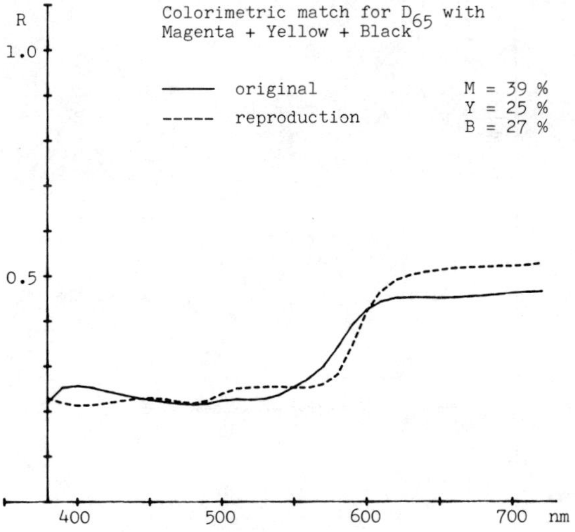

Fig. 14. DIN test color No. 1 (flesh tone).

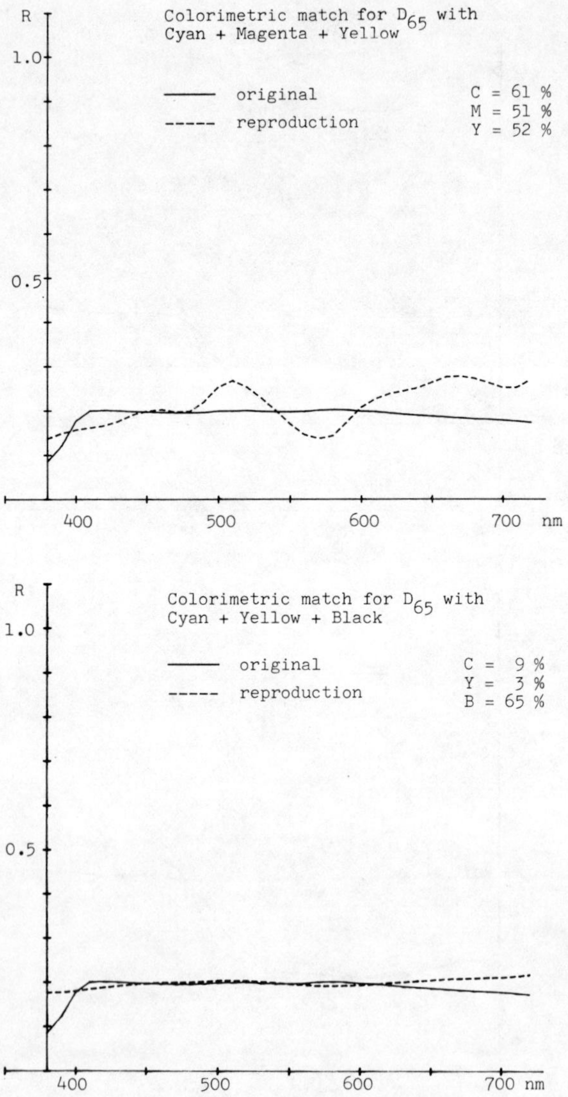

Fig. 15. DIN test color No. 16 (medium grey).

Table 5 and Figures 14 and 15 also permit comparison of the two different concepts of reproduction, i.e. with three colours and with two colours plus black. It is interesting to note that the reproduction with two colours plus black is considerably less sensitive to metamerism than the reproduction with three colours. This is a further argument for choosing a reproduction with only two colours plus black, i.e. for total undercolour removal.

REASONS FOR TOTAL UCR

In the last 25 years, the concept of total UCR has been proposed by many authors and recently it was even announced as a new colour space (Bruno[1]). Although UCR is not directly associated with colour separation, it has an indirect impact on colour separation as electronic scanners are better suited for this concept. To realize a total UCR with photographic separation techniques, it is necessary to apply special masks. This usually creates problems as the black printer is not colour-corrected and hence may contaminate the pure colours. On colour scanners, however, the concept of total UCR is feasible, although no commercial scanner is currently able to realize full UCR.

TABLE 6 Dot area to achieve an exact colourimetric reproduction with three colours and with two colours + black

No:	DIN test colour		C + M + Y	Black + 2 colours	Difference
1	Flesh tone	C	20	0	-25
		M	51	39	-12
		Y	43	25	-18
		B	0	27	
2	Light brown	C	38	0	-38
		M	40	14	-26
		Y	78	55	-23
		B	0	42	
4	Green	C	70	67	- 3
		M	7	0	- 7
		Y	76	73	- 3
		B	0	8	
5	Blue-green	C	64	56	- 8
		M	16	0	-16
		Y	34	24	-10
		B	0	19	
11	Saturated green	C	91	89	- 2
		M	12	0	-12
		Y	92	89	- 3
		B	0	11	
16	Medium grey	C	61	9	-52
		M	51	0	-51
		Y	52	3	-49
		B	0	65	

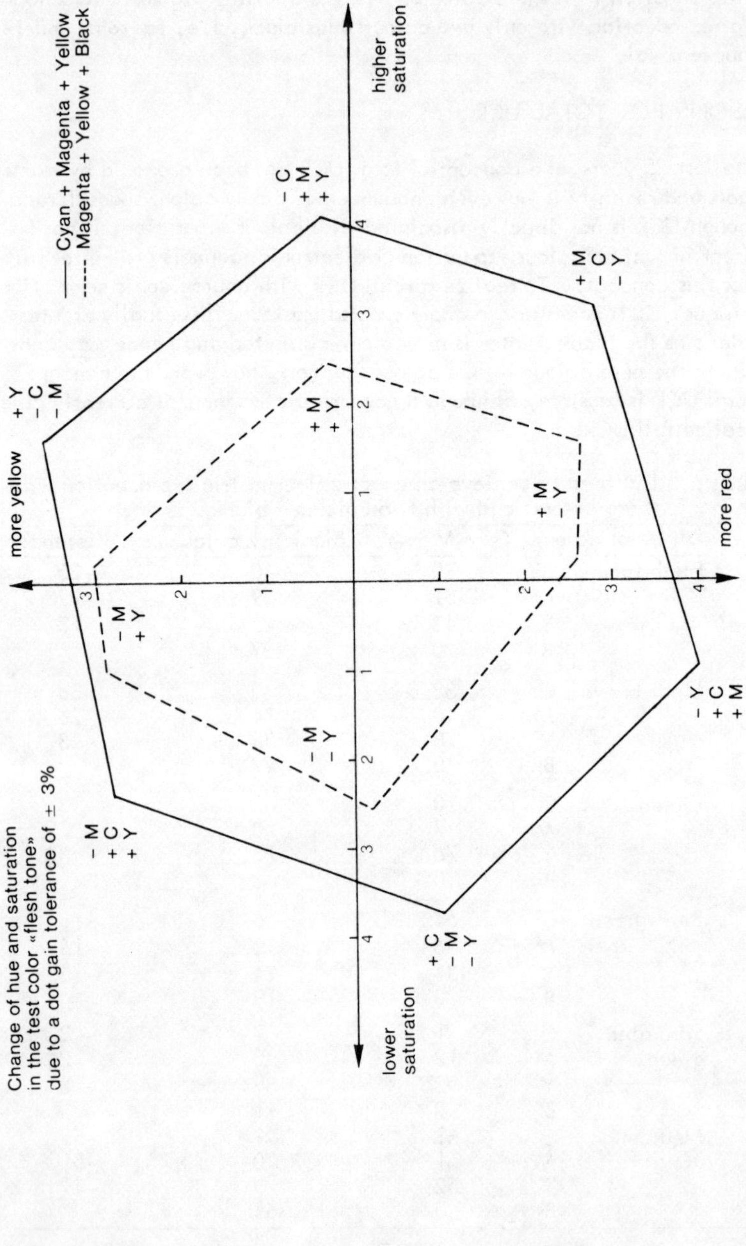
Change of hue and saturation
in the test color «flesh tone»
due to a dot gain tolerance of ± 3%
Cyan + Magenta + Yellow
Magenta + Yellow + Black
more yellow
more red
higher saturation
lower saturation
+ Y − C − M
− C + M + Y
+ M − C − Y
− Y + C + M
+ C − M − Y
− M + C + Y
+ M + Y
+ M − Y
− M − Y
− M + Y
1
2
3
4

Fig. 16.

Two advantages of total UCR are quantitatively illustrated in Table 6 and Figure 16:

(1) Table 6 shows the percentage of coloured inks which can be saved when the third colour is replaced by black.

(2) Figure 16 compares the change in hue and saturation within a specified dot gain tolerance, when no UCR and total UCR is applied.

In some cases the concept of total UCR may also create problems, the best known being the loss of density in the shadows. When nearly achromatic colours have to be reproduced, a drop-out of halftone dots may occur in the coloured inks (Scheuter[8]). All aspects considered, the merits of the concept of total UCR outweigh by far the disadvantages.

REFERENCES

1. M. Bruno, 'Brouwer Color Space', What's New(s) in Graphic Communications, No. 29 (1980).
2. W. T. Hanson, Jr. and W. L. Brewer, 'Subtractive Color Photography: The Role of Masks', J. Opt. Soc. Am., 44 (1954).
3. W. T. Hanson, Jr. and W. L. Brewer, 'Subtractive Color Photography: Spectral Sensitivities and Masks', J. Opt. Soc. Am., 45 (1955).
4. R. W. G. Hunt, 'The Reproduction of Colour', in The Reproduction of Colour in Photography, Printing and Television, Fountain Press (1975).
5. N. I. Korman and J. A. C. Yule, 'Digital computation of dot areas in a colour scanner', Proceedings 11th IARIGAI Conference (1971).
6. P. Kowaliski, 'Colour Reproduction Theory and Separation Filters', J. Phot. Sci., 12 (1964).
7. I. Pobbarovsky and M. Pearson, 'Computation of dot areas required to match a colorimetrically specified color using the modified Neugebauer equations', TAGA Proceedings (1972).
8. K. R. Scheuter, 'Etwas zum Farbendruck', Mitteilungen Forschungsgesellschaft Druckmaschinen, January (1978).
9. J. A. C. Yule, 'Principles of Color Reproduction', John Wiley & Sons (1967).
10. R. D. Warner, 'A three colour inter-image reflectance model', TAGA Proceedings (1980).

DISCUSSION

M. B. Lyne You have used the CIE tristumulus colour space. Is it by preference that you have chosen this colour space over the internationally accepted CIE LAB or LUV colour space which corresponds better to human perception?

K. Schläpfer To apply the Neugebauer equations we cannot but use the CIE tristimulus values. However, to assess colour differences, we also use the CIELAB colour space.

W. D. Schaeffer

(1) Are 35 point calculations (wavelength) essential for the calculation of your reflectance curves? For many years we have used 16 wavelengths.

(2) The degree of metamerism is sometimes related to the difference in areas between reflectance curves, e.g. Figures 14 and 15, or the number of times the curves cross. Do these measures agree with your values? The two selections of Figure 14 appear to involve the same number of contacts or crossings of the reflectance curves.

(3) Black is black, but what characterizes the black print or ink that you use? Is the black a blue-tone or another tone, since this frequently contributes to the subjective evaluation of blackness and neutrality?

K. Schläpfer

(1) 35 wavelengths give undoubtedly a better accuracy. Moreover, 35 wavelengths are calculated quicker on modern computers than were 16 wavelengths in former days. Therefore, there is no reason to go back to 16 wavelengths.

(2) There are indeed different methods of assessing metamerism. I think the one we used (difference between viewing at illuminant D65 and A) is more widely accepted. Moreover, it is directly related to visual perception.

(3) We have not investigated different types of black. Our black is defined by the spectral reflectance curve obtained on the chosen paper. This can be seen from our calculations made to match a neutral grey in the original; our black is indeed not neutral. It is, however, a very common black.

M. H. Bruno In the Brouwer colour space any black is used and the amounts of the other two colours are varied to compensate for the spectral characteristics of the black. The r, g, b values of the black are incorporated in the program and are used to correct the amounts of the other two colours.

LIMITS OF MICROSCOPIC PRINT QUALITY

P. Oittinen, The Finnish Pulp and Paper Research Institute, Helsinki, Finland.

ABSTRACT

The microscopic quality of printed halftone dots has been analysed by application of a system-analytical approach, with output modulation and signal-to-noise ratios being employed as the main quality criteria. The influence that solid density, gloss, optical and physical spreading phenomena and the noise of structural origin exercise on these are assessed with the aim of finding the weakest links.

Under the conditions of newspaper printing, it was found that physical ink behaviour and interactions with the paper structure are more important than optical effects such as light diffusion in the paper, and the lack of specular reflection of the print, which increases in significance when a high quality level is sought.

INTRODUCTION

It has proved feasible to apply a system-analytical approach in the study of image reproduction in printing (Saarelma and Oittinen[11], Oittinen and Saarelma[7], Saarelma[12]). This approach constitutes a practical method of analysis, which allows the influences of various factors that contribute to system behaviour to be investigated, and quantified in uniform terms. Such quantitative information is necessary for the assessment of system performance with a view to process control, along with process and material development.

Print quality is one aspect of the performance of the printing reproduction process. The way in which print quality is best defined is decided by the purpose for which information on print quality is being compiled. Essentially, the definition may be concerned with:

(1) the ratio of the output to input images,
(2) the difference between the output and input images,
(3) the output.

In the first of these cases, the criteria of quality primarily express the performance of the information transfer channel, i.e. the printing process, irrespective of the characteristics of the input image. This particularly applies if the process is linear in the system-analysis sense of the word. The quality criteria which quantify the difference between the output and the input images lay emphasis upon process fidelity (Pratt[10]). Such figures of quality depend upon the distribution of the signal power with different frequencies, i.e. the magnitude of small details in the image. This is because image transfer systems tend to attenuate the contrast of small details (Dainty and Shaw[2], Biberman[1]) as a consequence of spreading phenomena. One implication is that the overall deterioration becomes related to the magnitude of small details in the image. The quality criteria which are concerned solely with the output image are appropriate in the exploration of factors connected with the materials which form the image, the ink and the paper, or the structure of the image.

In principle, a print-quality criterion may express an average image property, or the relation of the average behaviour and its variations, the noise. If an increase occurs in the noise power relative to the signal power, increasing importance is attached to the inclusion of noise in the considerations of quality.

This paper is concerned with the output image, the print, and its quality on the microscopic scale. The image is characterized in terms of the continuous image signal, as defined by the halftone dot profiles. The aim is that of seeking the limits which restrict halftone dot reproduction on paper. The analysis serves as an application and extension of work published previously (Saarelma and Oittinen[11, 12]).

Halftone dots, however, do not carry any information on the object if they are not considered in relation to the size of the unit element of the screen, the screen cell. This is one way of stating that halftone printing is area-modulated. In system-analytical terminology, conversion of the microscopic halftone dot profiles into macroscopic halftone cell profiles is by its nature a filtering operation, in which the screen defines the filter function. It is thus obvious that the microscopic image properties set the ultimate limit to the image properties and print quality which are achievable. As a consequence, their analysis is necessary when the fundamental limits of printing are being sought. Knowledge of these limits allows location of the weakest link in the process; if the greatest advantage is to be gained, process and material development should be concentrated on the weakest link.

METHODS

Even if the optical print quality criteria were limited to the output image, this does not lead to a single definition of the quality. All the system-analytical criteria, however, are based upon a few variables, viz. the density, the noise, and a geometrical attribute which expresses the smallest detail that the process can record.

It should be noted that only indirectly is gloss a factor of system-analytical print quality. This is because in printing, variations in the object gloss are not transferable to the image, owing to the non-discriminating gloss level of the print. The function of gloss is that of influencing the formation of density in the form of a cut-off level, and the creation of variations in density. Print-through is not a print-quality factor, but forms a component of the noise on the reverse side of the point.

TABLE 1 OUTPUT IMAGE PROPERTIES

Property		Definition	Limits considered in this paper
Signal	D	Solid density	Gloss
	$D_{max}(f_s)$	Maximum density of halftone dot	
	$D_{min}(f_s)$	Minimum density of halftone dot	
	f_s	Screen frequency	
	M_{out}	$\frac{D_{max} - D_{min}}{D_{max} + D_{min}}$	Solid density, Optical spread, Physical spread
Noise	σ_D	RMS value of density variations	
	σ_θ	RMS value of the variations in relative halftone area	Paper structure
	D_{max}/D_{min}	Average potential of halftone dots to carry area modulation	
Signal-to-noise ratio	$(D_{max} - D_{min})/\sigma_D$	Potential of halftone dots to carry area modulation in the presence of density variations	
	θ/σ_θ	Potential of halftone dots to carry area modulation in the presence of variations in the relative halftone area	

The measures of print quality discussed in this paper are listed in Table 1. In an earlier study (Oittinen and Saarelma[12]), it was observed that the maximum density of halftone dots is not equal to the density of the solid area printed under the same conditions; moreover the minimum density was not zero. Instead, under the conditions of newspaper printing, halftone dot density formation developed to a large extent in agreement with the predictions of the linear image transfer theory. Thus, in view of the information-carrying capacity of halftone dots, it becomes of the utmost importance how the maximum and minimum halftone dot densities are related. In particular, dark halftones are critical, as their spreading most easily leads to a complete loss in carrying area modulation. Output modulation (Table 1), calculated for a given halftone percentage, is a measure of the potential of that tone in the preservation of its information-carrying capacity. This is discussed in the following in respect of a dark halftone as a function of the screen frequency.

Ultimately, however, the reproduction of halftone dots is governed by noise. The root mean square variations of the relative halftone area (θ), and of the maximum dot density (D_{max}) at each screen frequency, are used as measures of the noise. Its relative magnitude is evaluated by means of the ratios θ/σ_θ and $(D_{max} - D_{min})/\sigma_D$; these ratios express whether a halftone area reproduces any area modulation at a given screen frequency.

The influence of gloss on solid density has been investigated by the measurement of density from laboratory-scale prints, by means of the Elrepho photometer, with and without the gloss trap. The difference in measured reflectance, scaled into density, gives an expression of the influence upon density of the lack of gloss. Print gloss from the same samples was measured on the Hunter glossmeter.

To study the influence of the flow phenomena of the ink upon print gloss, prints were made with varnishes characterized by similar gloss levels. The viscosities were measured on the Ferranti-Shirley cone-and-plate viscometer.

RESULTS AND DISCUSSION

Solid density

In a linear density transfer process, the significance of the solid density with respect to the maximum density of the halftone dots is straightforward; the maximum dot density is formed as a product of the solid density, and a function which is defined by the halftone dot size, the screen frequency, and the spread function of the system. As a first approximation, it can be assumed that an image transfer process is linear. Process nonlinearity modifies the dependence of the maximum dot density upon the solid density, but in general it is true that a high solid density is a prerequisite for a high output modulation.

From a qualitative aspect, the factors which contribute to the solid density are well known. The nature of the limiting factor depends upon the level of the density achieved. This arises from the differences in type of influence exerted by the contributory factors. The absorption of light in the ink layer varies linearly with the layer thickness, while light scattering and the lack of specular surface reflectance act as cut-off levels of the density. These two different types of influence have been included in a density model for a solid printed area. In its basic form, the model can be given as (Tollenaar and Ernst [13]):

$$D = -\log_{10}\left[10^{-D_\infty} + (1 - 10^{-D_\infty})\, e^{-m(g_o)g_o}\right] \qquad (1)$$

Parameter D_∞ denotes the limiting cut-off density, and $m(g_o)$ the optical absorption efficiency (including the effect of light scattering) of an ink film of thickness g_o on paper. Under conditions in which the ink film thickness is inefficient in the production of density, the second additive term in Expression (1) largely controls the formation of density. With a rise in the level of density achieved, the cut-off level begins to play a significant role; to a large extent, this is controlled by the roughness of the top surface of the print, in other words, by a lack of gloss (Oittinen [8]). According to measurements (Oittinen [9]), the cut-off level set by light scattering in the ink is considerably higher than that set by deficient gloss. It can be concluded that, with a rise in the level of the solid density, the factor which becomes limiting for the solid density moves away from the pore structure of the paper, through the ink film and towards the top of the print surface.

An experimental exploration has been made of the quantitative influence of a given level of gloss upon the measured density. Figure 1 provides examples of the improvement in density resulting from increasing specular surface reflectance, as indicated by the gloss value. The significance of gloss is more marked at a higher level of density, consistent with the considerations given above. The dependence of the relations of Figure 1 upon the combination of ink and paper is attributable to a difference in the measuring conditions (detection and aperture angles) of the densitometer and the glossmeter employed.

Similar to the formation of the solid density, the factor which limits print gloss depends less on the paper with a rise in the level of gloss. This has been borne out by experiments in which the gloss is registered with a variation in the viscosity (Figure 2). The curves are characterized by an initial rise - interpreted as a diminution in the absorption of the ink into the paper - followed, on medium-glossy coated papers, by a maximum and a descending trend; the latter evidently arises from less flow-out of the ink, with an increase in the viscosity; a rougher print surface is obtained. The level of gloss in the region of the maximum is largely determined by the level of the surface smoothness of the paper. On uncoated paper, a continuously rising curve shape is prevalent at all viscosity levels. This provides an indication of the greater surface pore volume in comparison with coated papers. Figure 3 provides a schematic illustration of the type of behaviour generally observed.

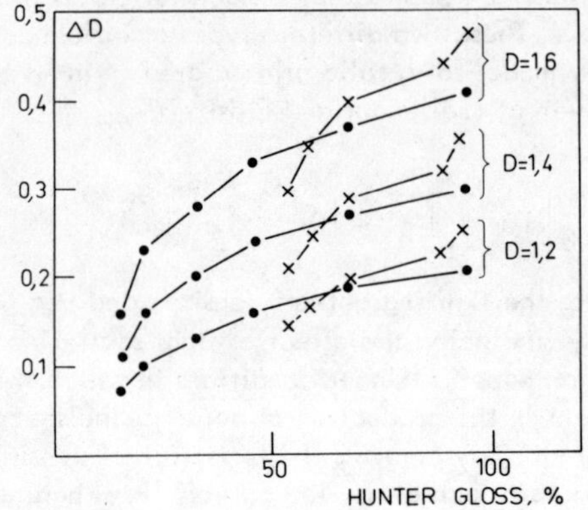

Fig. 1.

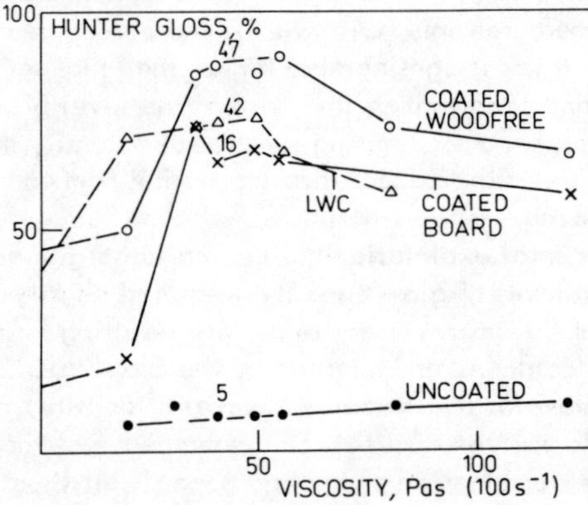

Fig. 2.

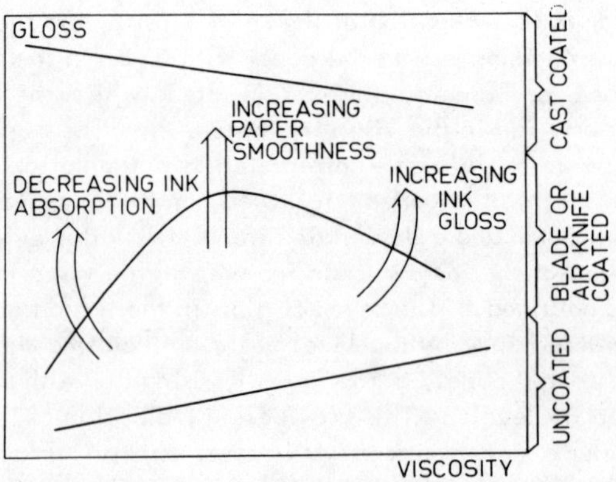

Fig. 3.

Halftone dot density

The density formation of halftone dots differs from that of solid areas by reason of optical and physical spreading phenomena which occur in the x, y-plane of the paper. On solid areas, optical light-spreading does not play any part, while the significance of the physical ink spreading is restricted to its effect on the coverage of the paper by the ink. This also is negligible when the roughness volume of the paper under compression is small in comparison with the thickness of ink film applied in printing.

The spreading phenomena have been included in a density model (Oittinen and Saarelma[7]) for the microscopic density of halftone dots. An extended version of the model is presented in Figure 4, with incorporation of the density cut-off level, and the influence of the physical absorption of the ink in the z-direction of the paper. The various sources of noise have not been included.

The paper constitutes the primary source of optical spreading in a print. Scattering of light in the paper structure makes it diffuse, from the point of the incidence, into the paper structure, in both the x, y-plane and the z-direction. On the assumption of isotropic light scattering, an optical spread function in the x, y-plane can be derived from the Kubelka-Munk theory. Differentiation of the expression for the reflectance of a layer of thickness x, measured against a black background, with respect to x, expresses the fraction of reflected light which travels a distance between x and x+dx from the point of incidence. Normalized in an appropriate manner, this differential is equivalent to the definition of a spread function $h_o(x)$ (Dainty and Shaw[2]):

$$h_o(x) = \frac{1}{R_\infty} \frac{dR_o(x)}{dx} \tag{2}$$

The expression obtained

$$h_o(x) = \frac{2bs\,(1 - R_\infty^2)\,e^{-2bsx}}{(1 - R_\infty^2 e^{-2bsx})^2} \tag{3}$$

can be approximated by means of an exponential spread function,

$$h_o = \frac{1}{2w_o}\, e^{-\left|\frac{x}{w_o}\right|} \tag{4}$$

Curves calculated from (3) and (4) are shown in Figure 5, with values for the light absorption and scattering coefficients of the paper typical of newsprint quality. The effective distance of light diffusion, which is equivalent to the mean of the exponential spread function, is of the order of 20 μm. This is of the same order as that determined experimentally (Yule et al.[16]). In principle, an increase in the scattering coefficient brings about a diminution in the distance of diffusion, whereas an

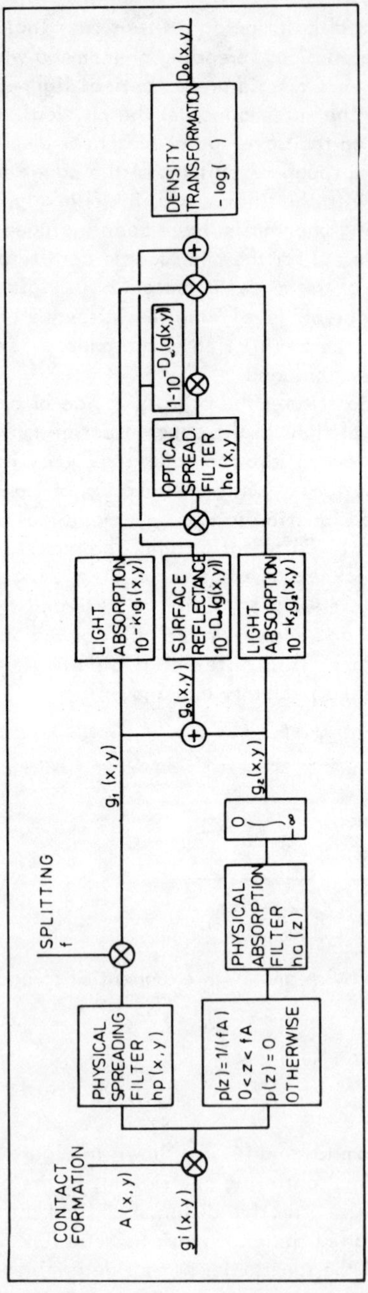
CONTACT FORMATION
A(x,y)
gi(x,y)
PHYSICAL SPREADING FILTER hp(x,y)
SPLITTING f
p(z)=1/(fA) 0<z<fA p(z)=0 OTHERWISE
PHYSICAL ABSORPTION FILTER ha(z)
g₁(x,y)
g₂(x,y)
g₀(x,y)
LIGHT ABSORPTION 10-k₁g₁(x,y)
SURFACE REFLECTANCE 10-D∞(g(x,y))
LIGHT ABSORPTION 10-k₂g₂(x,y)
OPTICAL SPREAD. FILTER h₀(x,y)
1-10-D∞(g(x,y))
DENSITY TRANSFORMATION -log( )
D₀(x,y)

Fig. 4.

increase in paper reflectance acts in the opposite direction.

Two types of physical spreading are distinguishable, i.e. spreading of the ink, and spreading of the oil from the ink. Both these spreading phenomena influence the halftone dot density profile through light absorption in the ink, that is, a linear physical spreading implies a linear density spreading in the absence of other factors contributing to the density.

Physical ink spreading can be approached as a starting point by means of a differential equation for flow between parallel plates (Dienes and Klemm[3]). Integration gives a formula which relates the increase in halftone dot diameter (Δa) to the ink viscosity (η), the applied pressure (p), the ink film thickness (g_i), and the time of application of the pressure (t):

$$\Delta a = \sqrt{\left(a^2 + \frac{8}{3}\,\frac{p g_i^{\,2} t}{\eta}\right)} - a \qquad (5)$$

Other starting points lead to slightly different forms (Mill et al.[6]). Expression (5) is comparable with the prediction of the linear image transfer theory; such a comparison is illustrated in Figure 6. Calculation of the increase in the dot diameter in a linear transfer system is based upon an exponential physical spread function (Expression (4)). The forms of the two curves are essentially the same. This demonstrates that under conditions in which dot spreading is controlled by viscosity, the phenomena can be described by the linear theory. On the other hand, spreading by viscous flow is inevitable in the nip, by reason of the application of pressure. As a consequence, physical ink spreading can constitute a limit to the print quality achievable.

Nevertheless, paper roughness can act as a physical constraint to spreading, through imperfect contact between the ink layer and the print. This is discussed later as a source of noise. Physical constraints also operate in the spreading of oil from the ink, as oil spreading depends on the surface pore volume. In essence, the pore volume functions in a similar way to a cut-off level, and can give rise to non-linearities in the density transfer, in the form of enhanced density for fine screens. The results of this effect have been observed under newspaper printing conditions (Saarelma[12])..

The operation of optical spreading in terms of reflectance, and of physical spreading in terms of density, implies in both cases a given magnitude of spread exerts a markedly different influence upon output modulation. For 75% halftone dots, Figure 7 illustrates that physical spreading is more detrimental than optical spreading. An optical spread of 20 μm approximately corresponds to a physical spread of about 5 μm. The difference lies mainly in the value of the maximum dot density, in which physical spreading brings about a pronounced diminution, while the influence of optical spreading is minor (Figure 8), or the reverse.

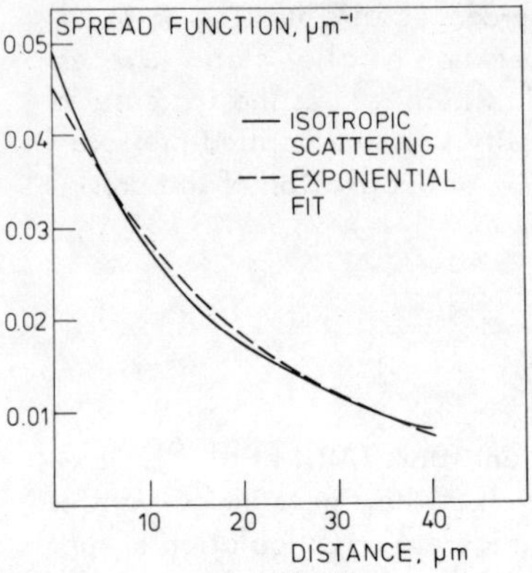

Fig. 5.

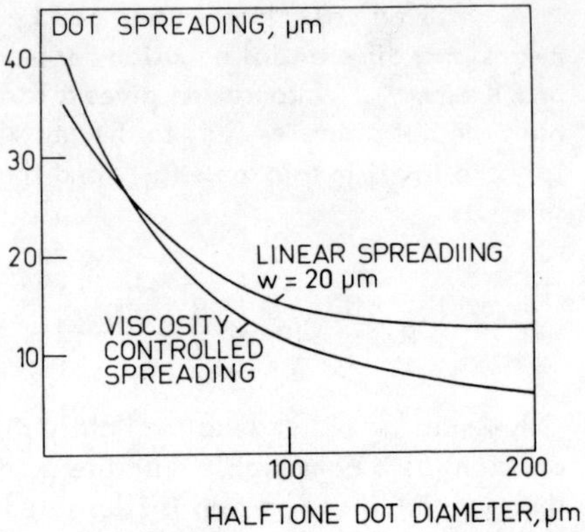

Fig. 6.

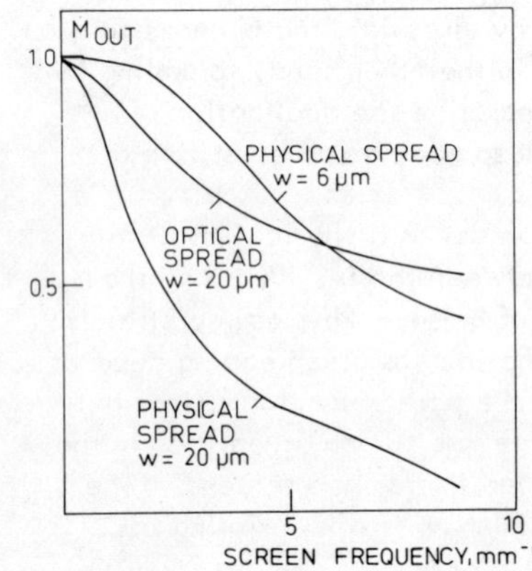

Fig. 7.

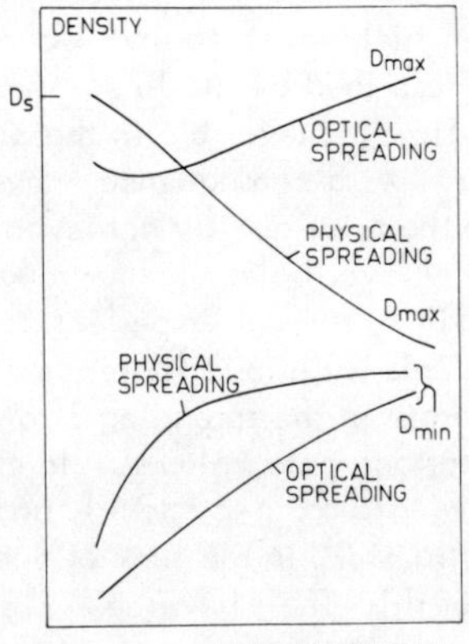

Fig. 8.

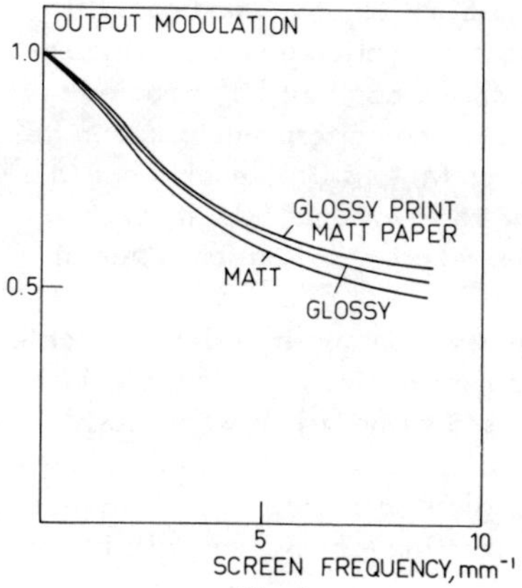

Fig. 9.

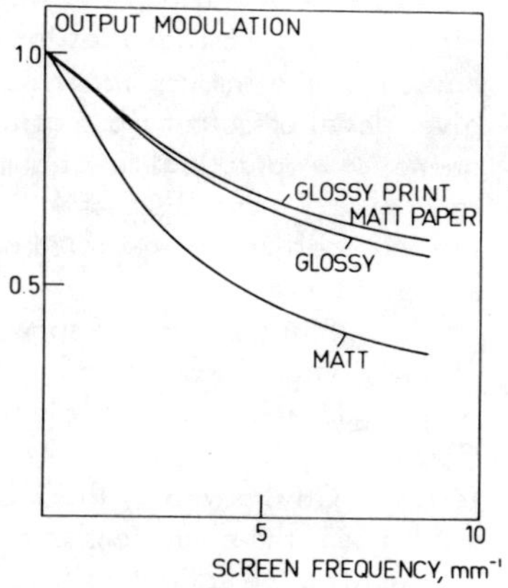

Fig. 10.

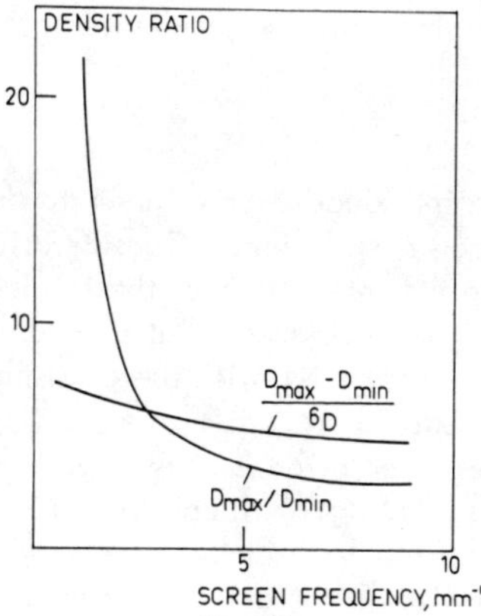

Fig. 11.

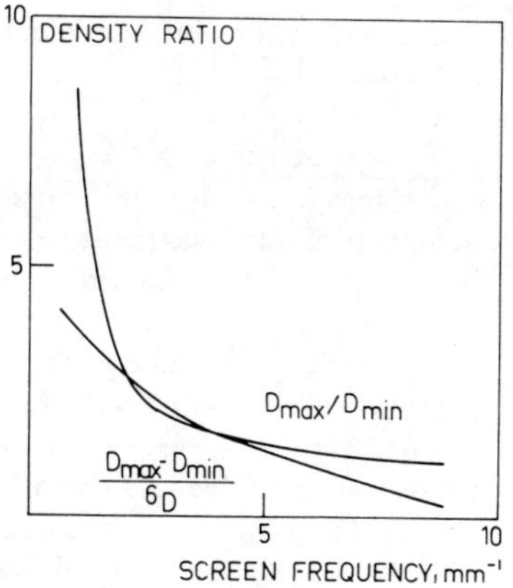

Fig. 12.

The magnitude of spread, assessed in terms of linear density transfer, under newspaper printing conditions (Oittinen and Saarelma[7]), was found to be in excess of 50 μm. Quite obviously, physical spreading restricts output modulation in newspaper printing, rather than optical spreading.

Figures 9 and 10 reproduce curves of output modulation at different levels of solid density, and print and paper gloss. The solid density level in Figure 9 corresponds to the conditions of newspaper printing (Lindqvist and Oittinen [5]), while the combination of ink and paper in Figure 10 is almost ideal as far as the light absorption and scattering behaviour of the ink are concerned. The curves can best be compared at a given level of output modulation in terms of the corresponding screen frequency. A practical level of reference of M_{out} is 0.6, equivalent to a ratio of D_{max} and D_{min} of 4. When the ratio exceeds this limit, the halftone cell can be regarded as reproducing area modulation in the absence of noise.

Figures 9 and 10 show that a glossy high quality print is capable of carrying area modulation up to screen frequencies which are considerably higher than that possible with a newspaper print if it were glossy. Conversely, if the high quality print is matt, its output modulation is inferior. On the whole, the influence of gloss upon output modulation is small under the conditions of newspaper printing - consistent with its role in regard to the solid density.

Figures 9 and 10 further illustrate that the best output modulation is achieved if the print is glossy, but made on a matt paper. A matt surface combats the rise in minimum density, and in this way enhances output modulation.

Influence of noise

Variations in density and halftone dot size introduce uncertainty into the potential of dark halftones to reproduce area modulation. Consideration is given below to noise which originates in the interaction of the ink layer with the paper structure. Nevertheless, it is well known that pronounced variations can also be caused by interaction of the ink with the printing process. Splitting patterns fall in this category.

Noise generated by ink and paper interaction is of two types. Paper roughness induces variations in dot spreading, with ensuing variation in dot size, while the roughness and pore volume of the paper are sources of variation in the thickness of ink film in the dots. The hypothesis can be made that the variation in dot size is more marked when the roughness profile of the paper is deep, on condition that the unevennesses in the plane of the paper surface are also large in size. This is equivalent to the assumption that the uncertainty in halftone dot diameter is the same as the typical roughness dimension of the paper, defined as a geometrical average:

$$\sigma_a = \sqrt{(\sigma_p r_p)} \qquad (6)$$

in which σ_p denotes the root mean square roughness of the paper, and r_p the average unevennesses. The relation between σ_a and the variation in the relative halftone area is given approximately by:

$$\sigma_\theta = \sqrt{\left(\frac{d\theta}{da}\right)}\ \sigma_a \tag{7}$$

For square dots, the signal-to-noise ratio of the relative halftone area becomes:

$$\frac{\theta}{\sigma_\theta} = \frac{\sqrt{\theta}}{2f_s\sqrt{(2\sigma_p r_p)}} \tag{8}$$

in which f_s denotes the screen frequency. According to Equation (8), the signal-to-noise ratio of the relative halftone area is inversely proportional to the screen frequency. If typical roughness values for newsprint are used ($\sigma_p = 4\,\mu m$, $r_p = 40\,\mu m$), it appears that the variation in dot size in a 75% tone is a limiting factor in reproduction at screen frequencies above 5 mm^{-1}.

The pore structure of paper can be regarded as being formed by a random deposition of fibres (Haglund et al. [4]), in which the position of the fibre centres follow Poisson statistics. Under such conditions, the distribution of the size of the gaps between the fibres, or the pigment particles, is exponential - according to the so-termed renewal theory (Trabka [14]). An implication is that the mean of the gap width is similar in magnitude to its root mean square variation. The constituents of paper have a micrometer scale size. As the size of the halftone dots and the spreading phenomena occur on the same scale, the variation which is induced by the structure becomes transformed into variations in the dot size without any filtration.

The theoretical reasoning which leads to the approximate result of the equal magnitudes of a structural variable and its variation is supported experimentally by printing results which relate the ink film thickness and the absorption of the ink into the pore structure. This phenomenon is reasonably well described by an exponential function (Walker and Fetsko [15]).

$$B = 1 - e^{-g_i/b} \tag{9}$$

Function B can be interpreted as the cumulative surface pore volume distribution expressed in terms of the ink film thickness (g_i) which interacts with the pore structure. The mean surface pore volume is given by b, while its variation also proves to be b. An assumption of process ergodicity allows of the deduction that at a given level of ink film thickness, the variation equals the ink film thickness which interacts with the surface pore volume. When the surface pore volume is small, the fraction of the ink film thickness which interacts and becomes absorbed is small, and consequently the variation remains low.

An expression of the type of Equation (7) can be applied for transformation of a variation in ink film thickness into a variation in density. The necessary differential dD/dg_i is obtained from a density model such as Equation (1). It is obvious that density is more sensitive to variations in the ink film thickness when the level of density markedly deviates from the cut-off level.

The significance of density variations to microscopic halftone dot quality can be estimated by comparison of the area modulation capacity in the absence, and in the presence of noise (cf. Table 1). If the area modulation capacity of dark halftones is controlled by spreading phenomena, the magnitude of D_{max}/D_{min} at a level of 4 corresponds to a smaller screen frequency; if $(D_{max} - D_{min})/\sigma_D$ is smaller, noise is the restricting factor.

Figures 11 and 12 reproduce curves calculated by means of typical newspaper printing values for the ink film thickness and density. The plot of Figure 11 suggests that optical spreading and noise induced by paper structure are reasonably well balanced, whereas the deterioration caused by structural noise becomes easily overwhelming with an increase in physical spreading. This is because physical spreading induces a rise in minimum dot density even with coarse screen frequencies (cf. Figure 8), whereas in the case of optical spreading the rise is more subtle.

When the quality of printing becomes superior to that of newspaper printing, variations in gloss are also introduced as a source of noise. Simultaneously, a general diminution occurs in the interactions between the ink layer and the paper structure. These emergent factors prevent generalized conclusions in regard to the significance of noise to system behaviour.

CONCLUSIONS

In this paper, the concept of microscopic density formation as an image transfer system has been applied to discuss some fundamental limits of print quality. The emphasis has been laid upon principles, although it is maintained that the criteria used are applicable when specific systems are assessed with respect to output quality. In particular, the approach appears to be well suited for material studies.

The result of these considerations is the crucial role played by print gloss in the search for a high optical quality level of print. However, gloss as a separate factor is not a means of achieving better microscopic quality in newspaper printing conditions, since other factors, such as physical ink spreading in the x, y-plane of the paper, and the interactions between the ink and the paper, with consequent variations in dot size and density, evidently set the quality level lower than does the lack of gloss. Conversely, optical spreading in the paper does not seem to constitute a critical factor.

Continuation of the work along the lines presented involves a more detailed analysis of signal and noise, and the analysis of criteria of macroscopic quality.

REFERENCES

1. Biberman, L. M. (Ed.) (1973), Perception of Visual Information, Plenum Press, New York.

2. Dainty, J. C. and Shaw, R. (1974), Image Science, Academic Press, London.

3. Dienes, G. H. and Klemm, H. F. (1946), 'Theory and Application of the Parallel Plate Plastometer', J. Appl. Physics 17: 458.

4. Haglund, L., Norman, B. and Wahren, D. (1974), 'Mass Distribution in Random Sheets - Theoretical Evaluation and Comparison with Real Sheets', Svensk Papperstidn, 77. 10:362-370.

5. Lindqvist, U. and Oittinen, P. (1980), A Study of Black Offset Newsinks, The Technical Research Centre of Finland, Otaniemi: (in Swedish).

6. Mill, C. C., Tritton, K. T., Birkenshaw, J. W. and Johnson, A. J. (1980), Improving the Correlation between a Litho Proof and a Production Print, PIRA Report PR 3(R)/1980.

7. Oittinen, P. and Saarelma, H. (1981), 'Density Formation of Halftone Dots', Tappi, 64:6.

8. Oittinen, P. (1980), 'Surface Reflection of Coated Papers and Prints', Advances in Printing Science and Technology, Vol. 15 (Ed. W. H. Banks), Pentech Press, London.

9. Oittinen, P. (Unpublished results).

10. Pratt, W. K. (1978), Digital Image Processing, Wiley-Interscience, New York.

11. Saarelma, H. and Oittinen, P. (1979), 'Introduction to Image Transfer and Picture Processing in Halftone Reproduction', Graphic Arts in Finland, 8. 2:9-17.

12. Saarelma, H. (1981), Colour Reproduction with Unconventional Screens, Advances in Printing Science and Technology, Vol. 16 (Ed. W. H. Banks), Pentech Press, London.

13. Tollenaar, D. and Ernst, P. A. H. (1971), 'Uneven Ink Transfer on Smooth Surfaces', Advances in Printing Science and Technology, Vol. 6 (Ed. W. H. Banks), Pergamon Press, London.

14. Trabka, E. A. (1971), 'Crowded Emulsions: Granularity Theory for Monolayers', J. Opt. Soc. Am. 61. 6:800-810.

15. Walker, W. C. and Fetsko, J. M. (1955) 'A Concept of Ink Transfer in Printing', Am. Ink Maker 33. 12:38-40, 42, 44, 69, 71.

16. Yule, J. A. C., Howe, D. J. and Altman, J. H. (1967),

'The Effect of the Spread-Function of Paper on Halftone Reproduction', Tappi 50. 7:337.

SYMBOLS

a	Halftone dot diameter
b	$\frac{1}{2}(R_\infty - R_\infty)$ (Expression (3)) mean surface pore volume (Expression (9))
f_s	Screen frequency
g_i	Ink film thickness offered for transfer
g_o	Ink film thickness on paper
h	Point spread function
m	Light absorption efficiency of ink on paper
o	As index denotes "optical"
p	Pressure, as index denotes "physical" or "paper"
r_p	Mean roughness width of paper
s	Scattering coefficient of paper
t	Time
w	Effective spread
x, y	Spatial coordinates
B	Cumulative pore volume distribution
D	Density
D_{max}	Maximum density of halftone dot
D_{min}	Minimum density of halftone dot
D_∞	Cut-off level of density
M_{out}	Output modulation (cf. Table 1)
$R_o(x)$	Reflectance of a layer of thickness x of paper
R_∞	Limiting reflectance of an infinitely thick layer of paper
θ	Relative halftone area
η	Viscosity
σ	Root mean square variation

DISCUSSION

L. O. Larsson The investigation presented here is, I think, a fine contribution to the study of an important but to a great extent unexplored field. We must appreciate that someone takes the trouble and makes the efforts. It is to be hoped that investigations of this type will contribute to a better understanding of the mechanisms of the creation of physical print quality. The results clearly show the paramount importance of the properties of the ink in printing on newsprint. If a study of this kind was supplemented by a corresponding investigation of the inks, and their required properties could be specified, it certainly would prove very useful.

In your paper you use the expressions 'roughness of the paper' and 'unevenness in the paper surface'. It is not quite clear if you mean different things or not. In case you do, what is the difference?

P. Oittinen The meaning is the same, but the term 'unevenness' has been used to refer to individual depressions at the paper surface rather than a measured value of roughness.

L. O. Larsson In some of your figures and in your discussion you refer to the curves using the terms 'physical spreading' and 'optical spreading'. Does not 'physical spreading' mean 'ink spreading' or 'effect of ink spreading'? What is your reason for the distinction? Physical as opposed to optical does not seem quite logical.

P. Oittinen It may be an unfortunate choice of terms. The term 'physical spreading' is used for ink spreading.

L. O. Larsson What is meant by 'maximum dot density'? Is it the average or the maximum density within a dot?

P. Oittinen Maximum dot density is calculated as the average density in the middle of the dot, well away from the edge.

L. O. Larsson What kind of measuring techniques have you used to assess the various entities on the microscopic level you are dealing with?

P. Oittinen For the measurements, a reflection microdensitometer is used.

L. O. Larsson You deal here with the print quality concept as a physical entity. How relevant do you think the microscopic print quality - in the terms you have used here - is to perceptual print quality?

P. Oittinen It is my understanding that as the optical quality level of the print rises, the microscopic phenomena gain in significance because the influence of the screen structure becomes less disturbing for perceived optical quality because the screen frequency increases.

M. H. Bruno Isn't a 75% screen tint too sensitive to press effects and difficult to keep open on an offset newspaper press to be a valid measure of print quality? Jorgensen at GATF developed a print quality strip using a 70% line tint at 45° and it was very sensitive to press conditions. It was found that on papers like newsprint a 50% tint was less sensitive to press effects and a more reliable measure of print quality. The screen ruling used was 150 l/in (60 l/cm). What screen ruling did you use in your studies?

P. Oittinen It is irrelevant to my argument whether the relative halftone area in the simulation is 70 or 75%. The conclusions are the same.

J. A. Bristow I gather that Figure 8 refers to 75% dots. Is the pattern the same for dots at other tone levels?

P. Oittinen Yes, the pattern is the same, but the critical limits of area modulation are shifted towards higher screen frequencies.

M. B. Lyne Dr Jordan of our Institute when asked to create an image analysis test for accutance of halftone dots printed on newsprint concluded that the optical spreading of the dots was sufficient to cause the dots to merge, making dot sharpness or accutance irrelevant for newsprint. Do your results support this conclusion?

P. Oittinen Clearly in such cases where the screen structure is visible, the halftone dot profiles have little influence on sharpness.

COLOUR REPRODUCTION WITH UNCONVENTIONAL SCREENS

Hannu Saarelma, Graphic Arts Research Institute, Finland

ABSTRACT

Studies have been made of the fundamental variables of halftone dot structure with regard to tone and colour formation. The essential variables proved to be frequency, i.e. the inverse of dot size and the degree of dot overlap. The inverse dot size exerts an effect through the MTF* of the printing process, while the influence of the degree of dot overlap finds its origin in the non-ideal spectral absorption of coloured printing inks.

INTRODUCTION

This paper reports some of the results obtained in more extensive work concerned with control of a digital reproduction process. Previous results are presented in Oittinen and Saarelma[3], and Saarelma and Oittinen[5].

A digital reproduction system (Figure 1) is controlled by means of mathematical algorithms which may be based solely upon measured relationships between input and output for the particular process to be controlled, or it may have fundamental knowledge of process mechanisms as a background. This fundamental knowledge makes the control model more flexible in application under varying process conditions. As is represented in Figure 1, picture processing should correct pre-reproduction and post-reproduction distortions. Post-reproduction distortions in printing reproduction are discussed below.

In digital reproduction, halftone dot structure deviates from that formed optically. Optical screening produces halftone dot patterns as plane cuts of a two dimensional, nearly sinusoidal surface (Saarelma[4]). The shape of the dot, as well as locations inside the screen cycle, are then related directly and solely to the fractional dot area. For this rea-

* Modulation transfer function

son, the fractional dot area, together with the screen ruling, gives sufficient information for defining the relation between the halftone dot structure and the tone and colour to be formed, when this relation has been established experimentally.

A digitally-produced halftone dot is formed of picture elements, i.e. pixels of finite size and quantity. A screen cycle is defined as a pixel matrix. Inside a matrix, pixels can be located freely and so form different dot structures. Digital screening algorithms have been discussed by Kekolahti[2]. It is obvious that for digitally produced screens, fractional dot area and screen ruling are no longer adequate for defining the pattern.

A more precise characterization of the pattern is derived either by observing the exact location of each pixel inside the matrix, and its influence upon image formation, or by the discovery of additional fundamental parameters, which can be used for both characterization of the pattern and the explanation of image formation.

The number of dot structure combinations for K pixels (one level of fractional dot area) in a matrix consisting of n elements is given by:

$$p = \frac{n(n-1)\ldots(n-k+1)}{k!} \quad (1)$$

in which p is the number of combinations. The total number of dot structure combination in colour reproduction is thus:

$$p = \left[\sum_{k=0}^{n} \frac{n(n-1)\ldots(n-k+1)}{k!}\right]^{t} \quad (2)$$

in which t is the number of inks to be printed.

The large number of combinations (Equations (1) and (2)) indicates that in most cases it is practically impossible to observe the exact location of each pixel. This demands the discovery of some fundamental screen structure parameters for relating the input, that is, the halftone film or printing plate, to the output, that is, the printed image.

METHODS

The density at a point in the area of print is:

$$\Delta D = -\log_{10}(I_o/I\Delta A) \quad (3)$$

in which D is the density, I_o the incident intensity, I the reflected intensity, and A is fractional area. If it is assumed that the print comprises a finite number of areas with equal reflection intensities, Equation (3) can be integrated as:

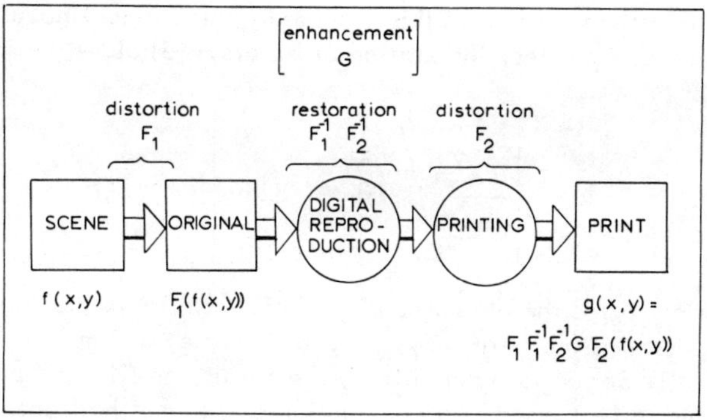

Fig. 1. Process control scheme for a digital reproduction process.

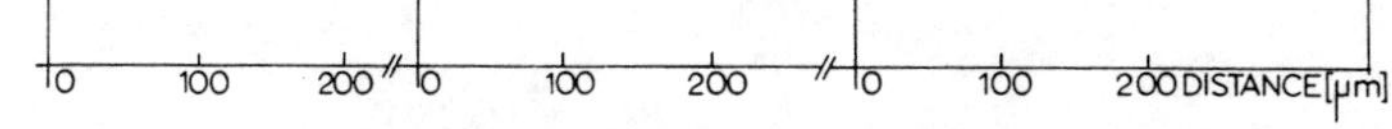

Fig. 2. Density profiles for different dot sizes at different frequencies. Dotted lines give solid and paper densities. Newsprint, line screen, black ink.

$$D_{int} = -\log_{10}\left[I_o / \sum_{i=1}^{m} I_i A_i \right]; \sum_{i=1}^{m} A_i = 1 \tag{4}$$

When the number of differently coloured areas is eight, Equation (4) corresponds to Neugebauer equations. However, it is known that the spread of light makes the boundaries of halftone dots diffuse[3, 6]. When it is assumed that the intensity of the reflected light is a continuous function of distance coordinates, integration of Equation (3) gives the integrated density of print as:

$$D_{int} = -\log_{10}\left[\frac{2u^2}{m^2} \iint_0^{m/\sqrt{2}u} 10^{-g(x,y)} \, dxdy \right] \tag{5}$$

in which g(x, y) is the output density profile of print as measured on a micro scale, u is the frequency of a screen cycle, and m is an integer.

Under certain conditions, the density profile of a print has been found to be formed as a convolution of a square halftone dot profile, f(x, y), and a point spread function, h(x, y)[3]:

$$g(x, y) = h(x, y) \circledast f(x, y) \tag{6}$$

It follows from Equation (6) that the printing process, when understood as an information transfer system, transfers density in a linear fashion[1]. The physical interpretation of the point spread function is that it re-distributes the density in the plane. This means that the sum of the density, measured on a micro scale, will not change as a function of image structure parameters, but the existing density 'flows' from high peaks into the neighbourhood. This brings about a fall in the sharpness of the print, and simultaneously contributes to the disappearance of individual halftone dots when the print is assessed visually.

On examination of Equations (5) and (6), the relation between the screened image and the integrated density on print can be written as:

$$D_{int} = -\log_{10}\left[\frac{2u^2}{m^2} \iint_0^{m/\sqrt{2}u} 10^{-h(x,y) \circledast f(x,y)} \, dxdy \right] \tag{7}$$

The point spread function of a linear system can be transformed into a modulation transfer function (MTF), or vice versa, by the employment of a Hankel transformation[1]. MTF gives the system response in the frequency domain, and it is physically defined as the transfer of a sinusoidal wave:

$$M(u,v) = M_{out}(u,v)/M_{in}(u,v) \text{ in which} \tag{8a}$$

$$M_{in} = \frac{f_{max} - f_{min}}{f_{max} + f_{min}} \qquad (8b)$$

$$M_{out} = \frac{g_{max} - g_{min}}{g_{max} + g_{min}} \qquad (8c)$$

Modulation transfer function (Equation 8a) is a measure of the quality of an image transfer system. The plain output modulation (Equation (8c)) is usable as a measure of image quality.

For a square wave, the contrast transfer function (CTF) can be measured and calculated from Equation (8) in a similar way to the MTF as a sinusoidal wave transfer. However, values of CTF are higher than those of MTF. A mathematical relationship exists between MTF and CTF [1].

Experimental details of the CTF of a print follow. The other halftone dot structure variable to be studied is the degree of dot overlap of different inks.

Test figures were plotted on a film with a computer-directed Optronics P1500 scanner, whose options were three square pixels of 25×25, 50×50, and $100 \times 100\ \mu m^2$. Films carrying test figures were exposed on offset plates and printed in a sheet-fed offset press. Integrated densities were measured with a GAM densitometer and density profiles on a micro scale with a Joyce-Loebl MK IIIC microdensitometer provided with a device for reflection measurements. In this measurement, a larger area is illuminated, and measurement made of the density of a smaller area determined by the slit. The width of the slit was 12μ m. The two densitometers were calibrated to display equal values.

RESULTS AND DISCUSSION

Figure 2 illustrates how the density spread modifies the halftone dot profile. The top density of halftone dots will diminish and the density between halftone dots will rise as a function of increasing frequency. Both these densities increase as a function of increasing fractional dot area (Figure 3(a)). Modulation (Equation (8c)), however, is invariant with respect to the relative dot area (Figure 3(b)). Modulations for coloured inks are shown in Figures 4(a), (b) and (c).

The integrated density of print is calculated for a non-linear system from Equation (5) and for a linear system from Formula (7). A reasonably accurate approximation for the integrated density is obtained by restricting consideration to the top and bottom densities of the halftone dots, instead of the whole dot profile. An example of this is shown in Figure 5.

The linearity of the system can be tested by studying the behaviour of the integral density profile on a micro scale, i.e.:

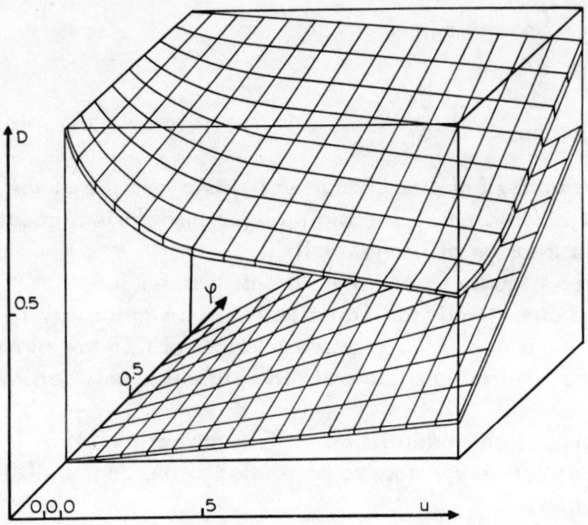

Fig. 3(a). *Maximum density of dot and density between halftone dots as a function of screen frequency and dot size. Newsprint, line screen, black ink.*

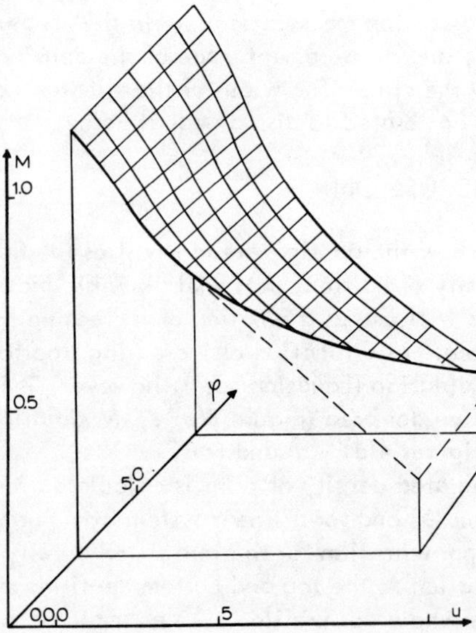

Fig. 3(b). *Output density modulation as a function of frequency and dot size. Newsprint, line screen, black ink.*

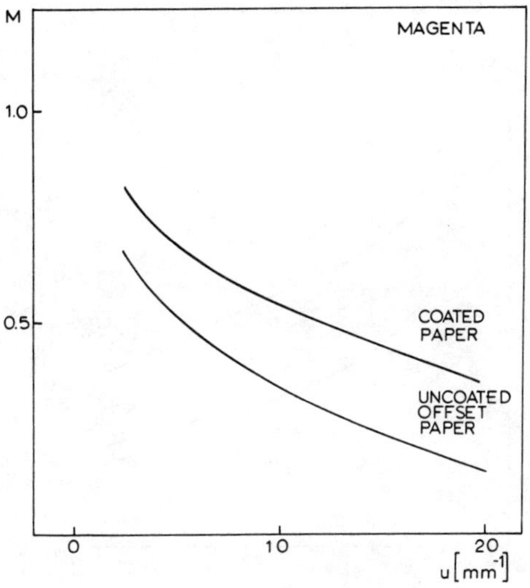

Fig. 4(a). Output modulation for magenta ink.

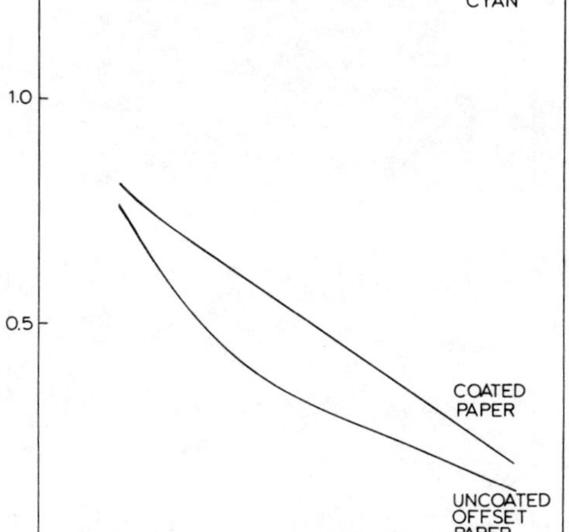

Fig. 4(b). Output modulation for cyan ink.

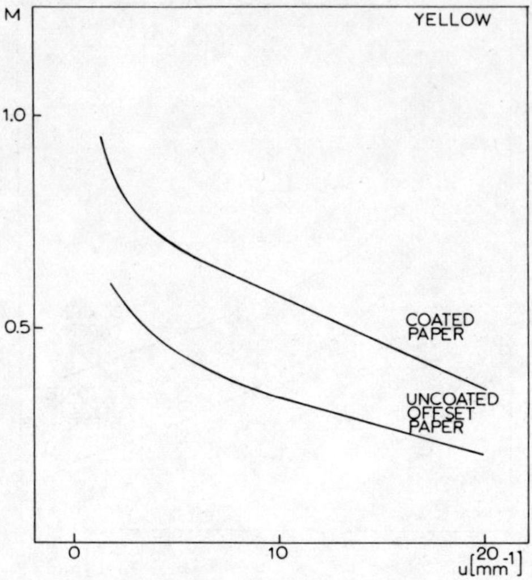

Fig. 4(c). Output modulation for yellow ink.

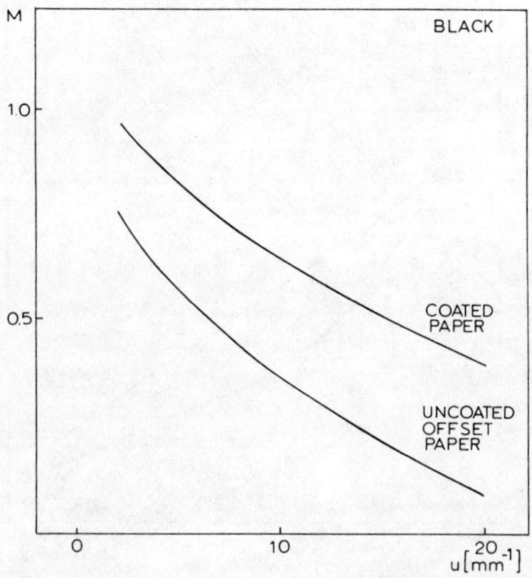

Fig. 4(d). Output modulation for black ink.

$$\overline{D} = \frac{2u^2}{m^2} \iint_0^{m/\sqrt{2}u} g(x,y)\, dxdy \qquad (9)$$

In a linear system $\overline{D}$ is invariable with respect to image structure. Figure 6 demonstrates that the system linearity is true at low frequencies but is lost at higher frequencies; the turning point depends on the inking level. Both physical ink spread and light spread at the edge induce linear density spread to the extent that density formation obeys Lambert-Beer's law. It is probably that light spread at higher frequencies induces the non-linearity observed.

When density spreads linearly, the integrated density of the print increases. This occurs because a constant alteration in density changes the reflected total intensity more in areas of low density than in areas of high density. On the micro-scale, density spread can be understood as a flow of density from the top of halftone dots into the area between the dots. Figure 7 illustrates the effectiveness of a constant density on the micro-scale as the originator of integrated density when the density is shared equally for different fractional areas.

When expressed by means of the MTF, integrated density will increase as a function of decreasing MT (Figure 8). For linear one-colour printing, dot frequency, expressed as the inverse of dot size, and the MTF of the process, give sufficient information for calculating the integrated output density.

In colour reproduction, the calculation of a subcolour, as was done for a single ink, will give inaccurate results. This is because the same sub-ink dots, superimposed, or side by side, will give different integrated densities. One well known contributory reason for this is that the sum of differently printed solid densities deviates from the density obtained when the inks are printed one upon the other.

The main reason, however, why the degree of dot overlap influences colour is the non-ideal spectral absorptions of the coloured printing inks; these make the absorption spectra overlap. Then, at a certain wavelength, many inks contribute to the absorption. When dots of different inks are superimposed, primary and secondary absorptions influence additively; the top density of the dots rises by reason of secondary absorption. When the dots of different inks are separated from this position, the secondary absorption increases the density between density peaks induced by the primary absorption. This affects the integrated density by the same mechanism as does density spread; integrated density increases with diminution in the degree of dot overlap.

The integrated density for a given wavelength can then be calculated from output densities of different inks on a micro-scale for this wavelength, and from the mutual position of the dots:

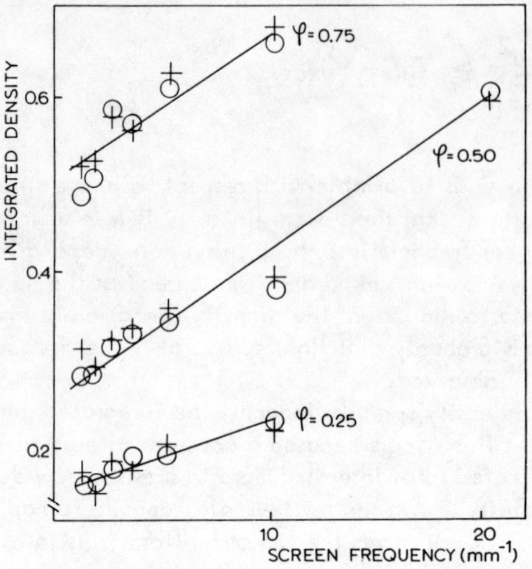

Fig. 5. Measured (+) and calculated (0) values for integrated density of print. Calculations are made using approximation Formula:

$$D_{int} = -\log_{10}[\varphi \cdot 10^{-D_{max}} + (1-\varphi)10^{-D_{min}}]$$

where φ is fractional dot size at density level

$$(D_{max} + D_{min})/2$$

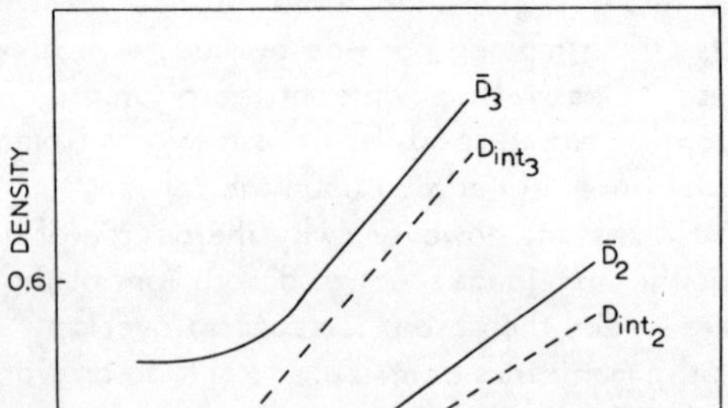

Fig. 6. Relation between D_{int} and $\bar{D}$ [Formula (9)] as a function of screen frequency for a $\varphi = 0.5$ dot size and for increasing inking levels (1 ... 3).

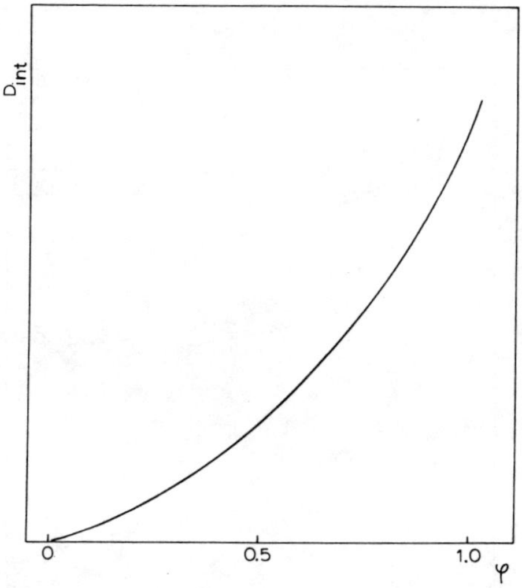

Fig. 7. Influence of density spread on integrated density. The curve is calculated from Formula:

$$D_{int} = -\log_{10}[\varphi + (1-\varphi)10^{-\bar{D}/\rho}]$$

in which $\bar{D}$ has been invariant.

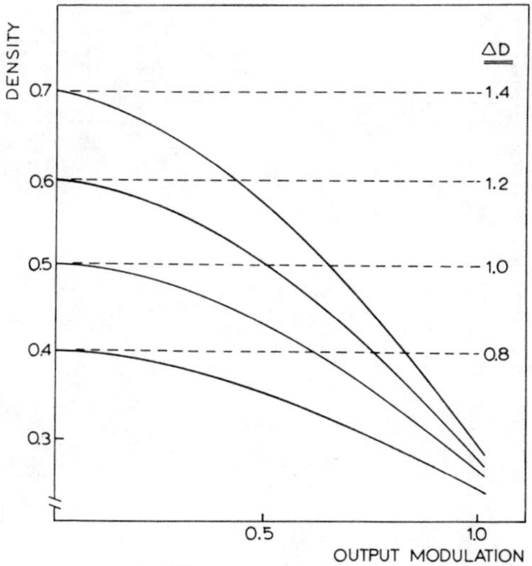

Fig. 8. Dependence between output modulation of print and integrated density for a $\varphi = 0.5$ dot size.

------------ $\bar{D}$ [Formula (9)]
———————— D_{int} [Formula (7)]
ΔD gives the density range of print

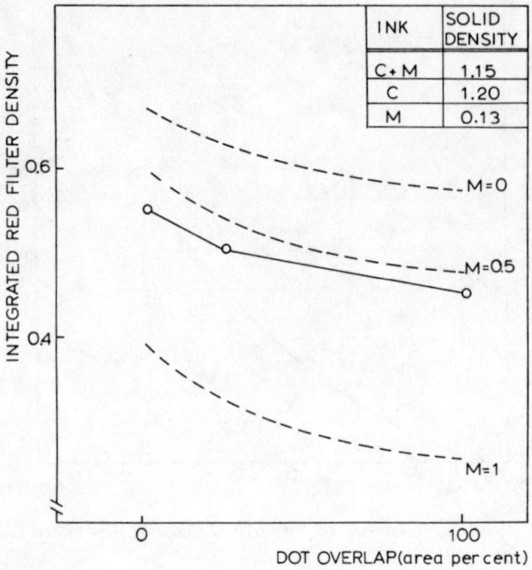

Fig. 9(a). Influence of dot overlap and modulation transfer (M) on integrated red filter densities of a $\rho_c = 0.5$, $\varphi_M = 0.5$ tone. Calculated values (dotted lines) and measured values (solid lines).

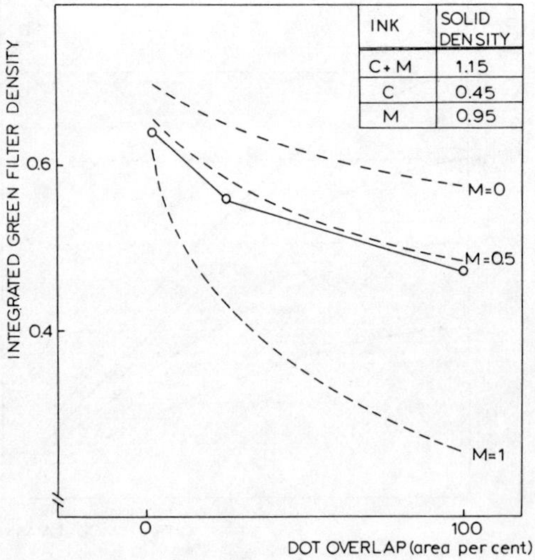

Fig. 9(b). Influence of dot overlap and modulation transfer (M) on integrated green filter densities of a $\varphi_c = 0.5$, $\varphi_M = 0.5$ tone. Calculated values (dotted lines) and measured values (solid lines).

$$D_{int} = -\log_{10}\left[\frac{1}{(x_2-x_1)(y_2-y_1)}\int_{x_1}^{x_2}\int_{y_1}^{y_2} 10^{-g_p(x,y)+\sum_i g_{si}(x+x_i,\, y+y_i)}\, dxdy\right] \quad (10)$$

in which indices p and s refer to the primary and secondary densities for the wavelength in question, i is the ink inducing the secondary absorption, and (x_i, y_i) gives the phase difference of the halftone dot patterns. In the case of a linear system, density profiles in output can be calculated from Equation (6). The influence of MTF and of dot overlap on colour formation can then be considered simultaneously.

Equation (10) is applicable also to separation filter densities. The measured filter densities from solid areas are then used as a basis for calculations. Figures 9(a) and (b) provide examples of the influence of MTF and dot overlap on filter densities. The Figures present both measured values and the values calculated from Equations (6) and (10). It is observable that the role of both MTF and the secondary absorption of printing inks is considerable in determining the colour.

CONCLUSIONS

The two fundamental variables of halftone dot structure which affect the tone and colour formation of the print are the frequency and the degree of dot overlap. When the frequency is increased, tones and colours become darker by reason of diminished modulation transfer. Dot overlap exercises an influence as a consequence of non-ideal spectral absorptions of the coloured inks. In conventional multicolour printing, this is then one reason contributing to the introduction of moire effects.

With a view to picture correction, the tone and colour can be predicted to a sufficient degree of accuracy by using the structure parameters of the halftone dot patterns. For a linear process, it can be predicted from the input of printing, i.e. halftone film or plate in relation with the MTF of the process. In the case of nonlinear printing, parameters of output, as ultimate densities for different dot sizes, are usable as a basis.

SYMBOLS

f	input
g	output
h	spread
i, m	integers
k	number of inked pixels in halftone dot matrix

n	number of pixels in halftone dot matrix
p	number of combinations
t	number of inks
u,v	spatial frequencies
x,y	distance coordinates
A	fractional area
D	density
$\bar{D}$	integral of density profile
D_{int}	integrated density
I	intensity
I_o	incident intensity
M	modulation transfer
M_{in}	input modulation
M_{out}	output modulation
ϕ	fractional inked area
$\otimes$	denotes convolution

REFERENCES

1. Dainty, J.C., Shaw, R., Image Science, Academic Press, London (1974)
2. Kekolahti, P., Evaluation of Digital Screening Methods for the Graphic Arts, 2nd Scandinavian Conference on Image analysis, Helsinki (1981)
3. Oittinen, P., Saarelma, H., Density of Halftone Dots, TAPPI, 64 (1981) 6. (in print).
4. Saarelma, H., Image Formation in Halftone Photography, Graphic Arts in Finland, 8 (1979) 1.
5. Saarelma, H., Oittinen, P., Introduction to Image Transfer and Picture processing in Halftone Reproduction, Graphic Arts in Finland, 8 (1979) 2.
6. Yule, J. A. C., Neilsen, W. J., The Penetration of Light into Paper and its Effect on Halftone Reproduction, TAGA Proceedings, (1951).

DISCUSSION

K. Schläpfer What do you mean by 'unconventional' screens in your paper?

H. Saarelma   I mean halftone dots generated by using picture element by picture element exposure instead of even exposure through glass or contact screen. The former allows for any structure for the halftone dot while the latter generates dots as plane cuts of a two-dimensional sinusoidal surface.

AN INVERSE SOLUTION OF THE NEUGEBAUER EQUATIONS FOR THE CALCULATION OF GRAY BALANCE REQUIREMENTS

H. B. Archer, Rochester Institute of Technology, U.S.A.

ABSTRACT

The Neugebauer Equations, as originally stated, yield three colorimetric reflectances, R, G and B of a three-color halftone overprint where the dot areas, c, m and y, and the colorimetric reflectances of the color solids and their overprints are known. When dot areas are the desired result of calculation, these equations must be solved for c, m and y by iterative techniques. Their algebraic inverse is not ordinarily obtainable.

This paper shows that for the special case of neutrals, an inverse solution of c, m and y is possible. For this solution the color reflectances of the substrate are all assumed to be 1.0 and for any neutral are assumed to be equal, i.e. for the substrate, R=G=B=1 and for any 3-color neutral printed on the substrate, R=G=B. The resulting equations are relatively simple to use and can be easily executed using a small programmable calculator.

The practical application of these new equations is demonstrated. Their accuracy is predicting the dot area requirements for gray balance is shown using densitometers with conventional broad band filters for measurement.

INTRODUCTION

The Neugebauer equations [2] have been used, with varying degrees of success, for both theoretical and practical applications in halftone color reproduction. As originally presented these equations are inadequate in that they fail to account for the effects of multiple internal reflections in the substrate. Additional errors result when the effects of dot gain are ignored.

Pobboravsky and Pearson [4] significantly improved the accuracy of the Neugebauer equations by converting all reflectances to effective values as follows:

$$R' = R^{1/n} \qquad (1)$$

The tri-stimulus values R, G and B of the ink solids and their overprints as well as the color to be reproduced were thus converted to effective values, R', G' and B', for use in the equations. The use of the 'n' factor in Equation (1) is analogous to its application in the Yule-Nielsen equation:

$$a = (1 - X)/(1 - Y) \qquad (2)$$

$$\text{where } X = 10^{-D_t/n}$$

$$Y = 10^{-D_s/n}$$

As stated here the Yule-Nielsen equation yields a fractional dot area for a measured halftone tint density, D_t and a solid density, D_s. The accuracy of the caiculated dot area (which we will call the 'apparent dot area') compared with the actual dot area on the paper depends upon the value of 'n' used.

Pobboravsky and Pearson also introduced a means of compensating for the effect of dot gain. Their method required a printed halftone scale for each of the primary colors. The apparent dot area of each step of the printed scales was then calculated using Equation (2) and then plotted against their corresponding film dot areas. This provided a relationship between the film and the press sheet for each of the colors. Calculated dot area values produced by the modified equations were thus converted to film dot areas for application. This technique proved to be the most valuable addition of all.

The objective of this present work is to present a simple method for calculating the gray balance requirements for a specific set of printing conditions, i.e. plate, press, inks and paper. The test data required for this method is readily available, e.g. that obtainable from a color control bar, using a conventional densitometer with broad band filters. The calculations required are relatively fast to perform and are well within the capabilities of a small programmable calculator.

AN INVERSE SOLUTION

The original Neugebauer equations can be written as follows:

$$R = f_1R_w + f_2R_c + f_3R_m + f_4R_y + f_5R_{cm} + f_6R_{cy} + f_7R_{my} + f_8R_{cmy} \qquad (3)$$

$$G = f_1G_w + f_2G_c + f_3G_m + f_4G_y + f_5G_{cm} + f_6G_{cy} + f_7G_{my} + f_8G_{cmy} \qquad (4)$$

$$B = f_1B_w + f_2B_c + f_3B_m + f_4B_y + f_5B_{cm} + f_6B_{cy} + f_7B_{my} + f_8B_{cmy} \qquad (5)$$

where the fractional dot areas, f_1 to f_8, created by the individual dot areas, c, m and y, are defined as follows:

$$f_1 = (1-c)(1-m)(1-y)$$
$$f_2 = c(1-m)(1-y)$$
$$f_3 = m(1-c)(1-y)$$
$$f_4 = y(1-c)(1-m)$$
$$f_5 = cm(1-y)$$
$$f_6 = cy(1-m)$$
$$f_7 = my(1-c)$$
$$f_8 = cmy$$

and the subscripts of the variables, R, G and B, have the following meaning:

TABLE 1

Subscript	Solid print reflectance of:
w	Substrate
c	Cyan Ink
m	Magenta Ink
y	Yellow Ink
cm	Cyan + Magenta Inks
cy	Cyan + Yellow Inks
my	Magenta + Yellow Inks

A simplification of Equations (3), (4) and (5) can be made in the case of neutrals provided the three color reflectances, R, G and B, which specify the reproduction color, are assumed to be equal.

By factoring out c, m and y and their products and by assuming $R_w=G_w=B_w=1.0$, Equations (3), (4) and (5) can be written as follows:

$$R = 1+cR_1+mR_2+yR_3+cmR_4+cyR_5+myR_6+cmyR_7 \quad (6)$$

$$G = 1+cG_1+mG_2+yG_3+cmG_4+cyG_5+myG_6+cmyG_7 \quad (7)$$

$$B = 1+cB_1+mB_2+yB_3+cmB_4+cyB_5+myB_6+cmyB_7 \quad (8)$$

where R_{1-7} are:

$$R_1 = R_c-1$$
$$R_2 = R_m-1$$
$$R_3 = R_y-1$$
$$R_4 = 1-R_c-R_m+R_{cm}$$
$$R_5 = 1-R_c-R_y+R_{cy}$$

$$R_6 = 1-R_m-R_y+R_{my}$$

$$R_7 = R_c+R_m+R_y-R_{cm}-R_{cy}-R_{my}+R_{cmy}-1$$

G_{1-7} and B_{1-7} taking the same form as R_{1-7}.

As specified, R=G=B for any neutral. Thus for R=G we have:

$$c(R_1-G_1)+m(R_2-G_2)+y(R_3-G_3)+cm(R_4-G_4)+cy(R_5-G_5)+ my(R_6-G_6)+cmy(R_7-G_7) = 0 \quad (9)$$

and for R=B we have:

$$c(R_1-B_1)+m(R_2-B_2)+y(R_3-B_3)+cm(R_4-B_4)+cy(R_5-B_5)+ my(R_6-B_6)+cmy(R_7-B_7) = 0 \quad (10)$$

For Equations (9) and (10) the variables R_{1-7}, G_{1-7} and B_{1-7} are constants for any single set of printing conditions. This leaves the variables c, m and y, one of which can be specified. Assuming that c, the cyan dot area is to be specified, Equations (9) and (10) can be written in terms of m and y as follows:

$$U_1+U_2m+U_3y+U_4my = 0 \quad (11)$$

$$V_1+V_2m+V_3y+V_4my = 0 \quad (12)$$

where U_{1-4} and V_{1-4} are:

$$U_1 = c(R_1-G_1) \qquad V_1 = c(R_1-B_1)$$

$$U_2 = c(R_4-G_4)+(R_2-G_2) \qquad V_2 = c(R_4-B_4)+(R_2-B_2)$$

$$U_3 = c(R_5-G_5)+(R_3-G_3) \qquad V_3 = c(R_5-B_5)+(R_3-B_3)$$

$$U_4 = c(R_7-G_7)+(R_6-G_6) \qquad V_4 = c(R_7-B_7)+(R_6-B_6)$$

Solving Equations (11) and (12) for m and y we have:

$$m = \left[-B-(B^2-4AC)^{\frac{1}{2}}\right]/2A \quad (13)$$

where A, B and C are:

$$A = U_4V_2-U_2V_4$$

$$B = U_4V_1+U_3V_2-U_2V_3-U_1V_4$$

$$C = U_3V_1-U_1V_3$$

and

$$y = -(U_1+U_2m)/(U_3+U_4m) \quad (14)$$

Thus where the cyan dot area, c, can be specified, the accompanying magenta and yellow dot areas, m and y, required to produce a neutral can be calculated using Equations (13) and (14). And with all three dot areas, c, m and y, the reflectance of the neutral may then be predicted using Equation (6), (7) or (8).

PROCEDURE AND APPLICATION

Density Measurement and Conversions

Proper calibration of the densitometer used for this procedure is important. The photometric calibration of the instrument should be checked using a filter which, in combination with the photodetector, approximates the luminosity function of the eye. The color filters used for measurement should have a relatively high colorimetric quality factor (Yule[5]). Filter measurements of a neutral scale should be linear, and preferably equal, with respect to the visual density measurements (photometric calibration) of the scale.

Zero Adjustment

The equations assume that $R_w+G_w=B_w=1.0$. The densitometer must therefore be zeroed on the white paper with each filter.

Slope Adjustment

When measuring a neutral on the paper the three filter densities should be equal and they should equal the visual density of that neutral. The procedure used by the author was to first check the slope of the densitometer with the check plaque, using the visual filter. Then zero the instrument on the white paper and measure the visual density of a mid-scale neutral (about 0.8 density) appearing on the paper. A black halftone image is satisfactory. A more standard neutral can be achieved by placing a low neutral density filter (about 0.4) on the paper for measurement. With each color filter, zero on the paper and then measure neutral. Each filter density measurement of the neutral should equal its visual density. For densitometers with individual slope control, this is an easy adjustment. For those instruments without individual slope control this adjustment can be made mathematically. In this case, as part of the calculator program, the densities of the neutral patch are input and used to calculate slope correction factors.

Density to Reflectance Conversion

All density values must be converted to reflectance values for use in the equations. The inclusion of the 'n' factor is most easily made at this stage. For example, for the red filter density, D_r, of the cyan solid, the red reflectance R_c is calculated as follows:

$$R_c = 10^{-D_r/n} \tag{15}$$

Reflectance to Density Conversion

From the cyan dot area, c, and the calculated magenta and yellow dot areas, m and y, the red reflectance, R, of the reproduction is calculated using Equation (6). The predicted neutral density, D_v, can then be obtained as follows:

$$D_v = -n\log_{10}R \tag{16}$$

Dot Gain Compensation

Equations (13) and (14) require a cyan dot area to be specified as it will be printed. The magenta and yellow dot areas that result from this calculation are likewise, printed dot areas. Since it is desirable to have results in terms of film dot area it is necessary to have some procedure for correcting for dot gain. One possible approach is the method employed by Pobboravsky and Pearson mentioned earlier. This method, however, requires that separate halftone scales be included as part of the test form.

In the interests of speed and simplicity the author chose to use Haller's mathematical model (1979) for dot gain compensation. Presented in terms of fractional dot area (i.e. 50% is 0.5) Haller's [1] model is as follows:

$$a = (1-L^{a_f})/(1-L) \tag{17}$$

where

$$L = \left[(1/a_{50})-1\right]^2 \tag{18}$$

a_f represents the measurable dot area on the film, a is the predicted dot area that will be printed and a_{50} is the apparent printed dot area resulting from a 50% film original which has been calculated using the Yule-Nielsen Equation (2).

The inverse of Equation (17) can be written as follows:

$$a_f = \log\left[1-a(1-L)\right]/\log L \tag{19}$$

General Procedure

The following general procedure was used for the results presented in this paper:

(1) Three color density measurements, D_r, D_g and D_b, are made of the seven color solids (see Table 1).

(2) Complementary filter density measurements are made of the color halftone tints produced from a 50% film original. A red filter density measurement is made of the cyan tint, a green filter measurement of the magenta tint and a blue filter measurement of the yellow tint.

(3) Upon input to the program, the 24 densities obtained in steps 1 and 2 are converted to the effective reflectance values using Equation (15).

(4) The halftone tint reflectances obtained from the 50% sample areas,

and their corresponding solid reflectances are used to calculate the apparent dot areas, a_{50}, for each color, using Equation (2). The values L_c, L_m and L_y are then calculated for each color using Equation (18).

(5) In order to conserve storage space, the accumulation of the values R_{1-7}, G_{1-7} and B_{1-7} can proceed as each density is input and converted to reflectance. Many of the registers can be used over again as the calculations proceed. The values R_{1-7}, however, should be retained since they will be needed to calculate the predicted neutral.

(6) With all preliminary calculations made, the program asks for a cyan dot area value to be designated. Alternatively the program can be designed to generate a series of cyan dot values.

(7) The cyan dot area input is assumed to be a film dot area, a_f, which is converted to a printed dot area, a, using Equation (17) and L_c.

(8) Step 7 provides printed dot area, c. The printed dot areas, m and y, can now be calculated using Equations (13) and (14). For output, the film dot areas for magenta and yellow are calculated using Equation (19) with L_m and L_y.

(9) The printed dot areas, c, m and y, are also used to calculate R, the red reflectance of the reproduction. R is then converted to density (Equation 16) which is the predicted neutral density.

EVALUATION TESTS

Selected printings of the R.I.T. TRAND Chart were used to evaluate this procedure. The TRAND Chart provides the needed test data required for calculation and also provides several identifiable neutrals for comparison with calculated results.

Neutrals were identified using the densitometer, where $D_r=D_g=D_b$, and then confirmed visually. Since the test form provides discrete areas for neutral selection rather than a continuous choice, it was often impossible to find a 'neutral' where $D_r=D_g=D_b$ exactly. At several different levels of cyan the 'best neutral' was selected. Results were then verified by plotting the cyan film dot areas, for each neutral vs. the required magenta and yellow dot areas. An example of this is shown in Figure 1. In this particular example the required magenta and yellow dot areas for any cyan dot area were equal. In other examples two curves resulted since the required amounts of magenta and yellow were different.

In Figure 1 it can be seen that three of the points follow a smooth functional relationship. Where the cyan is 45%, the plotted point is above the curve. In cases such as this, the value for magenta and/or yellow was adjusted so that it conformed to the curve. In the test examples that follow, observed magenta and yellow values thus

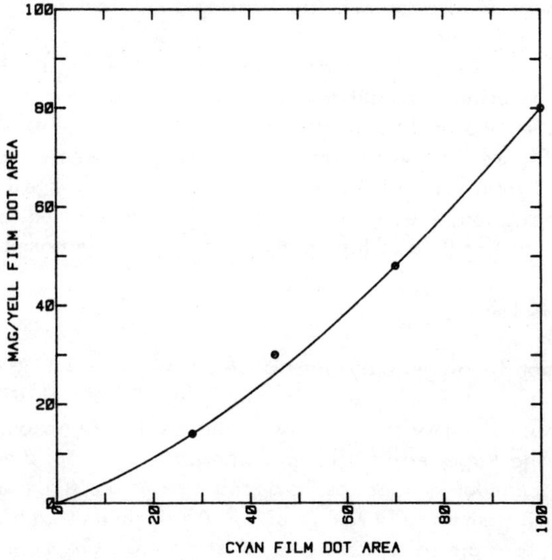

Fig. 1. *For each evaluation the observed dot area relationship for neutral was plotted as shown. Magenta and yellow dot areas not falling on the curve were thus corrected.*

adjusted have been marked with an asterisk (*).

Figures 2 to 6 are data sheets representing individual evaluation tests. Each data sheet identifie the conditions of the test, i.e. paper, press and densitometer used for measurement. The measurements made for neutral calibration are indicated. The value of 'n' used for density-reflectance conversions, is indicated. Density measurements, required for input, are tabulated as TEST DATA. TEST RESULTS are provided where for different specified cyan film dot areas the observed and calculated magenta and yellow film dot areas are indicated. The observed and calculated neutral densities are also given. In each case the difference between observed and calculated is given as the ERROR.

The density measurements for Figures 2, 3 and 4 were all made using the same densitometer. The printing conditions for these three tests are all different.

Figures 4, 5 and 6 are three tests made using the same printed test, i.e. the printing conditions are identical. Different densitometers were used for each of these tests.

Figure 7 shows the results of calculation where the value of 'n' was changed from 1.7 to 1.0 and also shows the result of eliminating the dot gain correction, i.e. where the printed dot is assumed to equal the film dot. The test data of Figure 4 was used for these comparisons.

CONCLUSIONS

An evaluation involving only three different sets of printing conditions and three different densitometers is insufficient for a definitive conclusion. As a preliminary test, however, the results are encouraging.

The larger errors all occur where the cyan dot area is 100%. This is understandable since results at this part of the scale are more sensitive to variations in solid ink density. Any non-uniformities of the printed test form are reflected here. Also neutrals are more difficult to identify at this part of the scale. If the errors, where cyan is maximum, are ignored then the maximum error in calculated dot size was only 6% and the maximum error in predicted neutral density was 0.06. These results are exceptionally good and well within acceptable tolerances.

The accuracy of this method depends upon the colorimetric qualities of the densitometer used for measurement. The choices of densitometers used in this experiment may have been exceptional. Greater error with some densitometers is quite possible.

The value of 'n' used in density-reflectance conversions has only a slight effect on results as shown in Figure 7. Where n=1.7, the results are slightly better. The choice of 1.7 is based on Pearson's findings [3] where 1.7 is indicated to be a good average for different papers, screen rulings, and dot sizes.

As Pobboravsky and Pearson found [4], correcting for dot gain effects has a profound effect. The accuracy of the predicted neutral

Neugebauer Gray Balance Test

Test No. 1 Date ______ Neutral Calibration

Paper Enamel Coated D_v .81 D_r .81

Ink ______ D_g .81 D_b .81

Press Sheet Fed

Densitometer Macbeth RD 514 "n" Factor 1.7

Test Data

	C_{50}	M_{50}	Y_{50}	C	M	Y	C+M	C+Y	M+Y	C+M+Y
D_r	.40	X	X	1.17	.10	.01	1.20	1.14	.12	1.21
D_g	X	.49	X	.38	1.14	.04	1.53	.44	1.16	1.44
D_b	X	X	.44	.08	.54	.92	.57	.93	1.46	1.45

Test Results

CYAN DOT AREA	MAGENTA DOT AREA			YELLOW DOT AREA			NEUTRAL		
	OBSERVED	CALCULATED	ERROR	OBSERVED	CALCULATED	ERROR	OBSERVED	CALCULATED	ERROR
16	6	7	+1	8	8	0	.14	.12	-.02
32	14	15	+1	17	17	0	.30	.26	-.04
50	25	26	+1	30	29	-1	.49	.43	-.06
70	42	41	-1	48	46	-2	.71	.67	-.04
100	80	79	-1	80	83	+3	1.17	1.20	+.03

Fig. 2.

Neugebauer Gray Balance Test

Test No. 2 Date ______

Paper Uncoated

Ink ______

Press Web

Densitometer Macbeth RD-514

Neutral Calibration

D_v .81 D_r .81

D_g .81 D_b .81

"n" Factor 1.7

Test Data

	C_{50}	M_{50}	Y_{50}	C	M	Y	C+M	C+Y	M+Y	C+M+Y
D_r	.37	X	X	.81	.07	0	.87	.89	.09	.91
D_g	X	.35	X	.34	.79	.02	.95	.38	.87	.97
D_b	X	X	.28	.07	.55	.62	.55	.67	1.00	.90

Test Results

CYAN DOT AREA	MAGENTA DOT AREA			YELLOW DOT AREA			NEUTRAL		
	OBSERVED	CALCULATED	ERROR	OBSERVED	CALCULATED	ERROR	OBSERVED	CALCULATED	ERROR
32	20	20	0	24	24	0	.31	.25	-.06
50	33*	34	+1	40	40	0	.46	.41	-.05
70	50	53	+3	58	60	+2	.65	.60	-.05
100	90	91	+1	90	108	+18	90	91	+.01

*Adjusted

Fig. 3.

Neugebauer Gray Balance Test

Test No. 3-1 Date ______

Paper Matt Coated

Ink ______

Press Web

Densitometer Macbeth RD-514

Neutral Calibration

D_v .81 D_r .81

D_g .81 D_b .81

"n" Factor 1.7

Test Data

	C_{50}	M_{50}	Y_{50}	C	M	Y	C+M	C+Y	M+Y	C+M+Y
D_r	.51	X	X	.95	.13	.03	.96	1.01	.14	1.03
D_g	X	.47	X	.47	.81	.07	1.01	.56	.84	1.11
D_b	X	X	.49	.15	.50	.75	.55	.82	1.06	1.10

Test Results

CYAN DOT AREA	MAGENTA DOT AREA			YELLOW DOT AREA			NEUTRAL		
	OBSERVED	CALCULATED	ERROR	OBSERVED	CALCULATED	ERROR	OBSERVED	CALCULATED	ERROR
28	14	14	0	14	16	+2	.29	.33	+.04
45	26*	24	-2	26*	27	+1	.49	.52	+.03
70	48	44	-4	48	48	0	.77	.77	0
100	80	77	-3	80	84	+4	1.03	1.02	-.01

*Adjusted

Fig. 4.

Neugebauer Gray Balance Test

Test No. 3-2 Date ______

Paper Matt Coated

Ink ______

Press Web

Densitometer Cosar

Neutral Calibration

D_v .82 D_r .82

D_g .82 D_b .82

"n" Factor 1.7

Test Data

	C_{50}	M_{50}	Y_{50}	C	M	Y	C+M	C+Y	M+Y	C+M+Y
D_r	.50	X	X	.94	.13	.03	.97	1.00	.15	1.03
D_g	X	.48	X	.48	.84	.07	1.03	.56	.87	1.13
D_b	X	X	.47	.16	.54	.70	.59	.79	1.06	1.10

Test Results

CYAN DOT AREA	MAGENTA DOT AREA			YELLOW DOT AREA			NEUTRAL		
	OBSERVED	CALCULATED	ERROR	OBSERVED	CALCULATED	ERROR	OBSERVED	CALCULATED	ERROR
28	14	13	-1	14	15	+1	.28	.32	+.04
45	26*	23	-3	26*	26	0	.50	.51	+.01
70	48	42	-6	48	47	-1	.77	.76	-.01
100	80	73	-7	80	87	+7	1.03	1.02	-.01

*Adjusted

Fig. 5.

Neugebauer Gray Balance Test

Test No. 3-3 Date ______ Neutral Calibration

Paper Matt Coated D_v 1.32 D_r 1.30

Ink ______ D_g 1.30 D_b 1.36

Press Web

Densitometer Macbeth TR-927 "n" Factor 1.7

Test Data

	C_{50}	M_{50}	Y_{50}	C	M	Y	C+M	C+Y	M+Y	C+M+Y
D_r	.54	X	X	.99	.13	.03	1.02	1.07	.15	1.09
D_g	X	.51	X	.43	.90	.09	1.08	.54	.94	1.17
D_b	X	X	.54	.16	.51	.85	.59	.93	1.15	1.19

Test Results

CYAN DOT AREA	MAGENTA DOT AREA			YELLOW DOT AREA			NEUTRAL		
	OBSERVED	CALCULATED	ERROR	OBSERVED	CALCULATED	ERROR	OBSERVED	CALCULATED	ERROR
28	14	15	+1	14	15	+1	.29	.35	+.06
45	26*	26	0	26*	26	0	.49	.55	+.06
70	48	47	-1	48	47	-1	.77	.82	+.05
100	80	81	+1	80	89	+9	1.03	1.11	+.08

*Adjusted

Fig. 6.

	Cyan Dot	Magenta Dot Error	Yellow Dot Error	Density Error
N=1.0 W/Dot Gain Correction	28	-3	+2	+.04
	45	-5	+2	+.03
	70	-7	-1	-.01
	100	-4	+1	0
N-1.7 W/Dot gain Correction	28	0	+2	+.04
	45	-2	+1	+.03
	70	-4	0	0
	100	-3	+4	-.01
N-1.7 W/No dot Gain Correction	28	0	+6	-.10
	45	0	+8	-.16
	70	0	+10	-.19
	100	+10	+15	-.01

Fig. 7. Comparison using the same test data (Fig. 4) with two different values of 'n' and with and without dot gain correction.

density is especially affected by such compensation (see Figure 7).

ACKNOWLEDGMENTS

The author wishes to thank Irving Pobboravsky for his help and support in this project.

REFERENCES

1. Haller, K., 'Mathematical Model for Screen Dot-Shapes and for Transfer Characteristic Curves', Advances in Printing Science and Technology, Ed. W. H. Banks, Vol. 15, Pentech Press, London (1980).

2. Neugebauer, H. E. J., 'Die Theoretischen Grundlagen des Mehrfarbendruckes', Z. wiss. Photogr., 36, pp 73-89 (1937).

3. Pearson, M., 'n Value for General Conditions', TAGA Proceedings, pp 415-25 (1980).

4. Pobboravsky, I. and Pearson, M., 'Computation of Dot Areas Required to Match a Colorimetrically Specified Color Using the Modified Neugebauer Equations', TAGA Proceedings, pp 65-77 (1972).

5. Yule, J. A. C., 'Principles of Color Reproduction', Wiley, p 138 (1979).

QUALITY CONTROL IN NEWSPAPER PRINTING

Åge Hansen, The Norwegian Research Institute for the Graphic Arts Industry, Oslo

ABSTRACT

The following main points are dealt with in this paper:

(1) Decision as to the optimum quality level.

(2) Organisation of the work.

(3) Specification of normal printing conditions, mechanical and ink-densities.

(4) Specification of the printing press reproduction characteristics for black-and-white halftones and also for 4-colour.

(5) Adjusting the foregoing processes, screen processing and colour separation, to the printing characteristics.

INTRODUCTION

This paper deals with a project carried out by INGRAF in cooperation with six Norwegian newspapers printed by offset. Most of the work was done in different production departments, and especially in the pressroom. Various problems were discussed, such as the weakest points in the print quality, and plans for improvement were worked out. We succeeded in increasing the interest in quality control, and in improving the cooperation between the different departments, with better overall quality as a result.

We have also received further inquiries from other newspapers about introducing quality control in the same manner.

The project should also result in a 'report' formulated as a guide for quality control in newspaper printing. The following is an abridged edition of this 'guide'.

QUALITY CONTROL AND THE CHOICE OF QUALITY LEVEL

The print quality of a newspaper and printed matter in general is not easy to evaluate by measurement and to quantify as is the case for products from most other industries. Print quality has still to be estimated from subjective judgments and comparative criteria. The following general definition of quality should also be applicable to newspaper print quality:

> 'The quality of a product is its ability to satisy the user of the product'. (The reader and the advertiser).

The aim of quality control need not necessarily be to obtain as high a quality as possible and thereby satisfy the user as much as possible. In newspaper printing it would involve a lot of expensive control work and a high wastage if perfection was the aim. It would also be necessary to invest in more expensive equipment and materials (paper, ink, etc.) The aim should be to satisfy the user within reasonable limits from consideration of the production costs since these will usually increase when the quality level is raised.

One must also consider the effect on revenue from increased sales and advertising arising from an increase in quality, e.g. using 4-colour pictures. It is usual for industrial products in general, to aim at an optimum quality level which gives maximum profit. This should also be the case for newspaper production.

The optimum quality of a newspaper can be defined as that quality where the net income from sale and advertisements is as high as possible. The newspaper should be produced at a level of quality in keeping with its cost. This is the aim of the quality control reported here.

Throughout the production run variations in the quality always occur. The top quality for a limited number of copies can be of interest if they are sent to the advertiser. But in general it is the lower limit (the waste limit) which should be controlled at an optimum level.

ORGANISATION OF THE WORK

Quality control in newspaper production is team work involving the following departments: editorial, photography, typesetting, processing, paste-up, colour-separation, plate making and printing.

The first thing to do when introducing systematic quality control, is to organise a team with participants from each department and with a qualified leader who can inspire his co-workers. This group should have regular meetings, say, weekly. The members must organise and manage the control program in their respective departments and co-ordinate the cooperation between them. If possible, the team should also include a 'neutral' extra mural consultant, such as from a graphic research institute.

All efforts of increasing or stabilising the print quality should be carried out as an integral part of the daily work, and not as an additional task. The driving force must be motivation, a real interest in obtaining a high quality product.

THE ACTIVITY OF THE QUALITY GROUP

The first thing to do is to discuss the optimum quality level for the newspaper, and compare it with the present quality level. Samples of the newspaper should be examined. What improvements are needed and how shall they be done? Both equipment, materials, procedures, and the need for more staff training should be considered. When discussing the print quality, it should be split up into different factors for separate judgments:

(1) The cleanliness of the pages (smearing, set-off, rub-off, print-through).

(2) The evenness of printed solids and screened areas.

(3) The text quality (sharpness).

(4) The ink density level (the contrast against the 'white' paper).

(5) Black and white halftone pictures.

(6) 4-colour pictures.

The team should decide what are the weakest points in the quality and in the production process, and consider what improvements are necessary.

VARIABLES THAT INFLUENCE THE QUALITY

They can be variations in material: paper, ink, dampening water, rubber blankets, photographic films and paper, chemicals, etc., and they can be technical faults in the production equipment. A certain degree of material control is thus necessary.

A program for preventive maintenance of the production equipment and machinery is imperative. But even if all this is good, the print quality can be highly variable depending on:

(1) The picture originals

(2) The phototype setting

(3) The screen processing

(4) The colour separation

(5) The plate making

(6) The printing conditions

It is of course important to demand certain requirements of the

SPECIFICATION OF THE PRINTING CONDITIONS

PRINTING PRESS: Goss Community, single ~~double~~ round, Cutoff = 579 mm

PRINTING PLATE: Howson-Algraphy 0.36 mm, Maraton.

INK, producer/type,

BLACK: Lorilleux

MAGENTA, CYAN, YELLOW: — " —

DAMPENING WATER, FOUNTAIN SOLUTION, type, concentration, pH : Vadic Wet, 3%, pH 5.5

OFFSET BLANKET: Mercury Compressible 1.70 mm

PAPER, producer, type, basis weight: --- Newsprint 45 g/m² 0.065 mm

MECHANICAL PRINTING CONDITIONS:

INK FORM ROLLER PRESSURE AGAINST PLATE CYLINDER, STRIPE OF CONTACT, WIDTH: 4.5 ± 1 mm

For the rest, the rollers in the ink- and dampening systems are adjusted in accordance with the press instruction manual.

		Type of printing unit: Blanket-to-blanket unit without cylinder bearers	Blanket-to-blanket unit with cylinder bearers in contact	"Satellite", 4 color unit, common impression cylinder, with steel surface ---- mm above bearer
PLATE HIGHT		= plate thickness 0.36 ± 0.02	Hight above bearers ± 0.02	Hight above bearers ± 0.02
BLANKET HIGHT		= blanket + packing when tentioned around cylinder 2.17 ± 0.03	Hight above bearers ± 0.03	Hight above bearers ± 0.03
Stripe of contact, width	Between blanket and plate.	10.5 ± 1.0	± 1.0	± 1.0
	Between blanket and paper.	11.5 ± 1.0	± 1.0	± 1.0
The distance between the cylinder steel surfaces, when pressure on.	Between plate- and blanket-cylinder.	2.30 ± 0.02		
	Between the blanket cylinders.	4.02 ± 0.02		

(All figures in millimeter, with ± tolerances.)

COLOR SEQUENCE: Black → Cyan → Magenta → Yellow

SOLID INK DENSITY: Type of densitometer: Cosar SOS 40

(Normal-density. Zero on "white" paper.)

BLACK	0.75	± 0.05 tolerance
CYAN	0.56	
MAGENTA	0.60	
YELLOW	0.56	

WASHING / SOLVENTS / PROCEDURE : Dayly: water - white spirit

Weekly: Varn ---

Fig. 1.

picture originals. They should all be suitable for reproduction on newsprint. The reflection copies should be consistent in density range, and the important details should lie in the best reproducible density area, i.e. in the middle tones.

Control strips should be used to test processing, exposure and developing to secure constant conditions in phototype setting and plate making.

We will, in the following, concentrate on the printing conditions, adjusting the screen processing, and colour separation for the printing press.

SETTING THE PRINTING CONDITIONS AND ADJUSTING THE PRE-PRESS REPRODUCTION TO THE PRINTING PRESS CHARACTERISTICS

Principles

(1) The press has to print in one definite way, and in the best way, with specified mechanical printing conditions and with specified ink film thickness on paper (the normal conditions).

(2) Camera work (negative film) and plate-making must be carried out at constant conditions (exposure and processing).

(3) Accommodation of the screening process and colour separation to the printing press (when printing at the fixed normal conditions).

Specification of the 'normal' printing conditions is illustrated by the example of the form shown in Figure 1. Special attention should be given to setting the pressure of the inking rollers and the offset blanket cylinder. The pressure should be sufficient, but not more than necessary.

In the case of the offset blanket, increasing the pressure (increased underpacking) can give increasing print evenness on the paper, but at the same time the pressure against the printing plate will also exaggerate dot gain and plate wear.

Solid ink density

As a measure for the ink film thickness, solid ink density for black and white printing should be chosen from the following criteria:

(1) Maximum density difference (absolute contrast) between printed solids and a 70 to 75% screened area. (Test element with same screen ruling as in the halftone pictures).

(2) Not so much ink that smearing, set-off or rub-off is a problem.

The solid ink densities for the cyan, magenta and yellow are chosen by the following criteria:

(1) The same as for black and white.

(2) The densities which have to be used when printing 4-colour advert-

isements where the colour separations or film negatives are supplied ready made from the advertising agency (Identical copies to many different printing companies).

To find these densities, small solid areas for density measurements should be placed at the bottom of the page, under the 4-colour picture in the printing direction. The ink densities are adjusted so that the print is in the best possible agreement with the supplied 'proof-print', (if any) or so that the colours 'look good'.

The average densities, for example, from ten 4-colour advertisements (preferably different advertisements), can be chosen and fixed as normal ink densities for cyan, magenta and yellow. The black density in 4-colour should be the same as for black and white printing.

Matching 'in-house' colour separations (editorial colour pictures) with ready made colour separations as received from the advertising agencies

When making 'in-house' colour separations, these should be accommodated to the printing press, when printing at the same 'normal' ink densities as explained in the foregoing. Editorial and advertising 4-colour pictures can then be printed side by side at the same ink density levels without problems.

ACCOMMODATION OF THE SCREENING PROCESS (THE DOT SIZES) TO THE NORMAL PRINTING CONDITIONS

Black and white

This can be done as illustrated by the data sheet, Figure 2 and Figure 4, for 'black and white halftone reproduction'. First the screentype and ruling has to be chosen. Then a grey scale is treated as a reflection picture original. The densities in the different steps are measured on the reflection original, on the negative film (where the dot sizes/percent dot area are estimated from Figure 3) and when printed on paper. The density figures are put into the form and then used to construct the curves in the diagram.

The curve in the upper right quadrant (Figure 4) represents the printing characteristic of the press and should be constant. The curve in the lower right quadrant represents the camera curve, from the screen positive (on the paste-up) to the negative film and should also be constant.

It is then easy to construct the lower left curve in Figure 4 (from reflection original to screen positive) which will result in the best tone rendering curve; in the upper left quadrant, for example, the straight line as indicated. The two constant curves to the right are then used as transfer curves. When the required lower left hand curve is constructed, variations in the screening process is carried out until the proper screening process conditions are found.

Printing test. Date:

Black and white halftones.

Solid ink density : 0.97
"White" paper density: 0.2

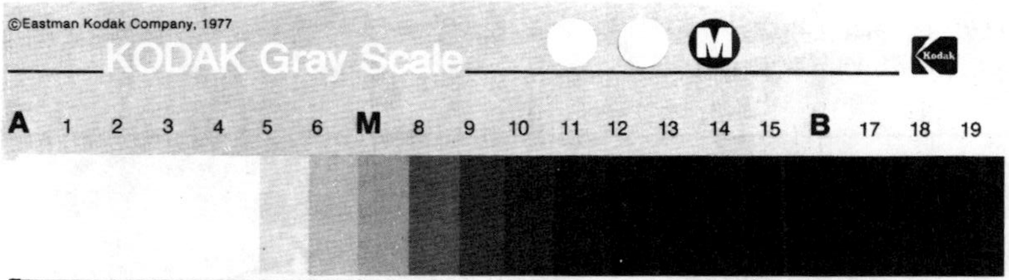

		A	1	2	3	4	5	6	M	8	9	10	11	12	13	14	15	B	17	18	19
Reflection-densitometer. Type: Cosar SOS 40	Reflection density of original	0.04	0.17	0.29	0.39	0.50	0.59	0.69	0.78	0.87	0.97	1.07	1.16	1.25	1.34	1.42	1.52	1.59	1.68	1.77	1.87
	Reflection density of print	0.26	0.30	0.37	0.43	0.46	0.53	0.60	0.63	0.66	0.69	0.72	0.75	0.78	0.81	0.82	0.84	0.86	0.87	0.89	0.90
	Reflection density of screen positive on fotographic paper	0.19	0.21	0.29	0.36	0.40	0.48	0.59	0.63	0.67	0.71	0.77	0.85	0.91	0.95	0.98	1.03	1.12	1.15	1.19	1.25
Transmission densitometer Type: MacBeth	Transmission-density on film negative	1.26	1.02	0.71	0.55	0.50	0.41	0.31	0.29	0.25	0.21	0.18	0.17	0.14	0.12	0.11	0.10	0.08	0.07	0.06	0.05
	Percent dot area % on negative film % The transparent dots which becomes printing dots on plate. From the dot size estimation curve/table	6	10	19	28	32	39	49	52	57	62	66	68	73	76	78	80	83	85	87	89

BLACK AND WHITE HALFTONE REPRODUCTION

REFLECTION ORIGINAL → SCREEN POSITIV → NEGATIVE FILM → (PLATE) → PRINT

Table, with figures to be used when drawing the curve-diagram

Fig. 2.

ESTIMATION OF THE DOT SIZE ON NEGATIVE FILM

Percent area of the transparent dots.

The transmission densitometer is zero'ed on transparent film.

The relationship between transmission density and percent dot area is given by the table and/or the curve.

Density	%	Density	%	Density	%	Density	%
0.00	100	0.17	68	0.34	46	0.64	23
0.01	98	0.18	66	0.35	45	0.68	21
0.02	96	0.19	65	0.36	44	0.72	19
0.03	93	0.20	64	0.37	43	0.76	18
0.04	91	0.21	62	0.38	42	0.80	16
0.05	89	0.22	61	0.39	41	0.84	15
0.06	87	0.23	59	0.40	40	0.88	13
0.07	85	0.24	58	0.42	38	0.92	12
0.08	83	0.25	57	0.44	37	0.96	11
0.09	81	0.26	55	0.46	35	1.00	10
0.10	80	0.27	54	0.48	33	1.10	8
0.11	78	0.28	53	0.50	32	1.25	6
0.12	76	0.29	52	0.52	30	1.50	3
0.13	74	0.30	50	0.54	29	1.75	2
0.14	73	0.31	49	0.56	28	2.00	1
0.15	71	0.32	48	0.58	26	(Black area of film, above 3.00)	
0.16	70	0.33	47	0.60	25		

DOT SIZE ON NEGATIVE FILM

Percent dot area of the transparent dots.

(which becomes printing dots on the plate.)

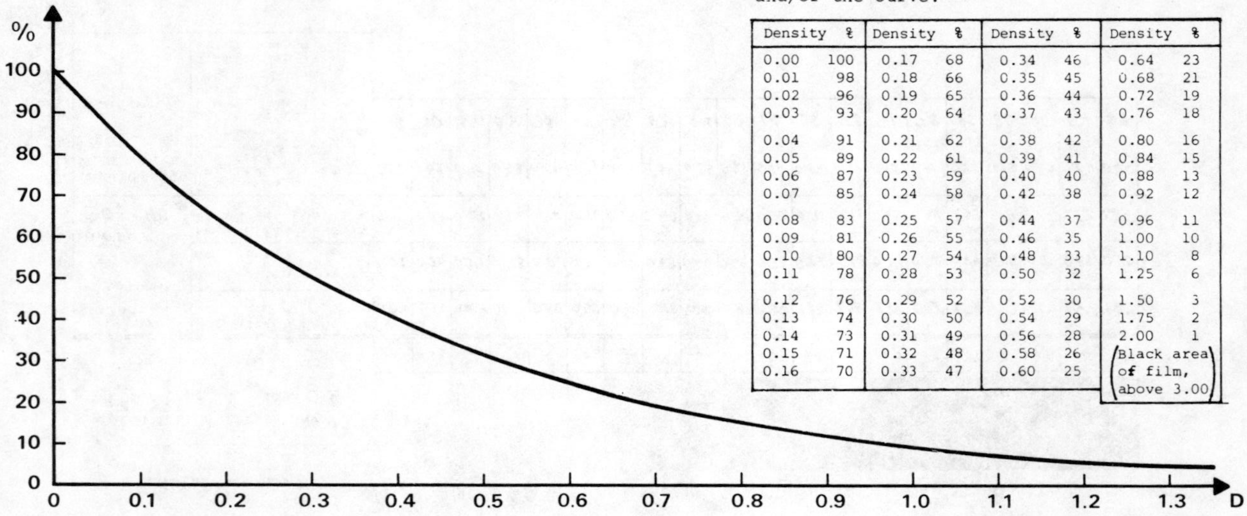

TRANSMISSION DENSITY OF HALFTONE NEGATIVE FILM.

(Zero on transparent film)

Fig. 3.

BLACK AND WHITE HALFTONE REPRODUCTION CURVES

Reflection original — screen positive — negative film — print

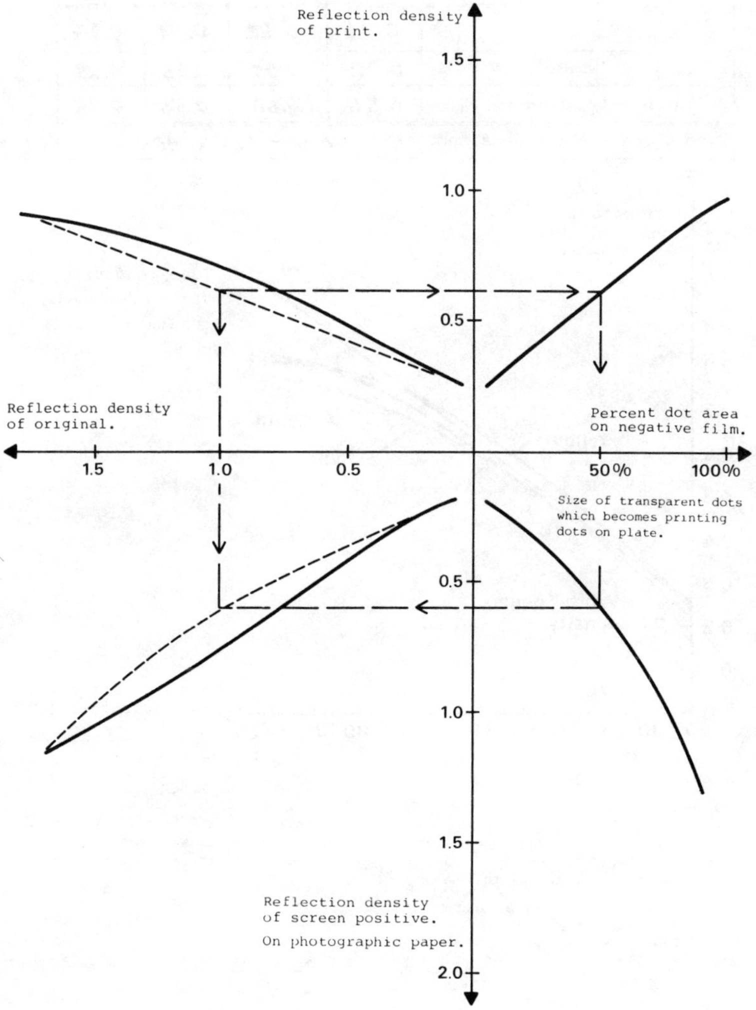

Fig. 4.

4 - COLOR

SPECIFICATION OF THE PRINTING PRESS REPRODUCTION CHACTERISTICS

(At the specified normal printing conditions.)

The normal ink densites are:	CYAN	MAGENTA	YELLOW	BLACK
Solid ink density Directly read from densitometer	0.78	0.82	0.81	0.97
Density on "white" paper	0.22	0.22	0.25	0.22
Solid density above paper	0.56	0.60	0.56	0.75
Type of reflection densitometer: Cosar SOS 40				

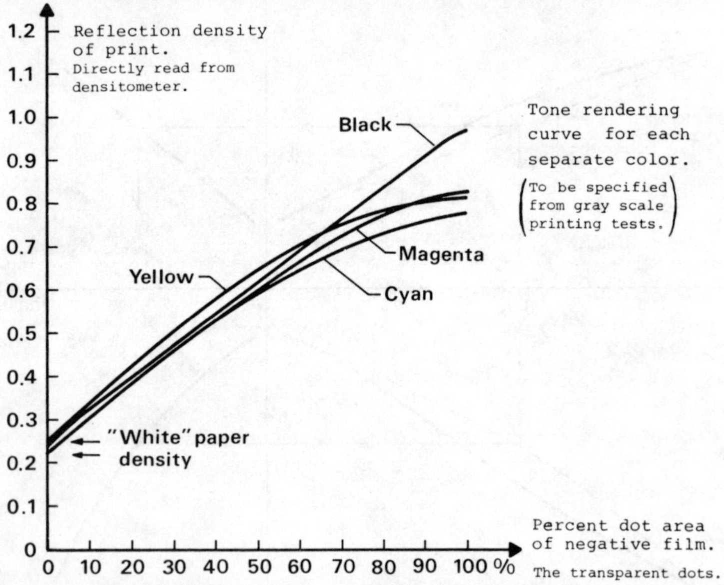

REQUIREMENTS FOR GRAY BALANCE (The relative dot areas)

To obtain neutral gray when printing CYAN, MAGENTA and YELLOW screened areas onto each other, with specified solid ink density (ink film thickness) and in specified color sequence, the dot sizes on the negative films (percent area of transparent dots) have to be adjusted relative to each other as specified in the following.

When the dot size of CYAN is about: the dot sizes of the two other colors must be larger or smaller by:		20%	50%	70%
(To be specified from gray balance printing test.)	MAGENTA	x %	x %	x %
	YELLOW	x %	x %	x %

Fig. 5.

4-Colour

The specification of the printing press reproduction characteristics is obtained as indicated in Figure 5. The colour separations/dot sizes on the negative films have to be accommodated to these characteristics.

DISCUSSION

W. D. Schaeffer You state in your paper that the black density in 4-colour should be the same as in black and white printing. Does this density refer to that of the black ink printer only as the cumulative four-colour print in the shadow areas?

We believe that the lack of shadow detail reproduction in newsprinting detracts from quality generally.

By appropriate choice of originals or by limiting the cumulative ink densities in shadow areas, improvement in quality reproduction can be obtained.

Åge Hansen The black density when printing 4-colour refers to the cumulative black printer. That is, the black printer solid areas above or below the 4-colour picture should have the same density and ink film thickness as do other black and white parts in the newspaper.

As to your comments, I agree. UCR-technique should be used to reduce the total ink film thickness in 4-colour shadows and original pictures with important details in shadows (and in highlights) should be omitted.

J. L. Poujade Practical experience shows that, in normal production conditions, quality variations due to start-ups are much wider than those caused by all the other factors, generally resulting in the biggest item in the waste statistics of any newspaper. In this field, the quality of the end product depends essentially on a very quick decision as to which copies are to be sorted out by the printing staff.

Is there, in your program, any specific training of printers in order that they can make adequate decisions to meet the requirements of the newspaper publisher and advertiser?

Åge Hansen The 'training' is participation in the quality group, with regular discussions of quality level and waste limits. Reactions from publishers and advertisers will be a prime guide.

INK TRANSFER EQUATIONS - PARAMETER ESTIMATION AND INTERPRETATION

P. J. Mangin, M. B. Lyne, D. H. Page and J. H. De Grâce, Pulp & Paper Research Institute of Canada

ABSTRACT

Ink transfer equations have been examined for goodness of fit to ink transfer data obtained for 29 different printing conditions. Solid letterpress plates and newsprint were used in this study. Fitting of the fraction of ink transferred from the plate to paper was found to result in an homogeneous residual error. The Walker-Fetsko equation was examined extensively and was found to have the best fit of all three-parameter equations. However, intercorrelation between estimates for the absorbency parameter and the split parameter, variation in the parameter estimates with the endpoint in ink weight on the printing plate, and the general inelasticity of the equation with respect to its parameters mitigate against interpretation of the parameters as general printing characteristics of paper. The addition of a fourth parameter to the Walker-Fetsko equation as proposed by Karttunen slightly improves the fit, but seriously increases intercorrelation among the parameter estimates. A fourth parameter power correction to the coverage factor in the Walker-Fetsko equation improves the fit significantly. However, it is concluded that modelling of the phenomena involved in ink transfer to paper is required if an ink transfer equation is to be created for purposes of describing the printing characteristics of paper.

INTRODUCTION

Perhaps the oldest and most elementary form of printing is the transfer of an ink consisting of a vehicle and pigment from a printing plate to porous paper when the two are pressed together. The idea of modelling this process mathematically in order better to understand ink transfer phenomena is not new, and yet only modest success has been achieved to date in elucidating the ink and paper factors that control the ink transfer process.

An S-shaped curve is characteristic of the change in transfer

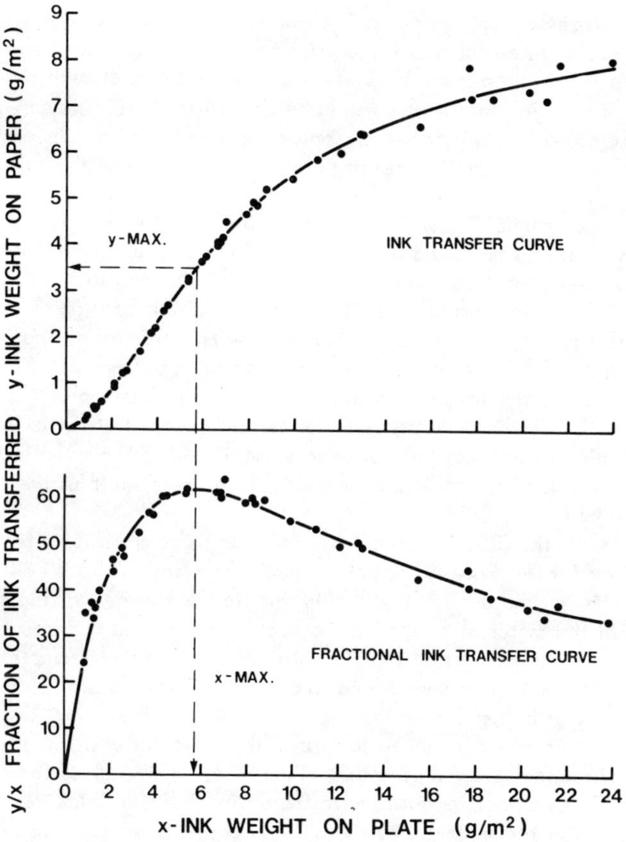

Fig. 1. Typical ink transfer and fractional ink transfer curves for newsprint printed at 4.6 m/s and 15.4 kN/m in a GFL rotary press.

of ink to paper as the amount of ink on the printing plate is varied. The transfer curve for a black news ink (14% carbon pigment in an oil vehicle) from the solid printing plates of a GFL rotary press to newsprint at a printing speed of 4.6 m/s and a line pressure of 15.4 KN/m is shown in Figure 1. Also shown are the same data expressed as the fraction of the ink on the printing plate that transferred to the paper.

As recognized by Walker and Fetsko[16], the S-shape of the transfer curve is due to several competing phenomena that control the transfer process. For example, if a perfectly smooth oil-receptive material were substituted for paper and printing was done at slow speed (i.e. less than 1 m/s) then the ink transfer curve would be a straight line and the fraction of ink transferred would be a constant 50%. Roughening the material would result in a loss of contact between the ink film and the material where surface grooves or pores were deeper than the thickness of the ink film.

At higher ink weights on the printing plate, progressively deeper pores would be bottomed as the thicker ink film was forced into the pores by the hydraulic pressure created in the printing nip. Thus, the fraction of ink transferred would be less than 50% at low ink weights on the plate (tending to zero percent at low ink film thickness) and would be greater than 50% once the surface pores had been bottomed. The latter effect is due to the immobilization of ink in the surface pores.

When the printing plate leaves the material at the exit of the printing nip the ink that has not been immobilized in the surface pores would split equally, leaving 50% of this 'free' ink film (plus the immobilized ink) with the printed material.

As the printing press is speeded-up to commercial rates the time taken for the material to pass through the printing nip is decreased to the order of one millisecond. On this time scale the wetting of the surface of the material by the ink may be imperfect and the flow behaviour of the ink during the ink film split can become very complex and irregular. These phenomena tend to depress the ink transfer curve, particularly at higher ink weights on the printing plate.

In summary, the initial dip in the ink transfer curve is due to imperfect contact between the thin ink film and the rough surface of paper. This contact increases with increasing ink film thickness causing the fraction of the ink transferred to the paper to increase. Ink is immobilized in the surface pores of the paper. The amount immobilized increases as the ink film thickness is increased up to a maximum value that depends on the surface porosity of the paper under printing compression. At normal printing speeds the split of the remaining 'free' ink film is less than 50%, and decreases with printing speed (Schaeffer, Fisch and Zettlemoyer[15]). The free ink split predominates at higher ink weights on the printing plate causing the overall fraction of ink transferred to diminish as higher ink weights on the printing plate are reached.

Many researchers have fitted the S-shaped ink transfer curve

TABLE 1 INK TRANSFER EQUATIONS FOR SOLID PRINTS

Authors	Equation	Parameters
Walker-Fetsko [16]	$y = A\left[bB + f(x - bB)\right]$ $A = 1 - e^{-kx}, \quad B = 1 - e^{\frac{-x}{b}}$	k, b, f
Rupp-Rieche [13]	$y = A\left[bB + f(x - bB)\right]$ $A = 1 - e^{-(kx)^2}, \quad B = 1 - e^{\frac{-x}{b}}$	k, b, f
Wultsch-Schubert [17]	$y = A\left[bB + f(x - bB)\right]$ $A = 1 - e^{-(kx)^{\frac{3}{2}}}, \quad B = 1 - e^{\frac{-x}{b}}$	k, b, f
Laraignou [8]	$y = A\left[b + f(x - b)\right]$ $A = \frac{x^2}{x^2 + k^2}$	k, b, f

TABLE 1 INK TRANSFER EQUATIONS FOR SOLID PRINTS (continued)

Authors	Equation	Parameters
Ichikawa et al [6]	$y = A\left[bB + f(x - bB)\right]$	f, b
	$A = \frac{\log e}{\sigma_1\sqrt{2\pi}}\int_0^x \frac{1}{x}\exp - \frac{(\log x - \mu 1)^2}{2\sigma_1^2}$, $B = \frac{\log e}{\sigma_2\sqrt{2\pi}}\int_0^x \frac{1}{x}\exp - \frac{(\log x - \mu_2)^2}{2\sigma_2^2}$	$\mu 1, \sigma_1$ μ_2, σ_2
Karttunen et al [7]	$y = Afx + (A - A_o)\, bB(1 - f)$	k, b, f, A_o
	$A = 1 - (1 - A_o)\exp - kx$, $B = 1 - \exp - x/b$	
Hultgren [4]	$y = \frac{y_{max}}{1 + \frac{1}{n}\left[\left(\frac{x_{max}}{x}\right)^{n_o} - 1\right]}$	y_{max}, x_{max} n_o
Bery [1]	$y_1 = Sx + \frac{Int\, x^2}{x_{max}^2}, \quad x \leqslant x_{max}$	x_{max}, S, Int
	$y_2 = Sx + Int, \quad x > x_{max}$	
nth order Polynomial	$y = a_o + a_1x + a_2x^2 + \ldots\ldots + a_nx^n$	$a_o, a_1, a_2, \ldots, a_n$

with equations having three or four parameters. Table 1 lists published ink transfer equations and Appendix 1 defines the symbols used. In the original Walker-Fetsko equation and subsequent modifications of it, the parameters are meant to represent the rate of increase in contact between the ink on the printing plate and the surface of the paper as the amount of ink on the plate is increased (k, or smoothness parameter), the maximum amount of ink immobilizable in the paper structure (b, or absorbency parameter) and the fraction of the free ink film which remains with the paper after the printing nip (f, or free ink film split).

As shown In Table 1, the Walker and Fetsko (WF) equation incorporates the k and b parameters in two exponential factors: A is the fraction of the paper surface covered with ink at ink level x on the plate; B is the fraction of the total amount of ink that could be immobilized at high ink levels which is actually immobilized at ink level x.

Rupp-Rieche[13] and Wultsch-Schubert[17] applied powers of 2 and 1.5, respectively, to the exponential function in the WF coverage factor A. The immobilization factor B remained unchanged.

Laraignou[8] (L) attempted to simplify the WF equation by letting the immobilization Factor B equal unity and replacing the exponential function in the coverage factor with the homographic function $x^2/(x^2+k^2)$.

Ichikawa et al[6] proposed a more complex equation in six parameters. The exponential functions of the WF equation were replaced with lognormal distribution functions meant to describe the distribution of contact between ink and fibres in a sheet of paper. The authors found physical interpretation of the mean and the standard deviation of these logarithmic functions to be difficult.

Karttunen et al[7] (K) modified the WF equation by adding a fourth parameter A_o, which they called the 'flattened fraction', meant to account for increased contact with the ink due to the flattening effect of printing pressure on the surface of paper.

Hultgren[4] (H) proposed a very simple equation based on three parameters - the amount of ink on the plate x_{max} and the amount of ink on the paper y_{max} at the point of maximum fractional transfer (See Figure 1), and a parameter n_o relating to the paper structure.

Bery[1, 2] (B) developed two models based on a theoretical approach to ink transfer. The ink transfer curve is divided into a quadratic portion before the point of maximum fractional ink transfer and a linear portion thereafter.

In this report the WF, L, K, and H equations, and Bery's second model (a three parameter, two equation model) have been fitted to 29 sets of ink transfer data using modern non-linear methods. Newsprint has been chosen as the test paper since its roughness, porosity and compressibility should generate considerable variation in parameters supposedly related to contact with an ink film under printing pressure, immobilization of ink in the paper structure, and ink film splitting from an irregular surface.

METHODS AND RESULTS

Experimental

Five different newsprint samples (See Appendix 2) representing a wide range in roughness and absorption characteristics have been printed with a solid plate in a GFL rotary letterpress press according to SCAN Standard P35:72. Three papers (A, B, C) have been printed over a wide range of speeds (1.1 m/s, 3.3 m/s, 4.6 m/s and 5.9 m/s) and pressures (10.7 kN/m, 15.4 kN/m and 17.85 kN/m), and papers D and S have been printed at 4.6 m/s and 15.4 kN/m. Approximately forty prints were made at different levels of ink on the plate for each transfer curve. The printing plate was deighed before and after inking and after printing. The weight of ink on the plate and the weight of ink transferred to paper (loss of ink on the printing plate) was calculated from these weighings.* Six data sets from the literature (listed in Appendix 2) were also analysed. These data sets had between 15 and 34 points per transfer curve. Thus a total of 29 ink transfer curves covering a broad range of printing conditions were used to examine the ink transfer equations.

Fitting Experimental Data

Marquardt's method [10, 11, 12], of curve fitting has been used to fit the ink transfer equations to the ink transfer data. This method is described in Appendix 3.

One of the basic requirements for proper use of regression analysis is homogeneity in the residual error. The difference between the observed value and that predicted by the equation should not be correlated with the level of the value. The ink transfer curve does not satisfy this condition. The scatter about the curve increases for increasing ink weights as shown in Figure 1. However, the fractional ink transfer curve of Figure 1 appears to have a homogeneous error. This is confirmed by fitting the WF equation to typical data. Figure 2, upper, shows the increase in residual error with ink weight for the ink transfer curve while Figure 2, lower, shows the homogeneity of residual error for the fractional ink transfer curve. For this figure and all the subsequent work of this report, the data were fitted to the fractional ink transfer curve using the 29 data sets.

The change in mean residual error with increasing ink weights for the fractional transfer fit reflects the poorness of fit of the model. Figure 3 shows the residual errors for the fractional transfer fit of the L equation. Clearly, the L equation has a poorer fit than the WF equation. As shown in Figure 4, the addition of a fourth parameter A_o to the WF equation gives the K equation a slightly better fit (the WF equation is equivalent to the K equation with $A_o = 0$) but the improvement hardly

*This data is available upon request to the Pulp and Paper Research Institute of Canada.

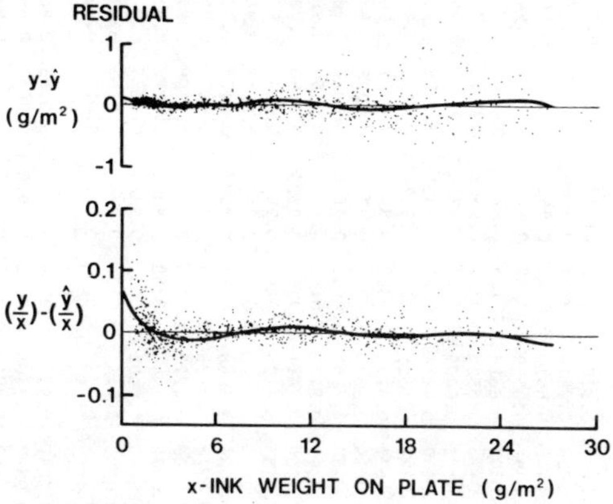

Fig. 2. *Distribution of the residual errors for the WF equation by conventional (upper) and fractional transfer fitting (lower) for the 23 ink transfer data sets obtained with the GFL rotary press.*

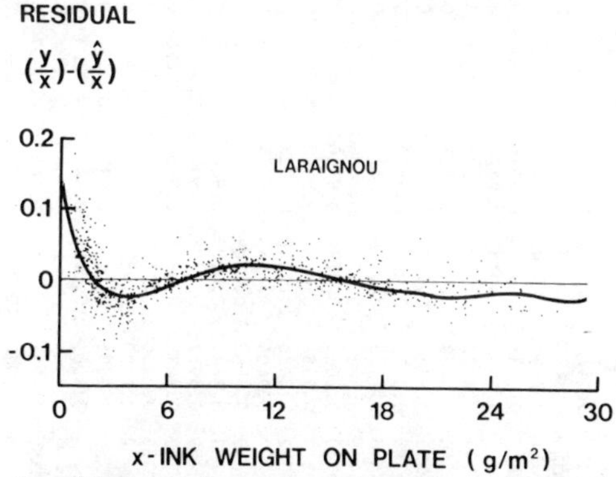

Fig. 3. *Distribution of the residual errors for the L equation by fractional transfer fitting for the 23 ink transfer data sets obtained with the GFL rotary press.*

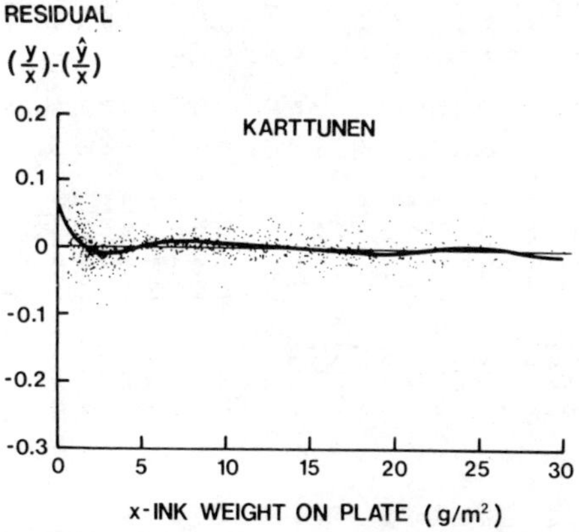

Fig. 4. *Distribution of the residual errors for the K equation by fractional transfer fitting for the 23 ink transfer data sets obtained with the GFL rotary press.*

TABLE 2

Sample and Printing Conditions			Equation WF SD	Equation WF r^2	Equation L SD	Equation L r^2	Equation K SD	Equation K r^2	Equation H SD	Equation H r^2	Equation B SD	Equation B r^2	Equation γ SE	Equation γ r^2	Poly(6th order) SE	Poly(6th order) r^2
Printed at PPRIC																
A,	4.6 m/s,	15.4 kN/m	0.0235	0.9753	0.0359	0.9424	0.0235	0.9759	0.0288	0.9608	0.0641	0.8208	0.0213	0.9801	0.0227	0.9795
A*,	4.6 m/s,	15.4 kN/m	0.0231	0.9563	0.0358	0.8497	0.0231	0.9575	0.0284	0.9334	0.0428	0.8544	0.0205	0.9666	0.0188	0.9744
B,	4.6 m/s,	15.4 kN/m	0.0251	0.9504	0.0211	0.9647	0.0205	0.9679	0.0247	0.9492	0.0447	0.8470	0.0212	0.9657	0.0217	0.9671
C,	4.6 m/s,	15.4 kN/m	0.0322	0.8975	0.0422	0.8237	0.0310	0.9076	0.0303	0.9040	0.0391	0.8529	0.0285	0.9220	0.0251	0.9443
C*	4.6 m/s,	15.4 kN/m	0.0135	0.9724	0.0191	0.9448	0.0137	0.9724	0.0155	0.9613	0.0307	0.8613	0.0136	0.9728	0.0130	0.9772
D,	4.6 m/s,	15.4 kN/m	0.0210	0.9557	0.0293	0.9139	0.0204	0.9595	0.0201	0.9569	0.0415	0.8324	0.0202	0.9603	0.0216	0.9587
D*,	4.6 m/s,	15.4 kN/m	0.0251	0.9317	0.0348	0.8689	0.0229	0.9451	0.0237	0.9360	0.0552	0.6809	0.0215	0.9513	0.0228	0.9505
S,	4.6 m/s,	15.4 kN/m	0.0218	0.9521	0.0255	0.9345	0.0221	0.9521	0.0272	0.9213	0.0339	0.8876	0.0221	0.9521	0.0199	0.9649
A,	4.6 m/s,	10.7 kN/m	0.0215	0.9347	0.0315	0.8602	0.0207	0.9409	0.0258	0.9018	0.0327	0.8528	0.0195	0.9476	0.0161	0.9673
A,	4.6 m/s,	17.9 kN/m	0.0359	0.8378	0.0448	0.7476	0.0349	0.8512	0.0469	0.7076	0.0627	0.5197	0.0307	0.8847	0.0297	0.9016
A,	1.1 m/s,	15.4 kN/m	0.0278	0.9584	0.0426	0.9020	0.0243	0.9688	0.0279	0.9559	0.0864	0.6062	0.0212	0.9762	0.0208	0.9788
A,	3.3 m/s,	15.4 kN/m	0.0197	0.9505	0.0323	0.8666	0.0194	0.9530	0.0257	0.9107	0.0480	0.7140	0.0151	0.9718	0.0129	0.9811
A,	5.9 m/s,	15.4 kN/m	0.0182	0.9664	0.0253	0.9351	0.0183	0.9668	0.0198	0.9577	0.0395	0.8458	0.0182	0.9673	0.0190	0.9674
C,	4.6 m/s,	10.7 kN/m	0.0235	0.8912	0.0312	0.8086	0.0226	0.9023	0.0278	0.8395	0.0368	0.7411	0.0201	0.9230	0.0190	0.9369
C,	4.6 m/s,	17.9 kN/m	0.0163	0.9604	0.0200	0.9404	0.0166	0.9604	0.0218	0.9252	0.0355	0.8179	0.0166	0.9604	0.0130	0.9777
C,	1.1 m/s,	15.4 kN/m	0.0330	0.9291	0.0449	0.8683	0.0309	0.9397	0.0266	0.9512	0.0600	0.7720	0.0287	0.9477	0.0248	0.9644
C,	3.3 m/s,	15.4 kN/m	0.0221	0.9286	0.0282	0.8832	0.0224	0.9287	0.0303	0.8582	0.0306	0.8668	0.0221	0.9307	0.0177	0.9591
C,	5.9 m/s,	15.4 kN/m	0.0205	0.9316	0.0259	0.8912	0.0208	0.9316	0.0252	0.8911	0.0278	0.8776	0.0199	0.9373	0.0171	0.9577
D,	4.6 m/s,	10.7 kN/m	0.0218	0.9335	0.0262	0.9038	0.0200	0.9454	0.0228	0.9234	0.0486	0.6773	0.0195	0.9482	0.0200	0.9499
D,	4.6 m/s,	17.9 kN/m	0.0259	0.9372	0.0286	0.9230	0.0257	0.9396	0.0264	0.9306	0.0475	0.7944	0.0258	0.9393	0.0271	0.9385
D,	1.1 m/s,	15.4 kN/m	0.0334	0.8650	0.0452	0.7520	0.0278	0.9086	0.0327	0.8628	0.0403	0.8082	0.0251	0.9260	0.0279	0.9162
D,	3.3 m/s,	15.4 kN/m	0.0164	0.9711	0.0215	0.9505	0.0159	0.9737	0.0234	0.9383	0.0500	0.7441	0.0154	0.9751	0.0161	0.9752
D,	5.9 m/s,	15.4 kN/m	0.0164	0.9457	0.0204	0.9161	0.0165	0.9465	0.0225	0.8919	0.0322	0.7965	0.0164	0.9472	0.0147	0.9614
CC-1	0.4 m/s,	35 kN/m	0.0333	0.8411	0.0351	0.8231	0.0311	0.8657	0.0307	0.8557	0.0261	0.9052	0.0309	0.8675	0.0163	0.9670
EE-1	0.8 m/s,	8.8 kN/m	0.0143	0.9822	0.0055	0.9972	0.0199	0.9882	0.0211	0.9540	0.0393	0.8708	0.0060	0.9970	0.0055	0.9980
FF-1	0.8 m/s,	17.5 kN/m	0.0177	0.9834	0.0173	0.9842	0.0170	0.9858	0.0227	0.9679	0.0463	0.8957	0.0183	0.9837	0.0105	0.9961
GG-1	0.8 m/s,	35 kN/m	0.0126	0.9881	0.0135	0.9862	0.0128	0.9883	0.0119	0.9880	0.0376	0.9017	0.0117	0.9902	0.0098	0.9943
QQ-1	0.8 m/s,	8.8 kN/m	0.0153	0.9846	0.0077	0.9961	0.0114	0.9921	0.0130	0.9875	0.0506	0.8438	0.0066	0.9974	0.0048	0.9989
MM-3	0.8 m/s,	17.5 kN/m	0.0545	0.7836	0.0456	0.8484	-	-	0.0494	0.7999	0.0795	0.5680	0.0295	0.9404	0.0147	0.9882

* Duplicate Experiment

Table 2. Fit of major ink transfer equations to 29 fractional ink transfer curves expressed as coefficient of determination (r^2) and the standard deviation of the residuals (SD).

justifies the incorporation of a fourth parameter. As shown in Figure 5, the residual error for the B equation is distinctly different and the overall fit is poorer than for the preceding equations. The fit is particularly poor at low ink weights and at the point of maximum ink transfer (interface between the linear and quadratic functions).

The coefficient of determination (r^2) for each of the ink transfer equations fitted to the individual ink transfer curves is listed in Table 2. A Wilcoxon ranked-sign test performed on the standard deviation of the residuals of the 29 sets of data (Table 2) ranks the equations from the best fit to the worst fit in the following order: K equation, WF equation, H equation, L equation and B equation.

Rupp and Rieche[13], and Wultsch and Schubert[17] modified the WF equation by applying powers of 2 and 1.5, respectively, to the exponential term in the coverage factor. In both cases the correlation was lower than for the WF equation. However, if this power is fitted to the data as an additional fourth parameter, γ, in the WF equation the fit is better than for the four-parameter K equation. As can be seen in Table 2, the coefficients of determination for the γ equation rank ahead of all the other ink transfer equations listed. The mean fitted value for γ was 0.9, with a standard deviation of 0.16. This is not far off the fixed value of γ (unity) in the WF equation.

The B equation shows a poor fit at low ink weights, at the point of maximum ink transfer, and after the maximum fractional transfer resulting in a lower coefficient of determination. The fitting procedure specified by Bery precludes the use of Marquardt's method of fitting the whole curve so a special fitting procedure was used for the B equation (see Appendix 3).

TABLE 3 Average Coefficient of Determination for Fractional Ink Transfer Fitting of Major Ink Transfer Equations and for the Polynomial Fitting of 29 Data Sets

Equation			Polynomial		
WF	r^2	0.9343	3rd order	r^2	0.8832
	SD	0.0491		SD	0.0654
L	r^2	0.8975	4th order	r^2	0.9514
	SD	0.0666		SD	0.0305
K	r^2	0.9470	5th order	r^2	0.9610
	SD	0.0347		SD	0.0245
H	r^2	0.9145	6th order	r^2	0.9653
	SD	0.0602		SD	0.0229
B	r^2	0.7951	7th order	r^2	0.9674
	SD	0.1013		SD	0.0221
γ	r^2	0.9527	8th order	r^2	0.9692
	SD	0.0300		SD	0.0216

Table 3 lists the average coefficient of determination for each of the major ink transfer equations fitted to the 29 sets of fractional ink transfer data. Also listed are the average coefficients of determination for 3rd to 8th order polynomials fitted to the 29 data sets. The coefficients of determination for the ink transfer equations are no better than that for an arbitrary 4th order polynomial which has five parameters. Also, if the standard deviation of the residual errors for, say, a 6th order polynomial (see Table 2) is taken as an indication of the true discrepancy between the data points and a smooth curve drawn through them, there remains a significant amount of residual error that could be explained by an improved ink transfer equation.

Intercorrelation between the Ink Transfer Parameter Estimates

The method for calculating the intercorrelation among the parameter estimates is explained in Appendix 3. Essentially, the covariances of the parameter estimates determine the intercorrelation which can be thought of as the degree of orientation of the confidence region about the parameter estimates in the multi-parameter space (see Figure 7 for a two-parameter example).

As shown in Table 3, the average coefficient of determination for the 29 data sets is 0.934 for the three-parameter WF equation and 0.947 for the four-parameter K equation. However, the intercorrelations among the parameters in the K equation (see Table 4) are all over 0.85 and highly significant. This may be compared with the intercorrelations between k and f, and between k and b in the WF equation which are 0.17 and -0.27, respectively, (not significant at the 0.05 probability level). This implies that the addition of the fourth parameter A_o to the WF equation causes the K equation to be over-determined (Hartley[3], Lipton and McGilchrist[9], Hunter and Reiner[5]).

TABLE 4 Average and Standard Deviation of Correlation Coefficient r among Estimates for Parameters k, b, f and A_o for 29 Data Sets

Equation		k vs f	k vs b	b vs f
WF	r	0.17	-0.27	-0.97
	SD	0.15	0.15	0.02
L	r	-0.90	0.97	-0.91
	SD	0.03	0.01	0.03
K	r	-0.89	0.96	-0.96
	SD	0.18	0.06	0.06
		A_o vs b	A_o vs k	A_o vs f
K	r	0.95	0.97	-0.86
	SD	0.07	0.04	0.14

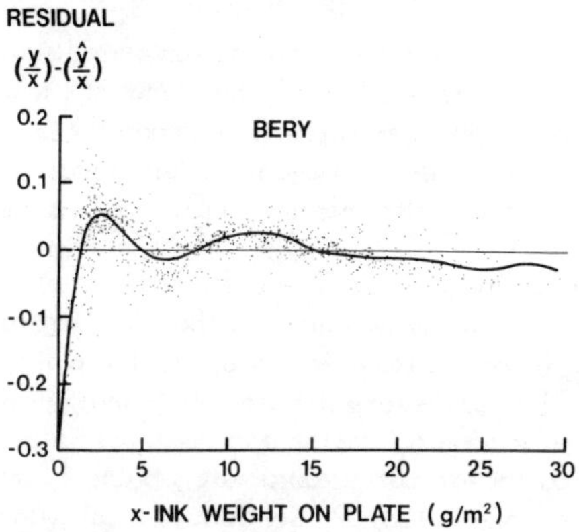

Fig. 5. Distribution of the residual errors for the B equation by fractional transfer fitting for the 23 ink transfer data sets obtained with the GFL rotary press.

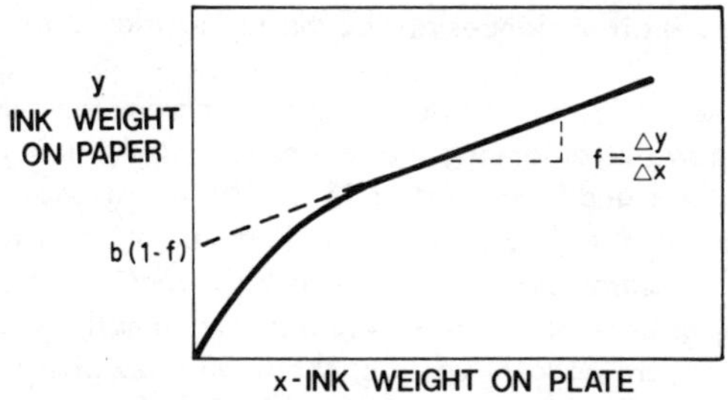

Fig. 6. Extrapolation method for calculation of parameters f and b.

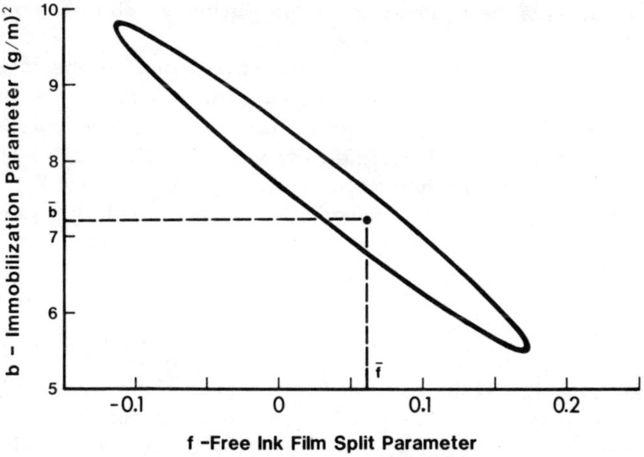

Fig. 7. Mean fitted values for b and f parameters in the WF equation for a newsprint sample printed at 4.6 m/s and 15.4 kN/m. Distorted ellipse is the 95 per cent confidence region for the parameter estimates.

The L equation, in which the exponential coverage term of the WF equation has been replaced with the homographic function $x^2/(x^2+k^2)$ suffers from a very high intercorrelation among the parameter estimates, as shown in Table 4. The coefficient of determination for the L equation fit to the 29 data sets is also inferior in all cases to the WF equation (see Table 2).

Considering the overall coefficient of determination and the intercorrelation among the parameters, the WF equation is superior to the L, K, and B equations. However, there is still a significant intercorrelation between the parameter estimates for b and f in the WF equation. The reason for this intercorrelation can be appreciated by examining the graphical method for estimating parameters b and f used by the earlier researchers (Walker and Fetsko[16]; Schaeffer, Fisch, and Zettlemoyer[15]; Zeleski, Schaeffer, and Zettlemoyer[18]). As shown in Figure 6, the WF equation reduces to a straight line with slope f and intercept b(1-f) at high ink weights on the plate. Since b and f are multiplied together in the intercept, their variances are bound to be related in the fitting process.

The difficulty of interpreting the parameters when there is a serious intercorrelation among the parameter estimates is illustrated in Figure 7 for the f and b parameters in the WF equation. The 95% confidence region forms a distorted ellipse about the mean fitted values for b and f in the b-f parameter space. This 95% confidence region includes a wide range of both parameters (eg. a negative split parameter and a high value of b are equally valid at the 0.05 probability level as the mean fitted values). In other words, this high degree of orientation of the confidence region for these parameter estimates indicates that neither parameter can be estimated independently of the other and that fitted combinations of the two parameters are possible which make no sense physically.

In Figure 7 the range of values for b and f in the 95% confidence region is large relative to the standard deviation of the fitted values for b and f for the 29 data sets (see Table 5). In Table 5 it can be seen that the mean standard errors for the fitted values of f and A_o are high relative to their fitted mean values in the WF, L, and K equations. This imprecision in the determination of parameter estimates seriously hampers any attempt to interpret the parameters in terms of paper or ink properties.

<u>Sensitivity of Ink Transfer Equations to Changes in the Parameters</u>

Mathematical calculation of the change in the fractional ink transfer caused by a shift in the k, b, and f parameters of the WF equation is described in Appendix 4. A shift equal to one standard deviation about the mean fitted values of k, b, and f for the 29 data sets is used in Figure 8 to show the sensitivity of the WF equation to changes in the parameters.

TABLE 5 Mean Value and Range for Ink Transfer Parameter Estimates for 29 Data Sets

	WF Equation	L Equation	K Equation
$\bar{k}$	0.333	4.807	0.401
SD	0.111	2.217	0.172
SE	0.014	0.108	0.012
$\bar{b}$	6.814	5.587	8.533
SD	2.972	1.970	3.263
SE	0.245	0.134	0.284
$\bar{f}$	0.090	0.175	0.060
SD	0.140	0.111	0.108
SE	0.016	0.019	0.016
$\bar{A}_o$	-	-	0.023
SD	-	-	0.353
SE	-	-	0.019

Bery Equation

$\bar{X}_{max}$	6.90	$\bar{S}$	0.257	$\overline{INT}$	2.615
SD	2.20	SD	0.072	SD	0.732
SE	0.17	SE	0.015	SE	0.191

SD Standard deviation of the mean value of each parameter for the 29 sets of data.

SE Mean of the standard errors of each determination of the parameter for the 29 data sets.

At low ink weights the smoothness parameter k predominates. An increase in the smoothness parameter by one standard deviation causes the fractional ink transfer to increase by 0.10 at 3 g/m^2 of ink on the printing plate. At commercial levels of ink on the printing plate (i.e. about 5 g/m^2) changes in the smoothness and the absorbency parameters k and b have an equivalent effect, and the split parameter f is beginning to influence the fractional transfer. At high ink weights, changes in the smoothness parameter have little effect and changes in the split parameter f begin to predominate.

One interpretation of this result for the WF equation is that under commercial conditions ink transfer can be improved with equal effect by increasing the smoothness or increasing the porosity of paper

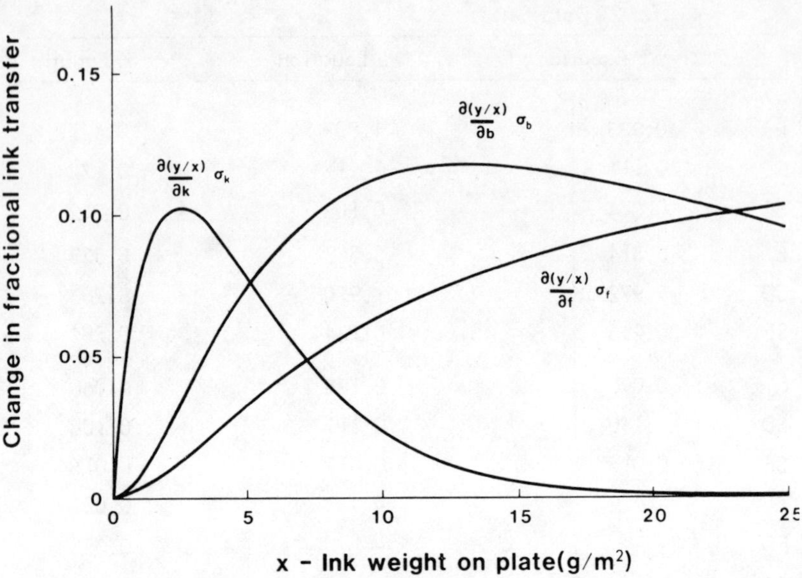

Fig. 8. Change in fractional ink transfer with a shift in parameters k, b, and f in the WF equation equal to one standard deviation about their mean fitted values for the 29 data sets.

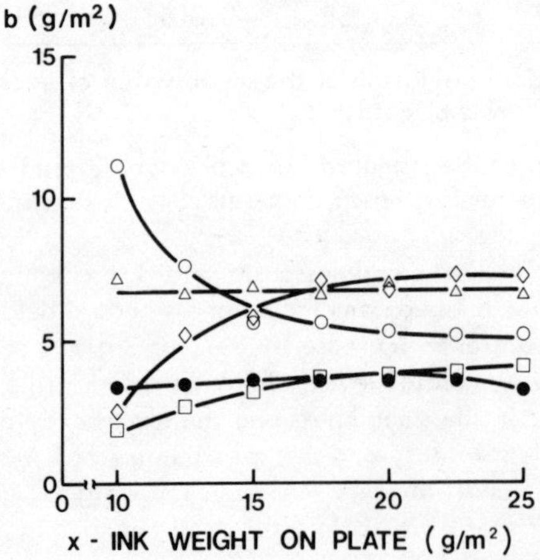

Fig. 9. Variation of the WF parameter b as a function of the end-point of the ink transfer curves for 5 newsprint samples printed at 4.6 m/s and 15.4 kN/m. Papers ◇-A, ○-B, △-C, □-D, and ●-S as listed in Appendix II.

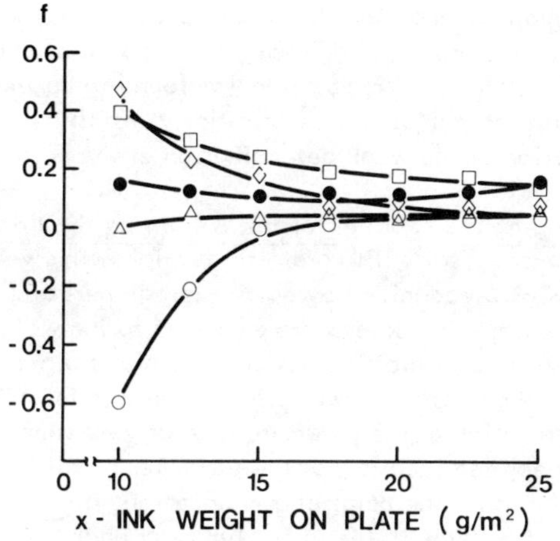

Fig. 10. Variation of the WF parameter f as a function of the end-point of the ink transfer curves for 5 newsprint samples printed at 4.6 m/s and 15.4 kN/m. Papers ◇-A, ○-B, △-C, □-D, and ●-S as listed in Appendix II.

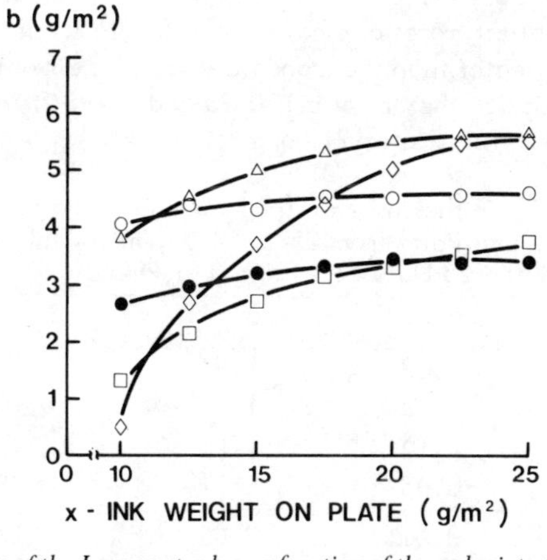

Fig. 11. Variation of the L parameter b as a function of the endpoint of the ink transfer curves for 5 newsprint samples printed at 4.6 m/s and 15.4 kN/m. Papers ◇-A, ○-B, △-C, □-D and ●-S as listed in Appendix II.

while the paper properties which influence the free ink film split cannot be neglected.

Finally, the elasticities of the WF equation to the three parameters (percentage change in ink transfer/percentage change in the parameter) have been calculated. Throughout the whole range of ink weights on the printing plate the WF equation was found to be inelastic with respect to the parameters k, b, and f (elasticity less than unity) which is another reflection of the weak determination of the parameter estimates.

Effect of Changing the Endpoint of Ink Weight on the Printing Plate

The predominance of the split parameter at higher ink weights on the printing plates was recognized by earlier researchers and was the chief reason why ink transfer studies were extended to ink weights that were far above those in commercial practice (vis. about 5 g/m^2 for letterpress). At higher ink weights transfer was primarily one of ink film splitting, making the estimation of the parameter f relatively simple. However, what has not been realized is that the estimates obtained for the parameter f (and therefore the parameter b since they are highly intercorrelated) vary considerably with the upper limit, or endpoint, of the ink weight on the plate. As shown in Figures 9 and 10 for the WF equation and Figures 11 and 12 for the L equation the estimates obtained for b and f not only vary with the endpoint ink weight, but the variation has a different character for the five newsprint samples.

The average shifts in the estimates for the ink transfer parameters in the WF, L, K, and B equations for the 29 data sets when the endpoint ink weight is reduced from 25 to 12.5 g/m^2 are listed in Table 6. These parameter estimates can have little generality if their variation with ink weight is greater than the standard error of their determination (e.g. for the WF equation the SE for b is 0.25 and the shift in b when the endpoint is reduced from 25 to 12.5 g/m^2 is 1.96).

TABLE 6 Shift in the Ink Transfer Parameters as a Function of Reducing the End Point from 25 to 12.5 g/m^2 of Ink on Printing Plate for the 23 Data Sets Printed at PPRIC

Equation		k	b	f	A_o
WF	$\overline{\Delta}$	0.02	1.96	0.16	-
	SD	0.02	1.41	0.10	-
L	$\overline{\Delta}$	1.12	1.83	0.17	-
	SD	0.65	1.11	0.08	-
K	$\overline{\Delta}$	0.04	2.05	0.17	0.11
	SD	0.04	1.55	0.14	0.10
		x_{max}	Slope	Intercept	
B	$\overline{\Delta}$	0	0.26	0.95	
	SD	-	0.07	0.48	

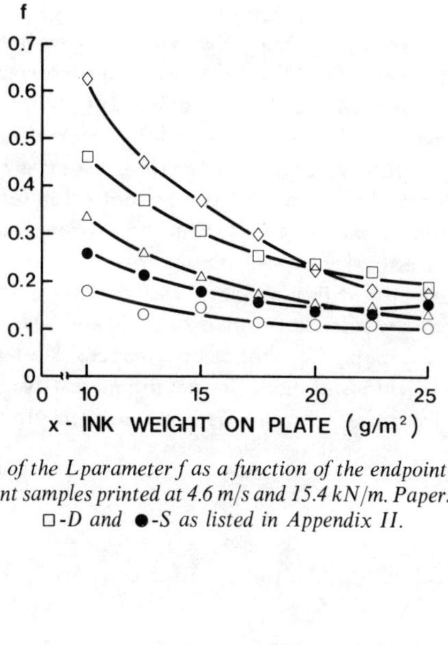

Fig. 12. Variation of the Lparameter f as a function of the endpoint of the ink transfer curves for 5 newsprint samples printed at 4.6 m/s and 15.4 kN/m. Papers ◇-A, ○-B, △-C, □-D and ●-S as listed in Appendix II.

CONCLUSIONS

The mathematical analysis of major ink transfer equations has shown some important but neglected aspects in the analysis of ink transfer curves.

By fitting the fraction of ink transferred instead of the amount of ink transferred, the residual errors become homogeneous over the whole range of ink weights on the printing plate. Fractional transfer fitting shows that the Walker-Fetsko equation is superior to the Laraignou and Bery equations, and that the addition of a fourth parameter in the Karttunen equation does not correct the basic lack of fit at low ink levels. Adjusting the coverage factor in the Walker-Fetsko equation as done by Rupp and Rieche, and Wultsch and Schubert worsens the fit, particularly at low ink levels. However, a fitted power correction to the exponential function in the coverage factor raises the correlation of the WF equation above that of the Karttunen equation.

Intercorrelation among the parameter estimates, and their change with the maximum ink weight make the interpretation of the parameter estimates in all ink transfer equations difficult and misleading. While the four-parameter equation involving a power correction in the coverage factor of the WF equation provides a better fit than all previously published ink transfer equations it does not offer further insight into the paper factors which control the ink transfer process. In order to arrive at an ink transfer equation whose parameters can be estimated independently of one another and independently of the range of ink weights on the printing plate, and whose parameters accurately reflect general printing characteristics of paper, the phenomena responsible for the transfer process must be identified and modelled mathematically. Without these models, fitting ink transfer curves is not likely to yield useful information for the printer or papermaker.

REFERENCES

1. Bery, Y., (1978), 'An Ink Transfer Equation', TAGA Proceedings, 172-191.

2. Bery, Y., (1979), 'Ink Transfer: Extended Application of the Weyerhaeuser Ink Transfer Equation', TAPPI, Vol. 62, No. 8, 89-91.

3. Hartley, H.O., (1961), 'The Modified Gauss-Newton Method for the Fitting of Non-Linear Regression Functions by Least-Squares', Technometrics, 3, No. 2, 269-280.

4. Hultgren, K., (1971), 'Calculation of Ink Transfer in Letterpress Printing', Proceedings Eleventh Conference IARIGAI, Canandaigua N.Y., U.S.A., 3-30.

5. Hunter, W. G. and Reiner, A. M., (1965), Designs for Discrimin-

ating Between Two Rival Models, Tech. Report 32, U. of Wisconsin, Depart. of Statist., Madison Wisconsin, Technometrics, 7, 307-323.

6. Ichikawa, I., Sato, K. and Ito, Y., (1962), 'A New Concept of Ink Transfer Equation', Res. Bull. Govern. Printing Bureau, Japan, No. 1

7. Karttunen, S., Kautto, H. and Oittinen, P., (1971), 'New Modifications of Ink Transfer Models', Proceedings Eleventh Conference IARIGAI, Canandaigua, N.Y., U.S.A., 33-52.

8. Laraignou, R., (1960), 'Le transfert de l'encre au papier, formules diverses, applications pratiques', ATIP Bull. No. 6, 217-225.

9. Lipton, S. and McGilchrist, C.A., (1964), 'The Derivation of Methods for Fitting Exponential Curves', Biometrica, 51, 504-508.

10. Marquardt, D.W., Bennet, G. and Burrell, E.J., (1961), 'Least-Squares Analysis of Electron Paramagnetic Resonance Spectra', J. Mol. Spectr., 7, No. 4, 269-279.

11. Marquardt, D.W., (1963), 'An Algorithm for Least-Squares Estimation of Non-Linear Parameters', J. Soc. Ind., Applied Math., Vol. 11, No. 2

12. Marquardt, D.W., (1966), 'Least-Squares Estimation of Non-linear Parameters', IBM Share Library No. 309401, NL1N2.

13. Rupp, E., and Rieche, L., (1959), 'Beiträge zur Bedrucharbeit von Papier und Folien', Inst. Grafische Technik, Leipzig.

14. SCAN P35:72, May (1972), 'Test Printing of Newsprint in the GFL Rotary Test Printing Press', Scandinavian Pulp, Paper and Board Testing Committee.

15. Schaeffer, W.D., Fisch, A. B. and Zettlemoyer, A.C., (1963), 'Transfer and Penetration Aspects of Ink Receptivity', TAPPI, June 1963, Vol. 46, No. 6, 359-375.

16. Walker, W.C. and Fetsko, J.M., (1955), 'A Concept of Ink Transfer During Printing', Am. Ink Maker, 33, No. 12, 38.

17. Wultsch, F. and Schubert, K., (1959), 'Untersuchungen vershiedener Papiersorten mit dem Fogra-Bedruchbarkeitsgerät', Das Papier, 13 (23/24), 600-607.

18. Zaleski, M., Schaeffer, W.D. and Zettlemoyer, A.C., (1969), 'An Extended Application of the Walker-Fetsko Ink Transfer Equation', Proceedings Tenth Conference IARIGAI, Krems, Austria, 175-184.

ACKNOWLEDGMENT

Emil Koller, Brenda Hilton and Cory Malin are thanked for the computer programming necessary for the analyses in this study. Emil Koller is acknowledged for many suggestions regarding curve fitting methods.

APPENDIX 1: Nomenclature

x	=	ink weight on plate (g/m^2 or μm)
y	=	ink weight on sample (g/m^2 or μm)
$Y = \frac{x}{y}$	=	fractional ink transfer
x_{max}	=	x coordinate of the peak in the fractional ink transfer curve
y_{max}	=	y coordinate of the peak in the fractional ink transfer curve corresponding to x_{max}
k	=	rate of coverage parameter (m^2/g or μm^{-1} in WF and K equations and g/m^2 or μm in L equation)
b	=	immobilization or absorption parameter (g/m^2 or μm)
f	=	free ink film split parameter
A	=	coverage factor
B	=	immobilization factor
μ_1 σ_1	=	mean and standard deviation of lognormal distribution of ink covering resistance
μ_2 σ_2	=	mean and standard deviation of lognormal distribution of ink absorption function
n_o	=	paper related parameter
A_o	=	flattened fraction
S	=	slope of linear portion of the transfer curve after x_{max}
Int	=	intercept of linear portion of the transfer curve after x_{max}
r^2	=	overall correlation (coefficient of determination)

APPENDIX 2 Papers Printed at PPRIC

	Sample	Origin	Basis Weight (g/m^2)	Apparent Density (kg/m^3)	Parker Print-Surf Roughness (μm) S10	Parker Print-Surf Roughness (μm) S20	Air Resistance Bendtsen (ml/min)
A	75% GWD +25% H.Y.S.	Eastern Canada	42.2	541	3.5	2.9	366
B	80% GWD +20% L.Y.S.	Eastern Canada	43.1	513	3.7	2.8	320
C	85% GWD +15% SBK	Western Canada	44.4	510	4.2	3.2	708
D	85% TMP +15% SBK	Eastern Canada	45.4	515	3.4	2.5	352
S*	100% TMP	Finland	41.4	545	3.2	2.3	-

* Supercalendered

Data from the Literature

Tappi Special Report 434: Transfer and Penetration Aspects of Ink Receptivity, August 1963, Tables III-CC-1, EE-1, FF-1, GG-1, QQ-1, MM-3.

APPENDIX 3: FITTING METHODS

In all non-linear techniques the model to be fitted by the data is:

$$E(y) = f(x, p_1, \ldots p_j, \ldots p_k) \tag{1}$$

where $E(y)$ = expected value for y (the ink weight on paper in g/m^2)

f = functional expression of the model

p_j = model parameters (the ink transfer parameters)

k = number of model parameters

x = independent variable (ink weight on the plate in g/m^2).

In fitting each of the models, the sum of the squares of the differences between the observed and the model predicted values is minimized:

$$\Phi = \sum_{i=1}^{n} (y_i - \hat{y}_i)^2 \tag{2}$$

y_i = observed value of ink weight on the sample

$\hat{y}_i$ = value of y_i predicted by the model

n = number of prints per curve.

When f is a linear function of the parameters p_j, the contours of a constant value of Φ are ellipsoids. A least-squares method can be used to find the optimum values for parameters where:

$$\frac{\partial < \Phi >}{\partial \delta_t} = 0 \tag{3}$$

δ_t = a correction vector for the parameter vector b, where

b = $(p_i, \ldots p_j, \ldots p_k)$ is the parameter vector.

$< >$ = predicted value from linear model versus non-linear model.

In the case of ink transfer models, the functions are far from linear and the contours of Φ are distorted accordingly. In the Gaussian method of non-linear fitting a Taylor series is used to expand the non-linear terms in a linear form:

$$<y(x_i, b + \delta_t)> = f(x_i, b) + \sum_{j=1}^{k} \left(\frac{\partial f_i}{\partial b_j}\right) (\delta_t)_j \tag{4}$$

or in a matrix form

$$<y> = f_o + P\,\delta_t \tag{5}$$

where f_o is a value of < y > corresponding to the initial or seed value of b, and

$$P^{(n,k)} = \left(\frac{\partial f_i}{\partial b_j}\right) \qquad \begin{matrix} i = 1,\ldots n \\ j = 1,\ldots k \end{matrix} \tag{6}$$

In Equation (5) δ_t is in a linear form and it is possible to determine its length by the least-squares method.

In the gradient method the negative gradient of Φ is chosen as the direction for the correction vector b.

$$\delta g = -\left(\frac{\partial\Phi}{\partial p_1},\ldots\frac{\partial\Phi}{\partial p_j},\ldots\frac{\partial\Phi}{\partial p_k}\right)^T \tag{7}$$

where $g = P^T P\ \delta_t$ (T is for matrix transposition) (8)

In both Gaussian and gradient approaches the stepsize has to be controlled carefully if convergence on a minimum value of Φ is to be achieved.

Marquardt's method[10, 11], of non-linear fitting has been used in the evaluation of the WF, L, H, K and γ ink transfer equations. It combines the Gaussian and gradient methods in a computer program which calculates the direction and length of the correction vector simultaneously. In Marquardt's method the vector is applied in controlled iterative steps to arrive at an optimum fit to the experimental ink transfer data. The computer program (Marquardt[12]) used in this study also provides options for fixing the values of parameters in order to force a constrained minimum for the function Φ. The analysis of interdependence among parameter-estimates is presented in a correlation matrix. The dependence between the parameter-estimates is represented by their 95% confidence contours in the parameter space. Practically, the matrix $P^T P$ (8) is transformed into the matrix of simple correlation coefficients among the derivatives.

Fitting the Bery Equation

Since the B equation involves two functions, a quadratic and a linear, it is not amenable to Marquardt's method of fitting. The sequence of steps in fitting the B equation is:

(1) Determination of x_{max} (point of maximum fractional transfer)

- polynomial fit of whole fractional transfer curve
- derivative tests for maximum
- local polynomial fit about maximum
- derivative test for x_{max}

(2) Linear regression of percent transfer after x_{max}

- slope and intercept estimates
- calculation of residual error

(3) Use of above slope and intercept to calculate quadratic portion of fractional transfer curve (before x_{max})

- calculation of quadratic
- calculation of residual error for quadratic portion (before x_{max})

(4) Pooling of residual errors for linear and quadratic portions of fractional transfer curve.

Since x_{max} is calculated separately from the slope and intercept in the B model, and since the slope and intercept of the linear portion of the curve are by definition independent, there can be no intercorrelation among the parameter estimates.

Coefficient of Determination

The goodness of fit has been calculated as follows:

$$r^2 = 1 - \frac{\Phi}{SSY}, \text{ with} \tag{9}$$

$$SSY = \sum_{i=1}^{n} (Y_i - \overline{Y})^2 \tag{10}$$

$$\overline{Y} = \sum_{i=1}^{n} \frac{Y_i}{n}, \text{ and} \tag{11}$$

$$Y_i = \frac{y_i}{x_i} \quad \text{(fraction of ink transferred from plate to paper)}$$

APPENDIX 4: SENSITIVITY ANALYSIS

The product of the partial derivative of the fractional ink transfer with respect to each parameter and the standard deviation of the mean value of the parameter estimates for the 29 sets of data describe the change of the fractional ink transfer due to a change of one standard deviation in the corresponding parameter. For the WF equation this gives:

$$\frac{\partial(\frac{y}{x})}{\partial k}\sigma_k = \left\{b(1-e^{-\frac{x}{b}}) + f\left[x - b(1-e^{-\frac{x}{b}})\right]\right\} e^{-kx}\,\sigma_k \qquad (12)$$

$$\frac{\partial(\frac{y}{x})}{\partial b}\sigma_b = (1-e^{-kx})(1-f)\left[(1-e^{-\frac{x}{b}}) - \frac{x e^{-\frac{x}{b}}}{b}\right]\frac{\sigma_b}{x} \qquad (13)$$

$$\frac{\partial(\frac{y}{x})}{\partial f}\sigma_f = (1-e^{-kx})\left[x - b(1-e^{-\frac{x}{b}})\right]\frac{\sigma_f}{x} \qquad (14)$$

Mean values for k, b, f and their standard deviations for the 29 data sets are given in Table 5.

THREE-EQUATION INK TRANSFER MODEL

Yuri Bery, Weyerhaeuser Company, Tacoma, Washington, U.S.A.

ABSTRACT

Eight sets of ink transfer data generated some twenty years ago at the National Printing Ink Research Institute along with eight sets of ink transfer data generated recently at the Pulp and Paper Research Institute of Canada have been analyzed.

When the rate of change of the percent ink transfer curve is examined, a maximum is discovered. This maximum, along with the second maximum in percent ink transfer curve, allows one to break down the ink transfer curve into three areas: before X_{max_1}, between X_{max_1} and X_{max_2}, and after X_{max_2}. Both X_{max_1} and X_{max_2} are interpreted as those critical levels of the amount of ink on the plate at which the nature of ink transfer process become progressively more complex. On the basis of these ideas, a three-equation ink transfer model has been developed. The model includes six parameters, all of which are readily available from experimental data. Interpretation of these parameters allows one to calculate some important paper and ink properties. The overall theory-to-experiment correlation for 16 sets of experimental ink transfer data averages 92.2%, leaving less than 8% of the total variation of the observed Y values unexplained by the postulated model. For that portion of the ink transfer curve which covers ink film thicknesses employed under realistic printing conditions, the theory-to-experiment correlation averages 94.6%, leaving only 5.4% of the total variation unexplained by the model.

INTRODUCTION

The usefulness of ink transfer equations has been challenged by P. J. Mangin[5]. In his presentation, the validity of the various models for newsprint was determined by obtaining a number of sets of ink transfer data and using Marquardt's method of fitting the curves. No model was found to be as good as an arbitrary low order polynomial. None of the models was found to correlate satisfactorily with experiments in the light ink coverage region. He concluded that useful physical interpretation of the ink transfer parameters requires a model built on known printing phenomena - not just curve fitting.

We agree that the poor agreement between theory-and-experiment in the light ink coverage region compromises the theory and that determination of the ink transfer parameters should be conceptual. In other words, the way these parameters are determined must follow from the same principles on which the model is built. In the light of this, the physical model is an ideal solution and the empirical model is only a satisfactory one.

There is a reasonable compromise between these two extremes, namely, a mechanistic approach to the given process. The constants of a mechanistic model are determined by manipulating the experimental data in a certain conceptual way; in contrast, the empirical model allows for manipulation with constants until the best fit is reached. In the empirical model, a variety of values for each constant can be obtained; therefore, the meaningfulness of these constants is diminished.

We do not think, however, that a polynomial of, for instance, the fifth order should be used as a criterion for evaluating the usefulness of the ink transfer models. We will discuss this statement later.

The ink transfer data for newsprint, generated at Pulp and Paper Research Institute of Canada (PPRIC), show that the classical linear approach to the transfer mechanism in the heavy ink coverage region provides only a rough approximation and neglects a distinctive curve for the dependence.

The data generated years ago at National Printing Ink Research Institute (NPIRI) created a more or less common impression that the linearity of the ink transfer curve in a heavy coverage region can be justified. The Weyerhaeuser two-equation ink transfer model (Bery, Y.[2, 3, 4]) represents an attempt at such justification. In the light of the new information, we conceptually reconsidered our approach. Below follows the derivation of a three-equation ink transfer model.

BASIC ASSUMPTIONS

(1) Printing surfaces, both paper and plate, require a certain amount of ink to satisfy their ability to adhere to the ink.

(2) This adhesion is related only to the facial interaction of the materials in physical contact.

(3) Regular ink transfer takes place when the amount of ink available exceeds the amount of ink required to satisfy the printing plate.

(4) A paper surface element is comprised of a plateau and a valley.

(5) The amount of ink immobilized by the plateau is proportional to

the amount of ink not affected by the ability of the plate to immobilize it.

(6) When there is enough ink on the plate to satisfy both plate and plateau of the paper element, ink flows from the high pressure zone (plate - plateau) to the low pressure zone (plate - valley).

(7) With further increases of the initial amount of ink on the plate, ink gradually fills up the volume of the valley until the bottom of the valley is reached.

(8) Further increase in the amount of ink available on the plate results in the penetration of ink into the bulk of the paper. However, this penetration is a strong function of time. Thus, under realistic printing conditions, a situation occurs when paper cannot take up all the ink available; at that point excessive ink becomes nothing more than a lubricant between two surfaces.

(9) According to classical lubrication theory, only a fraction of the ink on the plate will penetrate through the printing nip.

In summary, there are three different mechanisms controlling ink transfer; they are:

surface immobilization
valley filling
capillary penetration.

All three work along with the ink-film splitting mechanism. When there is not enough ink on the plate to satisfy the paper surface in direct contact with the ink, only the first mechanism is responsible for ink transfer.

The first two mechanisms control ink transfer when there is enough ink to flow into the valley. All three mechanisms are at work when there is more ink on the plate than can be carried through the nip by both plate and paper.

DERIVATION

The transfer process is illustrated in the scheme shown in Figure 1.

As Figure 1 shows, the walls and bottom of the valley hold a certain layer of ink with the same thickness, b_2. Our scheme implies that there are two critical values of X, namely, $X = X_{max_1}$ and $X = X_{max_2}$, designating a shift from mechanism to mechanism(s). The determination of these quantities is discussed later.

Now we will derive the corresponding equations for each of the three regions separately.

$b_1 < X < X_{max_1}$

According to the scheme, the total amount of ink transferred to the paper, Y'_1 is equal to the amount of ink immobilized on the plateau of the

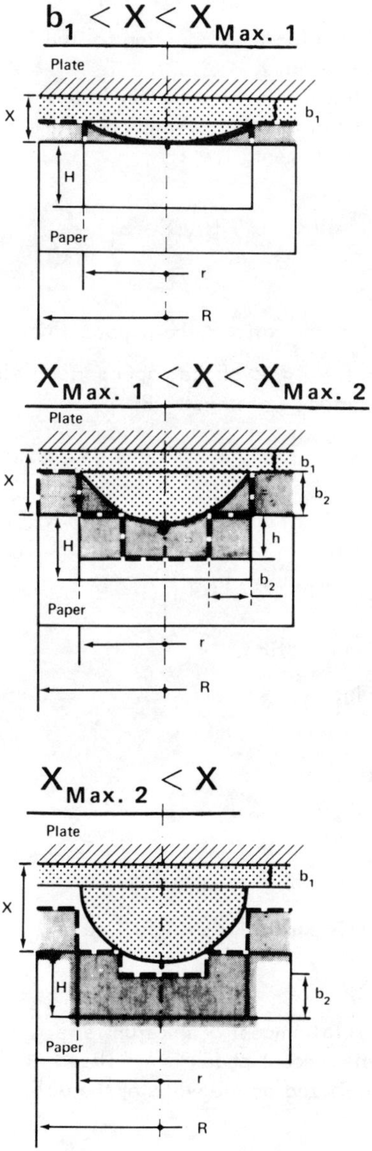

Fig. 1. Ink transfer scheme.

paper element plus a fraction of the remainder:

$$Y'_1 = (\pi R^2 - \pi r^2)(X - b_1) + f_1(X - b_1)\pi r^2 \tag{1}$$

where f_1 is the splitting factor accounting for the ink-film split between the plate and valley, when $X < X_{max_1}$.

In order to express this quantity in terms of the average thickness of the ink layer transferred to the paper element, we divide it over the area of the paper element, i.e.:

$$Y_1 = \frac{\pi R^2 - \pi r^2}{\pi R^2}(X - b_1) + f_1(X - b_1)\frac{\pi r^2}{\pi R^2} \tag{2}$$

The term $\frac{\pi R^2 - \pi r^2}{\pi R^2}$ is the ratio of the paper surface area in direct contact with ink to the total area of the paper surface element. We define this ratio as paper surface smoothness, P:

$$P = \frac{\pi R^2 - \pi r^2}{\pi R^2} \qquad P = 1 - \frac{r^2}{R^2} \tag{3}$$

Substituting Equation (3) in (2) and after rearrangement:

$$Y_1 = [P + f_1(1 - P)]X - [P + f_1(1 - P)]b_1 \tag{4}$$

This is a straight line equation:

$$Y_1 = S_1X + In_1 \tag{5}$$

where:

$$S_1 = P + f_1(1 - P) \tag{6}$$

is the slope and:

$$In_1 = - S_1b_1 \tag{7}$$

is the intercept of this particular straight line.

$X_{max_1} < X < X_{max_2}$

In this region, the total amount of ink transferred, Y'_2 is the sum of three quantities: the amount of ink immobilized on the plateau, the amount of ink immobilized on the walls of the valley, and a fraction of the remainder.

$$Y'_2 = (\pi R^2 - \pi r^2)b_2 + \left[\pi r^2 - \pi(r - b_2)^2\right]h + f_2\left[\pi r^2 b_2 + \pi(r - b_2)^2 h\right] \tag{8}$$

Here f_2 is the splitting factor accounting for the film split between the plate and valley, when $X_{max_1} < X < X_{max_2}$. Once again, expressing the amount of ink transferred in terms of the thickness of the ink layer, introducing the accepted definition of the paper surface smoothness, and rearranging the equation, we obtain:

$$Y_2 = \{P + f_2(1 - P)\}b_2 + \{1 - (1 - f_2)\frac{(r - b_2)^2}{r^2}\}(1 - P)h \qquad (9)$$

Equality of the ink volume before and after the pressure was applied is:

$$\pi R^2 X = \pi R^2 X_{max_1} + \pi r^2 h \qquad (10)$$

Solving for h:

$$h = \frac{X - X_{max_1}}{1 - P} \qquad (11)$$

Substituting Equation (11) in (9) and after rearrangement:

$$Y_2 = \{P + f_2(1 - P)\}b_2 - \{1 - (1 - f_2)\frac{(r - b_2)^2}{r^2}\}X_{max_1} +$$
$$\{1 - (1 - f_2)\frac{(r - b_2)^2}{r^2}\}X \qquad (12)$$

This is again a straight line equation:

$$Y_2 = S_2X + In_2 \qquad (13)$$

with slope:

$$S_2 = 1 - (1 - f_2)\frac{(r - b_2)^2}{r^2} \qquad (14)$$

and intercept:

$$In_2 = \{P + f_2(1 - P)\}b_2 - S_2X_{max_1} \qquad (15)$$

$X_{max_2} < X$

Here we have a situation in which the paper cannot take up all the ink available on the plate, and only a fraction of the ink penetrates the printing nip. As we showed elsewhere (Y. Bery[1]), the amount of ink that penetrates between the doctor blade and gravure cylinder is proportional to the square root of the ink on the cylinder surface before the doctoring.

In general, this is equally right for the situation under discussion. Here, the amount of ink transferred to the paper, Y_3', is a sum of four quantities: namely, the amount of ink immobilized on the plateau, the amount of ink immobilized on the walls and bottom of the valley, the amount of ink that has penetrated to the bulk of the paper, and a fraction of the remainder.

$$Y_3' = (\pi R^2 - \pi r^2)b_2 + \{\pi r^2 H - \pi(r - b_2)^2(H - b_2)\} +$$

$$\Delta Y_3'' + f_3 \{\pi R^2 \alpha\sqrt{X} - (\pi R^2 - \pi r^2)b_2 -$$

$$\{\pi r^2 H - \pi(r - b_2)^2(H - b_2)\} - \Delta Y_3'' - \pi R^2 b_1\} \qquad (16)$$

where:

- Y_3'' - depth of the ink penetration in the bulk of the paper.
- Δ - total area of the pores, through which this penetration materializes.
- f_3 - splitting factor accounting for the ink film split between the plate and valley, when $X > X_{max_2}$.
- α - hydrodynamic constant.

When X reaches a certain value, the following equality holds:

$$\pi R^2 \alpha\sqrt{X} = \pi R^2 X_{max_1} + \pi r^2 H + \Delta Y_3'' \qquad (17)$$

from which:

$$\frac{\Delta Y_3''}{\pi R^2} = \alpha\sqrt{X} - X_{max_1} - \frac{\pi r^2}{\pi R^2} H \qquad (18)$$

According to Equation (3) $\frac{\pi r^2}{\pi R^2} = (1 - P)$ (19)

According to Equation (11) and the ink transfer scheme, when $X = X_{max_2}$, then h = H.

Thus: $$H = \frac{X_{max_2} - X_{max_1}}{1 - P} \qquad (20)$$

Substituting Equations (18) and (19) in (17), we obtain:

$$\frac{\Delta Y_3''}{\pi R^2} = \alpha\sqrt{X} - X_{max_2} \qquad (21)$$

Our next assumption is that the value of the splitting factor, f_3, is determined by the corresponding areas of ink immobilization for both the plate and paper element. In this case:

$$f_3 = 1 - \frac{1}{2}\,\frac{\text{Immobilization area of the plate}}{\text{Immobilization area of the paper}} \tag{22}$$

When both printing surfaces (plate and paper) are identical, the splitting factor is equal to $\frac{1}{2}$, i.e. the amount of ink transferred to the paper is $\frac{1}{2}$ the amount of ink available on the plate.

$$\text{Immobilization area of the plate} = \pi r^2 + 2\pi r b_1$$

$$\text{Immobilization area of the paper} = \pi(r - b_2)^2 + 2\pi(r - b_2)H + 2\pi r b_2$$

Considering this and rearranging:

$$f_3 = 1 - \frac{1}{2}\,\frac{r^2 + 2rb_1}{(r^2 + b_2^2) + 2H(r - b_2)} \tag{23}$$

Substituting Equations (21) and (23) in (16), we arrive at the ink transfer equation for a heavy ink coverage region:

$$Y_3 = (1 - f_3)\left\{Pb_2 + \frac{1}{1 - f_2}\left[(S_2 - f_2)(X_{max_2} - X_{max_1}) + (1 - P)(1 - S_2)b_2\right] - X_{max_2}\right\} - f_3 b_1 + \alpha\sqrt{X} \tag{24}$$

or:

$$Y_3 = A + \alpha\sqrt{X} \tag{25}$$

Thus, Weyerhaeuser's three-equation ink transfer model contains six parameters, X_{max_1}, X_{max_2}, S_1, S_2, In_1, and In_2, as follows:

$$Y_1 = S_1 X + In_1 \quad \text{for } b_1 < X < X_{max_1}$$

$$Y_2 = S_2 X + In_2 \quad \text{for } X_{max_1} < X < X_{max_2}$$

$$Y_3 = A + \alpha/X \quad \text{for } X > X_{max_2} \tag{26}$$

The choice of the corresponding equation is thus determined by the initial amount of ink on the plate.

DETERMINATION OF INK TRANSFER PARAMETERS

All the parameters are readily available from experimental data. X_{max_2} is the X-coordinate of the peak in the percent ink transfer curve.

However, when the rate of change of the percent ink transfer curve is examined, another extremum (maximum) is discovered. X_{max_1} is the X-coordinate of this maximum (Figure 2).

S_1 and In_1 are obtained through linear regression of the experimental data for $X < X_{max_1}$ range. S_2, In_2 are calculated similarly for $X_{max_1} < X < X_{max_2}$ range.

Experimental Y values are expressed as a function of the experimental X values in the form of a fifth degree polynomial. The same is done for Y/X as a function of X.

The polynomial coefficients, thus obtained, are entered into the computer. We employ a simple program which calculates the maximum value of Y/X using an increment of 0.1. The solution is X_{max_2}. Using calculated values of Y and experimental values of X, we can calculate a $\frac{Y_i - Y_{i-1}}{X_i - X_{i-1}}$ ratio for all X values (beginning with i = 2) and again express them in the form of a fifth degree polynomial.

The second set of polynomial coefficients is entered into a computer loaded with the same program. The solution is X_{max_1}.

INTERPRETATION OF THE INK TRANSFER PARAMETERS

The thickness of the ink layer immobilized by the plate is readily available from expression (7):

$$b_1 = -\frac{In_1}{S_1} \tag{27}$$

According to the accepted ink transfer scheme, the thickness of the ink layer immobilized by the paper is:

$$b_2 = X_{max_1} - b_1 \tag{28}$$

The splitting factors f_1 and f_2 as well as the radius of the valley in the plane of plateau can be expressed as a function of paper smoothness, P, using corresponding Equations (6), (15) and (14), namely:

$$f_1 = \frac{S_1 - P}{1 - P} \tag{29}$$

$$f_2 = \frac{In_2 + S_2 \cdot X_{max_1} - Pb_2}{(1 - P)b_2} \tag{30}$$

$$r = \frac{b_2}{1 - \sqrt{\frac{1 - S_2}{1 - f_2}}} \tag{31}$$

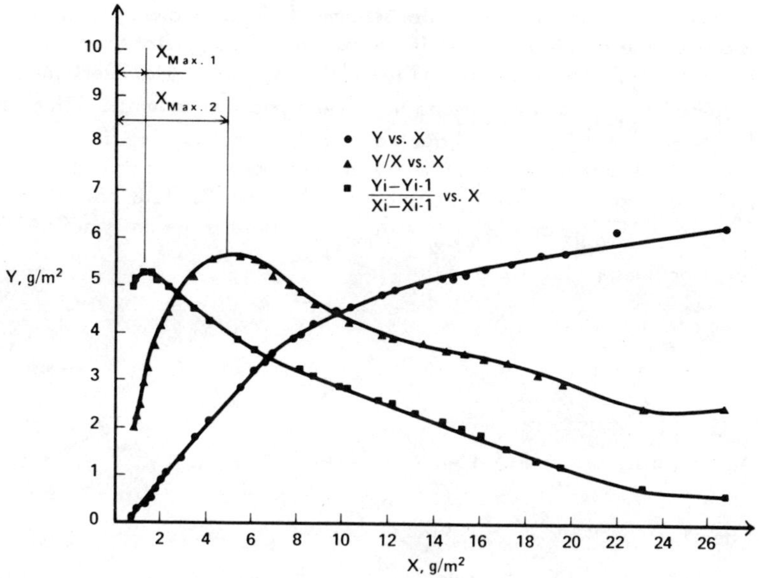

Fig. 2. Ink transfer parameters determination.

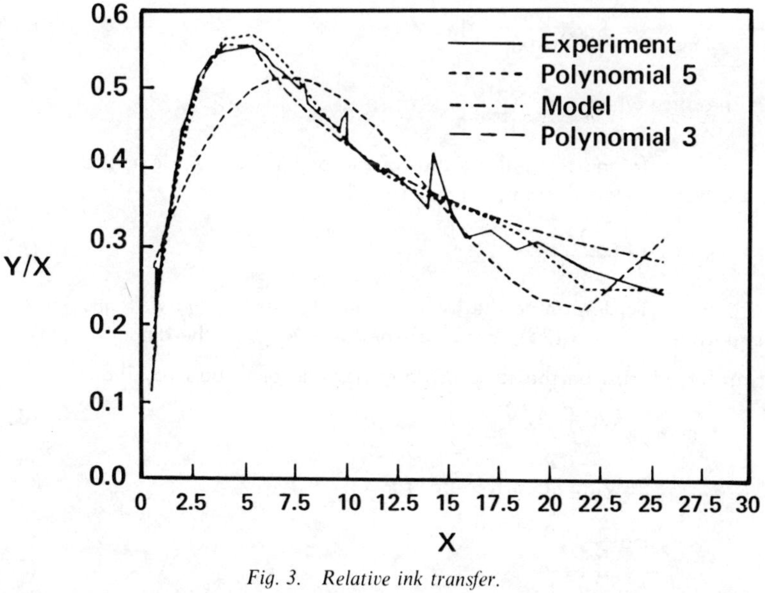

Fig. 3. Relative ink transfer.

Our final assumption concerns the nature of paper smoothness, P. According to the ink transfer scheme, P should increase when there is a decrease in the amount of ink transferred, Y_{max_2}, for some given increase in the amount of ink on the plate, X_{max_2}. However, the values of Y_{max_1} and X_{max_1} do not reflect paper smoothness. Thus, the last two values should be subtracted from the corresponding first two.

For the sake of similarity in the shapes of the two expressions for paper smoothness, we have to deal with squared differences of $(Y_{max_2} - Y_{max_1})$ and $(X_{max_2} - X_{max_1})$. Here is our basic definition for smoothness:

$$P = 1 - \frac{r^2}{R^2} \tag{3}$$

And our assumption: $$P = 1 - \frac{(Y_{max_2} - Y_{max_1})^2}{(X_{max_2} - X_{max_1})^2} \tag{32}$$

where:

$$Y_{max_1} = S_1 X_{max_1} + In_1 \tag{33}$$

is the amount of ink transferred when $X = X_{max_1}$.

$$Y_{max_2} = S_2 X_{max_2} + In_2 \tag{34}$$

is the same when $X = X_{max_2}$.

In addition, the radius of the paper surface element can be calculated as a solution for Equation (3):

$$R = r\sqrt{\left(\frac{1}{1 - P}\right)} \tag{35}$$

To determine the hydrodynamic constant, α, we equalize Equations (13) and (25), because when $X = X_{max_2}$, the amount of ink transferred must be the same using either one of those two, i.e.:

$$Y_{max_2} = A + \alpha\sqrt{X_{max_2}} \tag{36}$$

Solving Equation (36) for α:

$$\alpha = \frac{Y_{max_2} - A}{\sqrt{X_{max_2}}} \tag{37}$$

Thus, all the constants are determined.

CORRELATION OF THEORY AND EXPERIMENT

In order to test our approach, we used eight sets of experimental ink transfer data obtained at NPIRI and eight sets obtained at PPRIC.

The calculated values of the ink transfer parameters, along with their derivatives, are shown in Table A1 in the Appendix. Experimental X, Y, and Y/X values, along with theoretical values of Y and Y/X are shown in Table A2. Theoretical Y values were calculated using the present Weyerhaeuser ink transfer model. They were then compared with the experimental Y value, using Marquardt's method.

Table 1 compares experimental results with theory.

TABLE 1

	Coefficient of determination	
Sample ID	Whole curve	Light ink coverage
NPIRI		
111 - AA - 2	0.929	0.904
111 - BB - 2	0.902	0.974
111 - DD - 2	0.940	0.938
111 - FF - 2	0.889	1.000
111 - II - 2	0.955	0.872
111 - MM - 2(1)	0.926	0.964
111 - PP - 2	0.940	0.879
111 - QQ - 2	0.902	0.949
PPRIC		
B - STD	0.955	0.974
C - STD/2	0.926	0.951
D - STD/1	0.893	0.949
S - STD	0.893	0.981
A - LP	0.929	0.929
A - LS	0.901	0.954
A - HS	0.910	0.960
C - HP	0.958	0.961

DISCUSSION

The average coefficient of determination for the whole curve is 0.922; for the light ink coverage it is 0.946. According to Dr. Mangin, for four models he analyzed (the Weyerhaeuser two-equation model, Walker-Fetsko, Kartunnen, Laraignon) it was the other way around: the fit for the light ink coverage is some 10% worse than for the whole curve. We are talking about a range of ink film thickness (1 - 10 microns) actually employed under realistic printing conditions. Therefore, an improvement in correlation means something more than just a better fit.

We hope that independent measurements of the constants will prove the basic assumptions, on which the model is built, reflect the physical reality of the ink transfer process.

A typical percent ink transfer curve is shown in Figure 3. The reason we have included the polynomials is that the graphs clearly demonstrate the immense power of the least squares polynomial method to chase every single experimental point no matter how consistent or spread the data are. At the same time, any postulated model must 'prefer', so to speak, some data points over others, since a certain concept is utilized. Table 2 illustrates the point.

TABLE 2

	Coefficient of determination		
		Polynomial	
Sample ID	Model	Third order	Fifth order
B - STD	0.955	0.701	0.946
C - STD/2	0.926	0.907	0.976
D - STD/1	0.893	0.839	0.943
S - STD	0.893	0.711	0.936
A - LP	0.929	0.800	0.959
A - LS	0.901	0.948	0.978
A - HS	0.910	0.912	0.960
C - HP	0.958	0.868	0.971

A polynomial of the third order gives an average 8.6% worse correlation than does the model, and a polynomial of the fifth order improves the correlation by 3.6%. However, the first one yields a practically monotonous dependence, which is the basic concept behind all ink transfer models of which we are aware. Thus, if one likes to have a common denominator for the different models being compared, it should be the best curve fit rather than polynomial fit.

CONCLUSIONS

(1) When the rate of change of the percent ink transfer curve is examined, a maximum is discovered.

(2) This maximum, along with the second maximum in percent ink transfer curve, allows one to break down the ink transfer curve into three areas: before X_{max_1}, between X_{max_1} and X_{max_2}, and after X_{max_2}.

(3) Both X_{max_1} and X_{max_2} are interpreted as those critical levels of the amount of ink on the plate at which the nature of ink transfer process becomes progressively more complex.

(4) On the basis of these ideas, a three-equation ink transfer model has been developed.

(5) The model does not include empirical functions.

(6) The model includes six parameters, all of which are readily available from experimental data.

(7) Interpretation of these parameters allows one to calculate the:

(i) geometrical parameters of the average paper surface element;

(ii) surface immobilization capabilities for both plate and paper;

(iii) three ink splitting factors (different for the three zones of ink transfer);

(iv) hydrodynamic constant of the process which is characteristic for the given combination of printing conditions (pressure, speed, ink viscosity), geometry of the nip, and amount of ink on the plate;

(v) printing smoothness of the paper.

(8) The overall correlation between theory and experiment for 16 sets of experimental ink transfer data averages 92.2%, leaving less than 8% of the total variation of the observed Y values unexplained by the postulated model.

(9) For that portion of the ink transfer curve which covers ink film thicknesses employed under realistic printing conditions, the correlation averages 94.6%, leaving only 5.4% of the total variation unexplained by the model.

REFERENCES

1. Bery, Y., 'Gravure printing on non-absorbing materials', GRI Newsletter, No. 34, p. 56, June (1976).

2. Bery, Y., 'An Ink Transfer Equation', TAGA Proceedings (1978).

3. Bery, Y., 'Ink Transfer: Extended Application of the Weyerhaeuser Ink Transfer Equation', TAPPI, Vol. 62(8), p. 89 (1979).

4. Bery, Y., 'Ink Transfer: Another Interpretation of the Weyerhaeuser Equation', Proceedings of the TAPPI Printing and Reprography Conference (1980).

5. Mangin, P., 'A Discussion of the Usefulness of Ink Transfer Equations', Proceedings of the TAPPI Printing and Reprography Conference, (1980) (Abstract only).

Table A1.

INK TRANSFER PARAMETERS

Parameters	"NPIRI" Data TAPPI's Table							
	AA-2	BB-2	DD-2	FF-2	II-2	MM-2-1	PP-2	QQ-2
$Xmax_1$	5.925	5.861	5.940	4.180	5.052	3.500	6.240	6.630
$Xmax_2$	9.590	8.190	9.280	7.630	7.370	9.000	9.950	9.920
S_1	.472	.672	.436	.602	.505	.517	.583	.643
S_2	.670	.695	.597	.693	.525	.749	.571	.644
In_1	-.486	-.745	-.462	-.495	-.188	-.581	-.649	-.768
In_2	-1.236	-.827	-.861	-.687	-.309	-1.166	-.151	-.387
$Ymax_1$	2.311	3.194	2.128	2.021	2.363	1.229	2.989	3.495
$Ymax_2$	5.189	4.865	4.679	4.601	4.178	5.575	5.530	6.001
b_1	1.030	1.109	1.060	.822	.372	1.124	1.113	1.194
b_2	4.895	4.752	4.880	3.358	4.680	2.376	5.127	5.436
P	.383	.485	.417	.441	.387	.376	.531	.420
f_1	.144	.363	.033	.288	.192	.226	.111	.384
f_2	.284	.385	.229	.388	.401	.379	.287	.507
f_3	.652	.617	.638	.690	.565	.764	.654	.586
r	15.244	16.066	17.616	11.510	42.739	6.523	22.856	36.185
R	19.407	22.387	23.071	15.395	54.588	8.258	33.374	47.513
H	5.940	4.522	5.729	6.172	3.781	8.814	7.910	5.672
α	2.063	2.037	1.991	2.024	1.863	2.337	2.245	2.170
A	-1.201	-.965	-1.386	-.990	-.336	-1.435	-1.550	-.835

Table A1 (contd.)

INK TRANSFER PARAMETERS

Parameters	"PPRIC" Data Sample Identification							
	B-STD	C-STD/2	D-STD/1	S-STD	A-LP	A-LS	A-HS	C-HP
$Xmax_1$	1.303	2.800	1.342	2.330	3.331	3.230	2.643	2.100
$Xmax_2$	4.899	6.299	4.534	4.749	6.129	6.409	5.299	6.199
S_1	.561	.569	.581	.561	.609	.659	.628	.571
S_2	.670	.650	.659	.621	.692	.724	.697	.709
In_1	-.296	-.351	-.240	-.239	-.327	-.333	-.357	-.305
In_2	-.460	-.490	-.363	-.328	-.547	-.429	-.555	-.661
$Ymax_1$	.435	1.242	.540	1.068	1.702	1.796	1.303	.894
$Ymax_2$	2.822	3.604	2.625	2.621	3.694	4.211	3.138	3.734
b_1	.528	.619	.413	.426	.537	.505	.568	.534
b_2	.775	2.183	.929	1.904	2.794	2.725	2.075	1.566
P	.559	.544	.573	.588	.493	.423	.523	.520
f_1	.005	.055	.019	-.066	.229	.409	.220	.106
f_2	-.059	.143	-.028	-.001	.269	.481	.204	.018
f_3	.875	.781	.865	.775	.720	.706	.750	.833
r	1.754	6.049	2.191	4.950	7.959	10.074	5.418	3.437
R	2.641	8.958	3.353	7.712	11.178	13.262	7.845	4.961
H	8.154	7.673	7.475	5.871	5.519	5.51	5.568	8.54
α	1.590	1.858	1.511	1.543	1.935	2.144	1.786	1.831
A	-.697	-1.059	-.592	-.742	-1.097	-1.218	-.974	-.825

Table A2.

INK TRANSFER DATA
"NPIRI"

Table III-AA-2

Experimental			Calculated	
x micron	y micron	y/x	y micron	y/x
1.09	.120	.112	.028	.026
3.07	.787	.256	.963	.314
4.65	1.71	.368	1.709	.367
5.19	2.05	.395	1.964	.378
6.59	3.18	.483	3.179	.483
8.56	4.50	.525	4.499	.525
9.78	5.16	.527	5.251	.537
10.93	5.55	.507	5.619	.514
12.19	6.01	.494	6.002	.492
14.60	6.87	.471	6.682	.458
16.01	7.14	.446	7.054	.441
18.43	7.78	.422	7.655	.415
20.10	7.89	.392	8.048	.400
20.49	8.38	.409	8.230	.394
21.57	8.52	.395	8.380	.389

Table III-BB-2

Experimental			Calculated	
x micron	y micron	y/x	y micron	y/x
1.66	.427	.258	.371	.223
3.36	1.41	.420	1.513	.450
5.35	2.90	.542	2.850	.533
6.43	3.64	.566	3.641	.566
7.74	4.55	.587	4.549	.587
9.37	5.41	.577	5.270	.562
9.64	5.56	.576	5.360	.556
11.07	6.12	.553	5.812	.525
13.07	6.91	.528	6.399	.490
14.56	7.44	.511	6.808	.468
16.75	7.83	.467	7.372	.440
18.08	8.31	.459	7.696	.426
20.35	8.84	.434	8.224	.404
21.67	9.24	.426	8.517	.393
22.20	9.14	.412	8.633	.389

Table A2. (contd.)

INK TRANSFER DATA
"NPIRI"

Table III-DD-2

Experimental			Calculated	
x micron	y micron	y/x	y micron	y/x
1.09	.109	.100	.013	.012
2.02	.333	.165	.419	.207
3.28	.870	.265	.968	.295
4.49	1.58	.352	1.496	.333
6.06	2.76	.460	2.757	.455
6.77	3.17	.468	3.181	.470
8.95	4.48	.502	4.482	.501
9.61	4.81	.500	4.786	.498
11.95	5.58	.467	5.497	.460
13.25	6.05	.457	5.861	.442
15.50	6.73	.434	6.453	.416
17.86	6.98	.392	7.028	.394
18.84	7.37	.392	7.256	.385
20.09	7.54	.376	7.538	.375
21.59	7.55	.352	7.865	.364
22.90	7.71	.337	8.142	.356

Table III-FF-2

Experimental			Calculated	
x micron	y micron	y/x	y micron	y/x
1.32	.300	.232	.300	.232
2.98	1.30	.434	1.300	.434
5.01	2.78	.555	2.785	.556
6.02	3.49	.580	3.485	.579
6.83	4.04	.592	4.046	.592
7.80	4.59	.588	4.663	.598
9.54	5.52	.578	5.262	.552
10.83	6.01	.555	5.671	.524
11.38	6.32	.555	5.838	.513
12.63	6.76	.535	6.203	.491
14.45	7.40	.511	6.704	.464
16.03	7.90	.493	7.114	.444
16.73	8.09	.484	7.289	.436
19.28	8.46	.438	7.897	.410
21.04	8.83	.419	8.294	.394

Table III-II-2

Table A2. (contd.)

INK TRANSFER DATA "NPIRI"

Experimental			Calculated	
x micron	y micron	y/x	y micron	y/x
1.91	.645	.337	.782	.409
3.53	1.66	.469	1.600	.453
5.58	3.24	.580	3.239	.580
6.57	3.76	.572	3.758	.572
8.73	4.88	.560	5.169	.592
10.46	5.72	.547	5.689	.544
12.33	6.30	.511	6.206	.503
13.59	6.64	.489	6.532	.481
15.42	7.17	.465	6.980	.453
16.75	7.28	.434	7.289	.435
17.78	7.55	.424	7.520	.423
19.33	7.88	.408	7.855	.406
21.45	8.20	.382	8.292	.387
22.79	8.03	.352	8.558	.376

Table III-MM-2-1

Experimental			Calculated	
x micron	y micron	y/x	y micron	y/x
2.28	.597	.262	.598	.262
3.06	1.00	.328	1.001	.327
3.68	1.42	.385	1.590	.432
4.53	2.44	.538	2.227	.492
5.98	3.28	.549	3.313	.554
6.90	4.03	.585	4.002	.580
8.18	4.91	.600	4.961	.606
9.40	5.60	.595	5.730	.610
10.73	6.41	.598	6.220	.580
12.05	7.00	.580	6.677	.554
12.06	6.94	.576	6.681	.554
12.75	7.36	.577	6.910	.542
12.99	7.32	.564	6.988	.538
13.71	7.59	.552	7.218	.526
13.98	7.87	.563	7.303	.522
14.05	7.58	.537	7.325	.521
14.70	7.80	.530	7.525	.512
15.85	8.20	.518	7.869	.496
15.68	8.40	.536	7.819	.499
16.19	8.30	.517	7.968	.492
17.59	8.79	.499	8.366	.476
17.71	9.11	.514	8.400	.474
18.00	9.00	.501	8.480	.471
19.60	9.29	.473	8.911	.455
20.75	9.60	.462	9.211	.444
21.80	9.91	.452	9.477	.435
23.25	10.01	.431	9.834	.423
24.20	10.20	.420	10.062	.416
25.75	10.38	.402	10.424	.405
26.45	10.69	.406	10.584	.400
28.25	11.30	.400	10.986	.389
29.10	11.41	.392	11.172	.384
30.97	11.62	.376	11.571	.374

Table A2. (contd.)

INK TRANSFER DATA
"NPIRI"

Table III-PP-2

Experimental			Calculated	
x micron	y micron	y/x	y micron	y/x
2.80	.770	.277	.883	.351
4.23	1.66	.393	1.817	.430
4.83	2.16	.447	2.167	.449
5.92	2.98	.503	2.802	.473
7.06	3.83	.542	3.880	.550
8.21	4.63	.564	4.537	.553
9.85	5.44	.552	5.473	.556
11.18	6.20	.555	5.956	.533
12.13	6.63	.547	6.269	.517
13.15	6.84	.520	6.591	.501
13.27	7.00	.527	6.628	.499
14.26	7.31	.513	6.928	.486
14.75	7.46	.506	7.072	.479
14.89	7.69	.517	7.113	.478
16.20	7.72	.476	7.486	.462
18.04	8.27	.458	7.985	.443
18.64	8.52	.457	8.143	.437
19.34	8.62	.446	8.323	.430
19.88	8.72	.439	8.460	.426
21.22	9.11	.429	8.792	.414
22.25	9.15	.411	9.040	.406
22.46	9.25	.412	9.090	.405
23.28	9.40	.404	9.282	.399
25.41	9.30	.366	9.767	.384
26.25	9.71	.370	9.952	.379
28.03	9.97	.356	10.336	.369
28.42	9.93	.349	10.418	.367
28.73	10.12	.352	10.483	.365
31.02	10.44	.347	10.954	.353

Table A2. (contd.)

INK TRANSFER DATA
"NPIRI"

Table III-QQ-2

Experimental			Calculated	
x micron	y micron	y/x	y micron	y/x
1.54	.332	.216	.222	.144
2.83	.957	.338	1.052	.372
3.81	1.57	.414	1.682	.441
5.43	2.82	.520	2.723	.502
7.02	4.11	.586	4.134	.589
8.16	4.91	.601	4.868	.597
9.91	5.98	.604	5.995	.605
11.02	6.58	.597	6.369	.578
11.81	6.92	.586	6.622	.561
12.98	7.46	.577	6.983	.538
13.14	7.57	.576	7.031	.535
15.53	8.52	.549	7.717	.497
15.98	8.61	.539	7.840	.491
17.35	8.89	.514	8.204	.473
17.74	9.14	.515	8.305	.468
18.86	9.06	.482	8.589	.455
20.35	9.46	.474	8.954	.440
20.35	9.52	.466	8.954	.440
21.05	9.60	.456	9.121	.433
22.49	9.66	.429	9.456	.420
23.92	10.41	.435	9.778	.409
23.93	10.06	.418	9.780	.409
24.60	10.40	.423	9.928	.404
26.45	10.48	.395	10.325	.390
27.50	10.65	.387	10.545	.383
28.76	10.99	.382	10.802	.376
30.01	11.20	.372	11.053	.368
32.45	11.39	.351	11.526	.355

Table A2. (contd.)

INK TRANSFER DATA
"PPRIC"

Sample: B-STD

Experimental			Calculated	
x g/m^2	y g/m^2	y/x	y g/m^2	y/x
.672	.080	.119	.081	.121
.858	.149	.174	.185	.216
.863	.236	.273	.188	.218
1.003	.244	.243	.267	.266
1.163	.365	.314	.356	.306
1.407	.467	.332	.483	.343
1.695	.653	.385	.676	.399
2.007	.902	.449	.885	.441
2.240	1.040	.464	1.041	.465
2.768	1.429	.516	1.395	.504
3.373	1.825	.541	1.800	.534
4.076	2.234	.548	2.271	.557
5.414	3.007	.555	3.003	.555
6.141	3.325	.541	3.243	.528
6.478	3.425	.529	3.350	.517
6.945	3.617	.521	3.493	.503
7.779	3.894	.501	3.738	.481
7.981	4.048	.507	3.795	.476
8.160	3.937	.482	3.845	.471
8.668	4.087	.472	3.984	.460
9.779	4.394	.449	4.275	.437
9.862	4.580	.464	4.296	.436
10.138	4.794	.473	4.366	.431
10.143	4.396	.433	4.367	.431
10.298	4.441	.431	4.405	.428
11.643	4.644	.399	4.728	.406
12.065	4.791	.397	4.826	.400
12.119	4.858	.401	4.838	.399
13.325	4.997	.375	5.112	.384
14.156	4.982	.352	5.285	.373
14.353	6.043	.421	5.327	.371
15.310	5.313	.347	5.524	.361
16.041	5.075	.316	5.671	.354
17.230	5.576	.324	5.903	.343
18.486	5.561	.301	6.139	.332
19.516	6.065	.311	6.327	.324
21.877	6.067	.277	6.740	.308
25.645	6.318	.246	7.355	.287

Table A2 (contd.)

INK TRANSFER DATA
"PPRIC"

Sample: C-STD/2

Experimental			Calculated	
x g/m^2	y g/m^2	y/x	y g/m^2	y/x
1.289	.435	.337	.382	.296
1.557	.521	.335	.535	.344
1.916	.672	.351	.739	.386
2.180	.921	.422	.889	.408
2.244	.863	.385	.926	.413
2.266	.954	.421	.938	.414
2.660	1.200	.451	1.163	.437
2.698	1.194	.443	1.184	.439
3.055	1.488	.487	1.496	.490
3.323	1.648	.496	1.670	.503
3.492	1.801	.516	1.780	.510
3.998	2.106	.527	2.109	.528
5.483	3.094	.564	3.074	.561
5.492	3.152	.574	3.080	.561
6.104	3.412	.559	3.478	.570
6.703	3.805	.568	3.751	.560
6.992	3.853	.551	3.854	.551
7.580	4.177	.551	4.056	.535
8.015	4.387	.547	4.201	.524
8.071	4.433	.549	4.219	.523
8.445	4.566	.541	4.340	.514
9.021	4.755	.527	4.521	.501
9.712	4.999	.515	4.731	.487
10.032	5.075	.506	4.826	.481
11.036	5.691	.516	5.113	.463
11.382	5.780	.508	5.209	.458
11.901	5.559	.467	5.351	.450
12.100	5.624	.465	5.404	.447
13.775	6.167	.448	5.837	.424
13.967	6.177	.442	5.885	.421
14.208	6.041	.425	5.944	.418
16.618	6.614	.398	6.515	.392
16.984	6.726	.396	6.598	.388
17.111	6.726	.393	6.627	.387
18.930	6.778	.358	7.025	.371
19.291	6.783	.352	7.102	.368
20.971	7.215	.344	7.450	.355
23.732	7.488	.316	7.992	.337

Table A2. (contd.)

INK TRANSFER DATA
"PPRIC"

Sample: D-STD/1

Experimental			Calculated	
x g/m^2	y g/m^2	y/x	y g/m^2	y/x
.703	.143	.203	.168	.240
.986	.396	.402	.333	.338
1.027	.331	.322	.357	.347
1.090	.417	.383	.393	.361
1.235	.435	.352	.478	.387
1.241	.489	.394	.481	.388
1.420	.564	.397	.573	.403
1.524	.638	.419	.641	.421
1.799	.819	.455	.823	.457
2.058	1.008	.490	.993	.483
2.415	1.239	.513	1.228	.509
2.896	1.514	.523	1.545	.534
3.736	2.082	.557	2.099	.562
3.775	2.180	.577	2.125	.563
3.998	2.320	.580	2.272	.568
4.504	2.549	.566	2.605	.578
4.660	2.701	.580	2.670	.573
4.958	2.796	.564	2.772	.559
5.860	3.250	.555	3.066	.523
6.166	3.347	.543	3.160	.512
6.936	3.734	.538	3.387	.488
7.060	3.773	.534	3.423	.485
7.527	3.929	.522	3.553	.472
8.767	4.353	.497	3.882	.443
9.490	4.675	.493	4.063	.428
9.790	4.553	.465	4.136	.422
9.937	4.880	.491	4.171	.420
10[illegible]34	4.774	.437	4.404	.403
1[illegible]09	5.098	.414	4.709	.383
1[illegible]047	5.092	.390	4.866	.373
1[illegible]679	5.531	.377	5.197	.354
1[illegible].109	5.518	.365	5.281	.350
1[illegible].239	5.652	.328	5.682	.330
17.345	5.732	.330	5.701	.329
17.598	5.855	.333	5.747	.327
20.242	6.048	.299	6.206	.307

Table A2. (contd.)

INK TRANSFER DATA
"PPRIC"

Sample: S-STD

Experimental			Calculated	
x	y	y/x	y	y/x
g/m^2	g/m^2		g/m^2	
.807	.221	.271	.214	.265
.861	.251	.292	.244	.283
1.021	.322	.315	.334	.327
1.087	.361	.332	.371	.341
1.341	.525	.391	.513	.383
1.671	.685	.410	.698	.418
1.898	.835	.440	.826	.435
2.636	1.341	.509	1.309	.497
3.397	1.706	.502	1.782	.524
3.987	2.259	.567	2.148	.539
4.043	2.121	.525	2.183	.540
4.889	2.573	.526	2.670	.546
5.274	2.828	.536	2.802	.531
5.568	2.940	.528	2.899	.521
6.024	3.021	.501	3.045	.505
6.673	3.226	.483	3.244	.486
6.949	3.319	.478	3.325	.479
7.170	3.367	.470	3.390	.473
8.071	3.762	.466	3.642	.451
8.391	3.894	.464	3.728	.444
9.429	4.043	.429	3.996	.424
9.492	4.019	.423	4.012	.423
9.526	3.961	.416	4.020	.422
11.308	4.166	.368	4.447	.393
12.287	4.647	.378	4.667	.380
13.607	4.780	.351	4.950	.364
14.508	4.573	.315	5.135	.354
15.064	4.934	.328	5.247	.348
15.397	4.521	.294	5.313	.345
17.256	5.745	.333	5.668	.328
17.529	4.940	.282	5.718	.326
18.584	4.932	.265	5.910	.318
19.905	5.931	.298	6.142	.309
20.365	6.487	.319	6.221	.305
22.329	5.315	.238	6.549	.293
24.612	8.173	.332	6.913	.281
27.129	6.123	.226	7.295	.269

Table A2. (contd.)

INK TRANSFER DATA
"PPRIC"

Sample: A-LP

Experimental			Calculated	
x g/m^2	y g/m^2	y/x	y g/m^2	y/x
1.046	.337	.322	.310	.296
1.152	.437	.379	.375	.325
1.232	.422	.343	.423	.344
1.369	.544	.397	.507	.370
1.563	.551	.353	.625	.400
1.667	.688	.413	.688	.413
1.727	.672	.389	.725	.420
1.847	.763	.413	.798	.432
1.914	.813	.425	.839	.438
2.300	1.066	.463	1.074	.467
2.739	1.384	.505	1.341	.490
2.882	1.453	.504	1.428	.496
3.667	1.974	.538	1.991	.543
3.674	2.000	.544	1.995	.543
4.015	2.249	.560	2.231	.556
4.424	2.499	.565	2.514	.568
5.271	3.100	.588	3.101	.588
6.988	4.220	.604	4.018	.575
7.250	4.032	.556	4.113	.567
7.392	4.525	.612	4.164	.563
8.197	4.599	.561	4.443	.542
8.837	4.750	.538	4.655	.527
9.438	5.057	.536	4.848	.514
9.721	5.126	.527	4.936	.508
11.274	5.766	.511	5.400	.479
12.067	5.771	.478	5.625	.466
12.125	5.810	.479	5.641	.465
12.285	5.881	.479	5.685	.463
12.876	5.790	.450	5.846	.454
13.455	5.996	.446	6.001	.446
13.996	6.376	.456	6.142	.439
14.100	6.324	.449	6.169	.438
14.700	6.290	.428	6.322	.430
14.757	6.532	.443	6.336	.429
15.574	6.521	.419	6.539	.420
15.637	6.625	.424	6.555	.419
16.573	6.458	.390	6.780	.409
18.620	6.763	.363	7.253	.390
19.619	7.036	.359	7.489	.380
21.801	8.292	.380	7.938	.364
23.399	9.040	.386	8.263	.353
25.961	7.875	.303	8.762	.338
27.661	8.167	.295	9.080	.328

Table A2. (contd.)

INK TRANSFER DATA
"PPRIC"

Sample: A-LS

Experimental			Calculated	
x g/m^2	y g/m^2	y/x	y g/m^2	y/x
1.380	.610	.442	.576	.417
1.598	.735	.460	.720	.451
1.909	.902	.472	.928	.486
1.946	.949	.488	.952	.489
2.216	1.098	.495	1.127	.509
2.447	1.247	.510	1.280	.523
2.549	1.345	.528	1.347	.528
3.168	1.795	.567	1.755	.554
3.379	1.931	.571	2.017	.597
3.660	2.309	.631	2.221	.607
3.853	2.380	.618	2.361	.613
3.952	2.344	.593	2.432	.615
4.540	2.975	.655	2.858	.630
4.664	2.928	.628	2.948	.632
5.840	3.777	.647	3.799	.651
6.022	3.909	.649	3.931	.653
7.031	4.456	.634	4.467	.635
7.609	4.873	.640	4.696	.617
8.225	5.220	.635	4.931	.600
8.767	5.503	.628	5.130	.585
9.144	5.617	.614	5.265	.576
9.243	5.552	.601	5.300	.573
9.485	5.907	.623	5.385	.568
9.767	6.128	.627	5.482	.561
11.347	6.387	.563	6.004	.529
12.184	6.735	.553	6.266	.514
12.322	7.053	.572	6.308	.512
14.593	7.555	.518	6.972	.467
14.932	7.325	.491	7.067	.473
15.085	7.539	.500	7.109	.471
16.052	7.708	.480	7.372	.459
16.082	7.604	.473	7.380	.459
16.822	7.754	.461	7.576	.450
16.899	7.669	.454	7.596	.449
18.830	7.788	.414	8.086	.429
18.647	7.875	.422	8.040	.431
20.690	8.095	.391	8.534	.412
21.805	8.493	.389	8.794	.403
23.836	8.716	.366	9.249	.388

Table A2. (contd.)

INK TRANSFER DATA
"PPRIC"

Sample: A-HS

Experimental			Calculated	
x g/m^2	y g/m^2	y/x	y g/m^2	y/x
.850	.192	.226	.177	.208
1.008	.266	.264	.276	.274
1.213	.467	.385	.405	.334
1.420	.521	.367	.535	.377
1.490	.590	.396	.579	.389
1.704	.640	.376	.713	.418
2.056	.934	.454	.934	.454
2.084	.873	.419	.952	.457
2.484	1.185	.477	1.203	.484
2.512	1.317	.524	1.221	.486
3.113	1.656	.532	1.615	.519
3.472	1.909	.550	1.865	.537
3.764	2.076	.552	2.074	.551
4.141	2.320	.560	2.336	.564
4.173	2.471	.592	2.359	.565
4.837	2.837	.587	2.816	.582
5.008	2.843	.568	2.936	.586
5.180	3.070	.593	3.055	.590
5.284	3.127	.592	3.128	.592
5.792	3.321	.573	3.324	.574
6.143	3.624	.590	3.453	.562
6.595	3.872	.587	3.613	.548
6.675	3.814	.571	3.640	.545
7.568	4.220	.558	3.939	.520
8.002	4.342	.543	4.078	.510
8.158	4.571	.560	4.127	.506
8.880	4.867	.548	4.348	.490
9.020	4.759	.528	4.390	.487
9.771	4.990	.511	4.609	.472
10.333	5.340	.517	4.767	.461
11.122	5.487	.493	4.982	.448
12.753	5.877	.461	5.404	.424
12.889	5.529	.429	5.438	.422
13.505	5.840	.432	5.589	.414
15.535	6.407	.412	6.065	.390
15.544	5.909	.380	6.067	.390
15.604	6.268	.402	6.081	.390
17.676	6.182	.350	6.535	.370
17.680	6.588	.373	6.536	.370

Table A2. (contd.)

INK TRANSFER DATA
"PPRIC"

Sample: C-HP

Experimental			Calculated	
x g/m^2	y g/m^2	y/x	y g/m^2	y/x
1.654	.629	.380	.639	.386
1.838	.768	.418	.744	.405
1.968	.804	.409	.819	.416
2.186	.934	.427	.889	.407
2.270	.986	.434	.948	.418
2.521	1.120	.444	1.126	.447
2.677	1.183	.442	1.237	.462
2.787	1.319	.473	1.315	.472
2.992	1.393	.466	1.460	.488
3.377	1.743	.516	1.734	.513
3.905	2.110	.540	2.108	.540
4.313	2.420	.561	2.397	.556
5.174	3.010	.582	3.007	.581
6.318	3.676	.582	3.777	.598
6.460	3.734	.578	3.829	.593
7.505	4.260	.568	4.191	.558
8.056	4.541	.564	4.372	.543
8.882	4.746	.534	4.632	.522
9.100	4.889	.537	4.698	.516
9.775	5.148	.527	4.900	.501
9.935	5.176	.521	4.946	.498
10.521	5.250	.499	5.114	.486
11.274	5.604	.497	5.323	.472
11.600	5.431	.468	5.411	.466
12.007	5.821	.485	5.520	.460
12.612	5.788	.459	5.677	.450
14.045	5.849	.416	6.037	.430
15.457	6.180	.400	6.374	.412
15.847	6.136	.387	6.464	.408
16.839	7.219	.429	6.689	.397
17.224	6.419	.373	6.774	.393
17.821	7.380	.414	6.905	.387
19.829	7.420	.374	7.328	.370
20.445	7.395	.362	7.454	.365
22.618	7.833	.346	7.883	.349
23.687	7.786	.329	8.086	.341
24.527	7.860	.320	8.243	.336
25.533	7.643	.299	8.427	.330

RHEOLOGICAL PROPERTIES OF NEWS INKS AND SURFACE STRENGTH TEST LIQUIDS

J. S. Aspler; M. B. Lyne, Pulp and Paper Research Institute of Canada
Pointe Claire, Quebec, Canada
J. M. Dealy and G. C. Pangalos, McGill University, Montreal,
Quebec, Canada

ABSTRACT

The rheological properties of black letterpress and lithographic news inks, and surface strength test liquids such as polyisobutene fluids and tack-graded inks are examined in a Rheometrics mechanical spectrometer. While the ink vehicles are Newtonian liquids the addition of carbon black pigment makes the inks pseudoplastic (decreasing viscosity with increasing shear rate) and mildly thixotropic (reduced viscosity with time at constant shear rate). Pitch added to the ink to improve the wetting of the carbon pigment tends to reduce viscosity at low shear rates by acting to disperse pigment agglomerates. Elastic behaviour is exhibited when synthetic polymer resins are added to the ink. The elasticity of lithographic and tack-graded inks does not appear to be the cause of ink tack since polyisobutene surface strength test oils exhibit high tack, but are inelastic Newtonian liquids. Results also suggest that while viscosity and ink tack may be correlated for inks with chemically similar vehicles there is no general causal relationship between viscosity and tack.

INTRODUCTION

Traditionally, the rheological properties of inks have been described in pragmatic terms such as length, yield, misting, tack, and viscosity (as measured in a falling rod viscometer). While these terms may aid the ink manufacturer in understanding the requirements of the printer, they are not suited to a fundamental investigation of the factors which govern the behaviour of ink in a printing press. In part, this is due to the terms being ill-defined, as exemplified by the fact that where test methods exist the results are specific to the type of apparatus used. Each of these terms actually describes a combination of well-defined rheological properties.

In this study the basic rheological properties of various news inks and surface strength testing liquids for paper have been measured in a mechanical spectrometer. The effect of ink formulation on these properties is investigated and the relationship between these properties and tack is explored.

The black news inks used in North America have simple formulations - at least, superficially. They set by the penetration of an oil vehicle into the porous surface of paper leaving carbon pigment filtered out near the surface. Besides these two components a news ink may contain petroleum pitch to improve the wetting of the carbon pigment by the oil vehicle, and polymer resins added to impart properties such as better transfer in the printing press. News inks are low cost, high volume inks whose major component, the oil vehicle, is manufactured with a minimum degree of refining. The vehicle is a complex blend of aromatic, paraffinic and napthenic oils with wide distributions of molecular weight. Similarly, the carbon pigment has a primary particle size of approximately 0.02 μm, but agglomerates range in size up to 100 μm in diameter.

The formulation of the news inks used in this study has been varied by the stepwise addition of the above components, and by choosing two oil vehicles with different mean molecular weights. Results of similar research on European sheet-fed offset inks can be found in the work of Oittinen[1].

Basic Rheology of Fluids

Fundamental rheological properties of fluids are described in standard texts[2, 3] and only a brief review will be given.

As shown in Figure 1, if a material is confined between a stationary plate of area A and a parallel plate moving with velocity V at distance h, the shear rate $\dot{\gamma}$ is:

$$\dot{\gamma} = \frac{dV}{dh} \tag{1}$$

The shear stress is equal to the force that resists this deformation per unit area:

$$\tau = F/A \tag{2}$$

The viscosity η is defined by:

$$\eta = \tau/\dot{\gamma} \tag{3}$$

If the viscosity depends on shear rate or time, or both, the fluid is said to be non-Newtonian. Some common types of flow behaviour are shown in Figure 2. In this figure, the asymptotic slope of the curve defining pseudoplasticity can be referred to as the apparent plastic viscosity. The resulting viscosity-shear rate dependence is shown in Figure 3. Pseudoplasticity is a common property of inks, and is believed to result from the change in particle-fluid or particle-particle interactions with increased rate of shear such as the progressive reduction in the size of particle agglomerates. Thixotropy, the drop in viscosity with time at constant rate

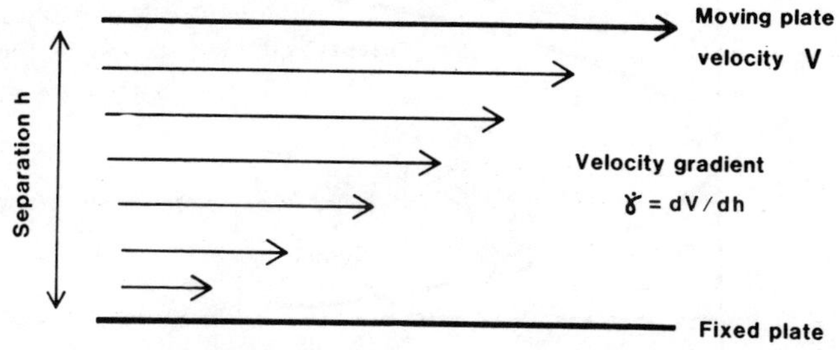

Fig. 1. Shear profile obtained in a viscometry experiment.

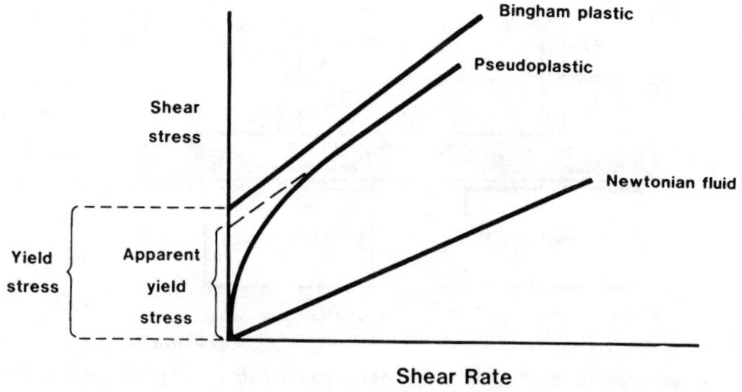

Fig. 2. Shear stress-shear rate behaviour for different classes of fluids.

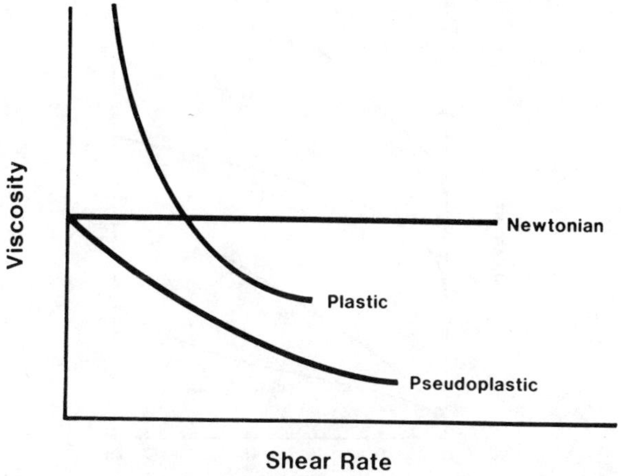

Fig. 3. Viscosity-shear rate behaviour for different classes of fluids.

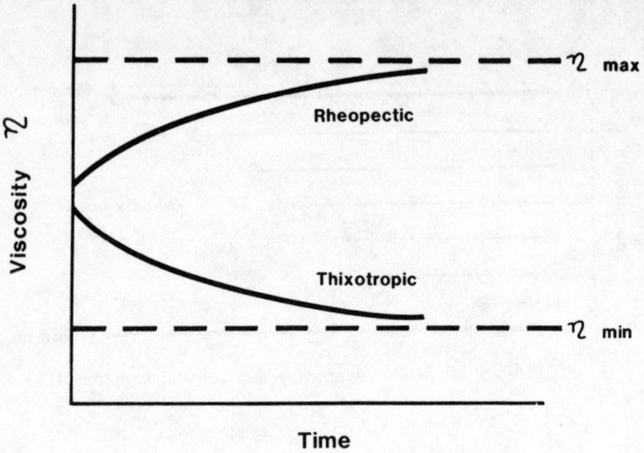

Fig. 4. Time dependence of viscosity.

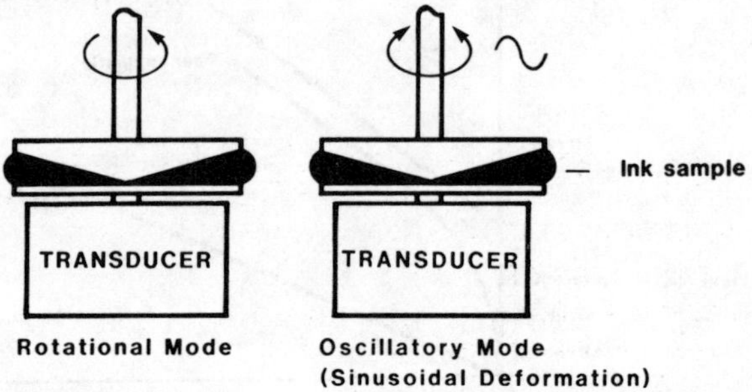

Fig. 5. Schematic of the cone and plate fixture in the Mechanical Spectrometer operating in the rotational mode for viscosity measurement at left, and operating in the oscillatory mode for the measurement of dynamic viscosity and storage modulus at right.

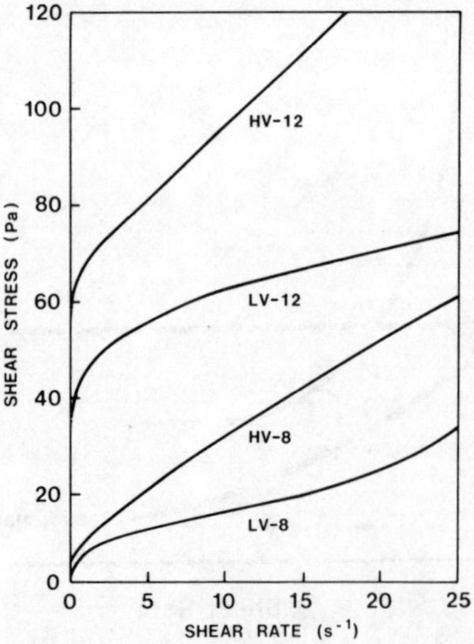

Fig. 6. Shear stress-shear rate dependence for model letterpress inks.

of shear, is thought to result from similar effects. This is illustrated in Figure 4.

Another type of fluid behaviour is viscoelasticity, which occurs when the fluid response includes both viscous dissipation and elastic storage of the work of shearing. This behaviour depends on both the composition of the mixture and the rate of shearing. Under conditions of sinusoidal deformation of frequency ω, the shear strain is:

$$\gamma(t) = \gamma_o \sin(\omega t) \tag{4}$$

and the shear rate is:

$$\dot{\gamma}(t) = \dot{\gamma}_o \cos(\omega t) \tag{5}$$

For small strain amplitude, the stress function is given by the sum of the elastic and viscous components:

$$\tau(t) = G' \gamma_o \sin(\omega t) + n' \dot{\gamma}_o \cos(\omega t) \tag{6}$$

$G'(\omega)$ is the storage modulus function related to the in-phase (elastic) component of the stress, and $n'(\omega)$ is the dynamic viscosity function, related to the out-of-phase (viscous) component. The dynamic viscosity n' is equal to G''/ω, where G'' is the loss modulus. The familiar loss tangent, tan δ, is equal to G''/G', where δ is the phase shift between stress and strain.

EXPERIMENTAL

Various inks, oils, and model materials were examined on a Rheometrics mechanical spectrometer. A cone and plate arrangement was used for viscosity determination (Figure 5). All measurements were made at 24°C.

The instrument also permits sinusoidal oscillation of the cone about its axis of rotation over several decades of frequency. Shear forces transmitted to the plate are recorded by a torque-sensing transducer. The phase relationship between applied strain and the measured stress is then used to calculate the dynamic viscosity and storage modulus, as above. The viscosity is measured over a range of shear rates, and dynamic viscosity and storage modulus can be obtained over a range of frequencies.

All the materials examined are shown in Table 1. The oils and carbon black suspensions that can be considered as model letterpress inks were supplied by Canada Printing Ink. Concentrations are given as weight percent of the components. The oils are heavy mineral oil ink vehicles, with carbon atoms distributed in the approximate order: aromatic 20%; paraffinic 40%, and napthenic (saturated cycloparaffinic) 40%. Other materials included an IPI "tack graded ink", SCAN standard test ink, a commercial lithographic ink, a medium-viscosity polyisobutene pick test oil from IGT, and a silicone (polydimethylsiloxane) fluid.

In addition, ink samples made by the stepwise addition of

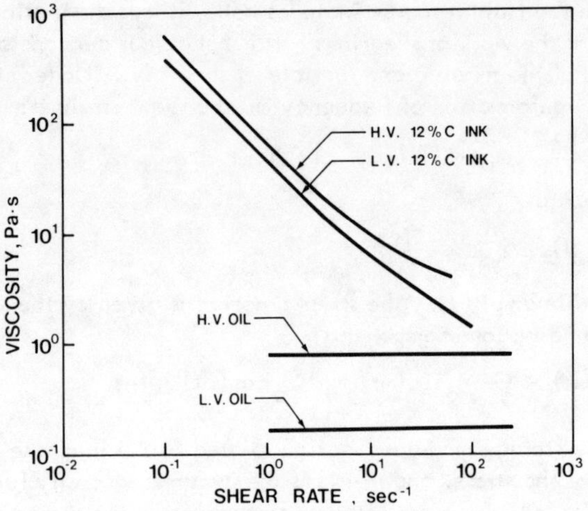

Fig. 7. Viscosity-shear rate dependence for model letterpress inks.

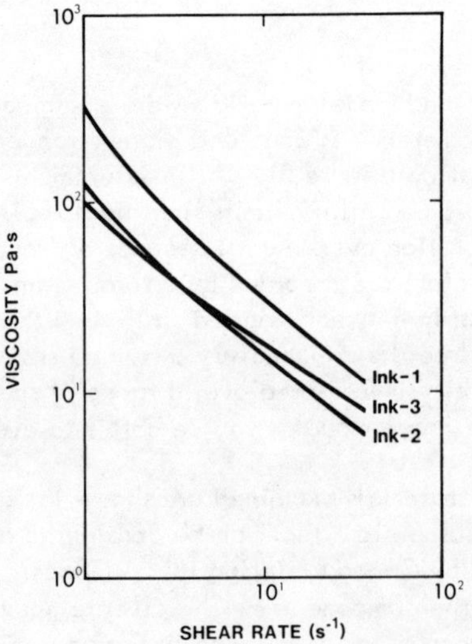

Fig. 8. Viscosity-shear rate dependence for three ink samples prepared by stepwise addition of materials.

materials to the same base were examined (Ink-1, Ink-2, Ink-3).

TABLE 1 SUMMARY OF MATERIALS USED

Sample, Ink vehicle	Description
LV oil	Light mineral oil
LV-8	Light mineral oil + 8% carbon
LV-12	Light mineral oil + 12% carbon
HV oil	Heavy mineral oil
HV-8	Heavy mineral oil + 8% carbon
HV-12	Heavy mineral oil + 12% carbon
IGT oil	'Medium viscosity' polyisobutene fluid, as per standard NEN 3095, ISO/DIS 3782 and 3783
GFL standard test ink	SCAN standard test method P35:72
Silicone fluid	Poly (dimethyl siloxane), Dow Corning 200, viscosity = 60 Pa·s
IPI tack graded ink	Y49506 8/24/73 BA No: 658521
Commercial lithographic offset ink	-
Ink-1	81.6 parts mineral oil plus 18.4 parts carbon
Ink-2	As above plus 3.2 parts pitch
Ink-3	47.5 parts mineral oil, 18.4 parts carbon black, 3.2 parts pitch plus 30.8 parts polyindene resin solution

RESULTS

Rheology of letterpress and lithographic news inks

Figures 6 and 7 show the rheological properties of the mineral oils and simple carbon black suspensions. The pure LV and HV oils are Newtonian fluids with viscosities of 0.175 Pa s (1.75 poise) and 0.80 Pa s (8.0 poise), respectively. The carbon black suspensions are all non-Newtonian, with fluid LV-8 showing a classical pseudoplastic behaviour. It is difficult to give a definitive classification for the remaining fluids, although these too appear to be pseudoplastic. A slight time dependence of the viscosity (not shown) indicates that these materials are also mildly thixotropic.

Figure 7 shows the equilibrium viscosities plotted vs shear rate on a logarithmic scale for the 12% carbon black suspension. At shear rates of $10^2\ s^{-1}$ the viscosities appear to be approaching the limiting values for the viscosities of the oil vehicles, while at intermediate shear rate the behaviour is of the form:

$$\eta = k\dot{\gamma}^{n-1} \tag{7}$$

At very low shear rate, a definite limiting value would be expected. Thus, it can be concluded that at low shear rate, the carbon black content governs the behaviour, while at high shear rate the vehicle oil governs.

Figure 8 shows the viscosity - shear rate dependence for the three samples designated Ink-1, Ink-2 and Ink-3, prepared by the stepwise addition of ink components to the same base. The exact formulations are given in Table 1. All three are highly non-Newtonian, and exhibit approximate power law behaviour. The simple carbon black - oil suspension Ink-1 showed the highest viscosity. The addition of the hydrocarbon pitch wetting agent caused a significant drop in viscosity (Ink-2). The addition of "varnish" (polymer resin solution) caused only a slight change in the viscosity, but affected the dynamic (viscoelastic) properties (Ink-3). Yield stresses were obtained that decreased in the order Ink-1 > Ink-2 > Ink-3. Explicit evidence of the viscoelastic nature of the inks was obtained by oscillatory shear. The curves of storage and loss moduli versus frequency indicated significant elasticity, but the shapes of the curves were quite different from those typical of solutions of synthetic polymers.

Rheology of surface strength test liquids

As stated in the introduction, "tack" is an important but poorly defined concept. Tack can best be described as the splitting resistance, or tensile force involved in the splitting of an ink film at the exit of a printing nip. However, tack measuring devices generally measure some combination of properties not directly related to the phenomenon of interest[4]. For example, the "Inkometer" measures the tendency of one roller to follow another under the influence of a splitting ink film. Since the tack forces are created on the outgoing side of the nip between the rollers, a component of this force should act to make one roller follow the other. However, the transmission of shear through the liquid should also cause one roller to follow another. This may explain why, for inks with similar formulations, a statistical correlation exists between the "Inkometer number" and the ink viscosity[5]. Tack measured on similar instruments has also been related to elasticity, shortness (defined as the ratio of apparent yield stress to apparent plastic viscosity) and surface tension.

The tack forces generated in a printing press nip must be borne by the surface of paper. Paper is tested for surface strength using tacky liquids like tack-graded inks or polyisobutene oils. These liquids and a commercial lithographic ink have been examined in an attempt to separate and identify the various factors contributing to tack. These materials are listed in Table 1. A high viscosity silicone fluid which exhibits very low tack is included for reference.

The IGT polyisobutene oil and the silicone fluid are both New-

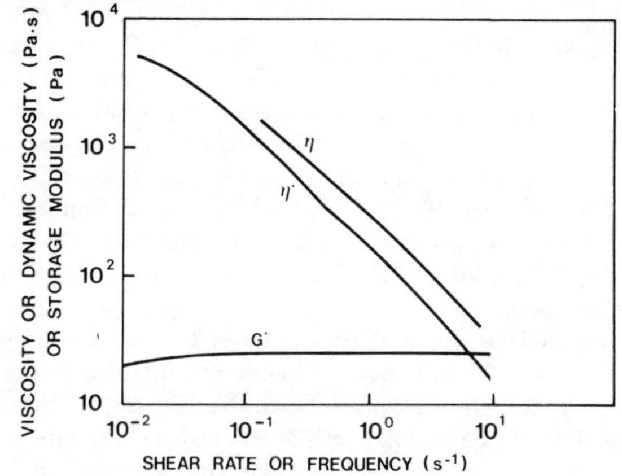

Fig. 9. Rheological properties of SCAN standard test ink.

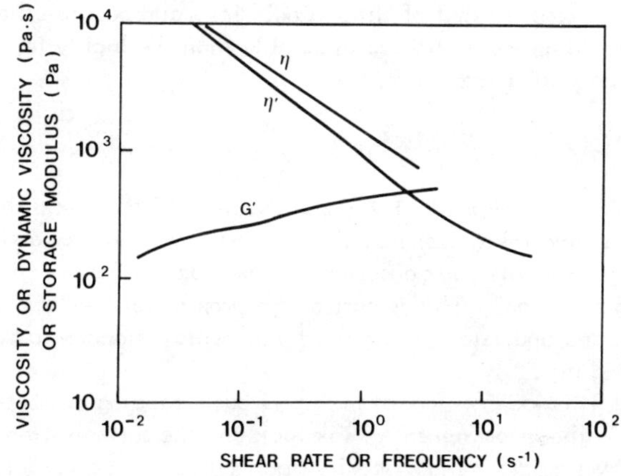

Fig. 10. Rheological properties of IPI tack-graded ink.

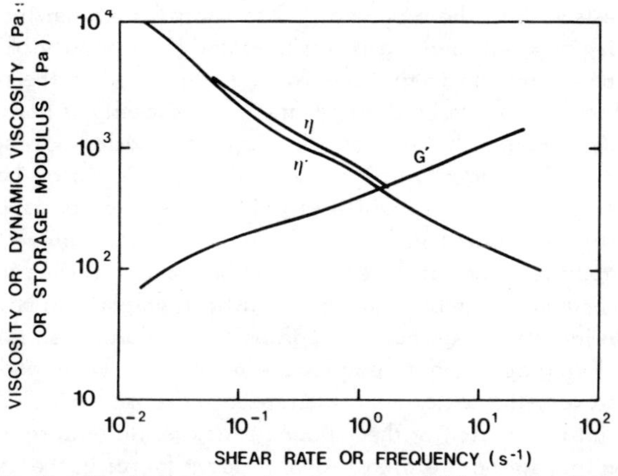

Fig. 11. Rheological properties of commercial lithographic offset ink.

tonian fluids with similar viscosities (50 Pas and 60 Pas, respectively). The IGT oil showed a high tack in an IGT surface strength test manifested by severe picking of the surface of paper at low printing speeds. The silicone fluid showed no sign of picking the surface of paper even at the highest printing speeds in the IGT A1C2 tester.

The GFL and IPI inks are slightly elastic (Figures 9 and 10) while the lithographic ink is very elastic (Figure 11). The values of the loss and storage modulae are similar to those found by Oittinen [1] for an equivalent European ink.

Polyindene is a synthetic polymer that is commonly added to lithographic inks. When dissolved in heavy mineral oil it is particularly tacky, causing picking at slow speeds in the IGT test. These polyindene solutions exhibit viscoelasticity unlike the IGT polyisobutene oils. Thus, these solutions in different concentrations may be better choices for surface strength testing liquids than the polyisobutene oils because their rheology is closer to that of lithographic ink, and because their tack properties should be more temperature stable than the highly temperature sensitive polyisobutene oils [6].

DISCUSSION

It is clear from the complexity of the rheology of news inks that tests in the printing and ink industries, like viscosity measured at a single shear rate, are inadequate for measuring the rheological behaviour of an ink in a printing situation. This is part of the problem of the lack of a proper definition and understanding of the fundamental phenomena occurring in the printing nip.

This lack of understanding of basic rheological behaviour is apparent in the measurement of ink tack and the surface strength of paper using tacky liquids. The viscosity contribution to the tendency of one roller to follow another in tack measuring devices is ignored while surface strength tests such as the proposed ISO standard [7] consider the product of the velocity at which picking is initiated and the viscosity of the test liquid to be equivalent to the tack force experienced by the surface of paper. Further, it can be demonstrated that viscosity is not always a factor in determining the tack of a liquid by comparing polyisobutene oil with silicone oil of equal or higher viscosity. Silicone oil has virtually no tack, evidenced by its inability to pick the surface of paper at even the highest speeds in the IGT tester. Bassemir [8] has pointed out that the viscosity-tack relationship is tenuous and breaks down when ink vehicle chemistry or ink 'shortness' change. Elasticity might also be considered a factor in ink tack since both lithographic ink and concentrated solutions of its tackifying component, polyindene resin, are elastic and cause picking. However, the tacky polyisobutene oils are inelastic. The lower tack and surface tension of the silicone fluids would tend to support the assumption that surface tension is a significant factor in the generation of cohesive tack forces. However, the surface tensions of polyisobutene liquids of vastly differing tacks were found to have the same surface tension (as measured by pendant drop photography). Stress relaxation phenomena in these liquids may also play a role in tack.

These studies and those in the literature suggest that the tack forces generated during the splitting of ink films are caused by several competing and inter-related phenomena. Furthermore, there are several factors not considered in this study that bear on the rheology and tack of news inks. Aqueous fountain solution emulsified by the ink in the lithographic process has a measurable effect on rheological properties and tack[8, 9]. The final steps in the ink film splitting - cavitation and filamentation - should also be taken into account as they have been associated with the property of tack. Temperature effects must also be considered since an increase in temperature results in a lowering of viscosity, tack, and picking of the surface of paper by ink and surface strength test liquids[6, 10]. In this regard a miniature force transducer is being mounted in the face of a printing cylinder in a thermostated lithographic printing press at the Pulp and Paper Research Institute of Canada. This should enable the measurement of tack forces under realistic conditions of shear and emulsification of fountain solution, and may resolve whether the maximum tack force is reached before cavitation and filamentation of the ink. The role of the surface chemistry and roughness of paper in the tack forces generated in the printing nip will also be studied.

CONCLUSIONS

The first step in understanding the relationship between ink formulations and behaviour under press conditions is a systematic approach to the measurement of rheological and other properties of inks in terms which have a fundamental scientific meaning. As it stands, the terminology used in the industry is highly confusing, with concepts such as 'tack', 'length', and 'holdback' being used at the same time as fundamental properties such as viscosity. Even a basic property such as viscosity can be easily abused when measured at only one shear rate or at an undefined shear rate such as with the falling rod viscometer.

Model letterpress inks were found to be both pseudoplastic and thixotropic, but inelastic, fluids. A GFL standard test ink and an IPI tack graded ink were slightly elastic, while a commercial lithographic ink was highly elastic. Polyisobutene surfaces strength testing oils are Newtonian liquids, and so do not have the rheological properties of real inks, and also exhibit unsatisfactorily high variation in viscosity and tack with temperature. The creation of stable surface strength test fluids that are of greater relevance to printing situations is being considered.

Specialized instrumentation is needed in order to measure the various aspects of ink rheology. In this regard, a device designed to measure ink tack directly as the force created by the splitting of an ink film in a printing nip is being developed at P.P.R.I.C.

REFERENCES

1. Oittinen, P., 'Fundamental Rheological Properties and Tack of Printing Inks and Their Influence On Ink Behaviour In a Printing Nip', D. Tech. Thesis, Helsinki University of Technology (1976).
2. Ferry, J.D., 'Viscoelastic Properties of Polymers', 2nd edn., Wiley, New York (1970).
3. Zettlemoyer, A.C. and Myers, R.R., 'The Rheology of Printing Inks', in 'Rheology-Theory and Applications', Vol. III (F.R. Eirich, Ed.), Academic Press, New York (1960).
4. Hellinckx, L. and Mewis, J., 'Rheological Behaviour of Pigment Dispersions as Related to Roller Passage', Rheologica Acta 8, 519 (1969); Mewis, J. and Dobbles, F., 'Nip Flow and Tack of Printing Inks', To be published.
5. Rosted, C.O., 'Rheology and Printing Properties', J. Oil Col. Chem. Assoc., 54, 420 (1971).
6. Lyne, M.B., 'Testing the Surface Strength of Paper', Tappi, 63 (11), 31 (1980).
7. Draft International Standard ISO/DIS 3783, 'Paper-Determination of Resistance to Picking - Accelerating Speed Method Using the IGT Tester (Electric Model)'.
8. Bassemir, R.W., 'The Physical Chemistry of Lithographic Inks', American Ink Maker, February (1981).
9. Surland, A., 'A Laboratory Test Method for Prediction of Lithographic Ink Performance', T.A.G.A., 32nd Conference (1980).
10. Snider, E.H. and Smith, R.E., 'Is It The Newsprint, The Ink, or The Printing Process?', Pulp and Paper Canada 81 (11), 97 (1980); Canadian Patent No. 1020807 (Nov. 15, 1977).

DISCUSSION

W. Hansen Rheological properties including tack are normally measured in the laboratory at 25°C (77°F) or 32°C (90°F). On a news press the temperature will be about 45°C (110-120°F) and the rheological properties will be completely different. Have you any comments on the relationship between laboratory measurements and practical experiences?

J. Aspler Both the mechanical spectrometer and our research printing press allow temperature control. In the next stage of our work, the rheological measurements in the laboratory and force and other measurements on the press will be done at the same temperature.

J. M. Fetsko Did your cone and plate have temperature control? Probably not. We inserted a thermistor during runs and found that temperature increases with oils were minor but surprisingly large with inks.

J. Aspler At 23°C, the only temperature control is that provided by an air-conditioned laboratory. However, a thermocouple mounted in our device showed that a constant temperature was in fact maintained.

W. Schaeffer (1) May I suggest that since you are interested in the rheology of colloidal systems, you describe not only the composition but also the sequence of additions and their form because these have an influence.

(2) Did you encounter with your inks any difficulties on the rheometer in maintaining contact between the cone and plate and also edge effects?

(3) In any definition of tack with other industrial use of this term, if you wish to restrict the definition to the tensile forces, then Voet's approach should be useful even though parallel plates rather than a roller system was used by him. If, on the other hand, you wish to include the instrumental evaluation inkometer, wouldn't the analysis of C. C. Mill be useful?

(4) Finally, Equation 6 of your paper is for small strain amplitudes and quite different from that encountered in printing.

J. Aspler (1) Thank you for this comment. We will discuss this with our ink supplier.

(2) The effects that you mention mark the upper limit of the measurements.

(3) It is our hope that the force transducer mounted in the impression cylinder of our offset press will enable us to measure directly forces in the nip. Once this is done, we intend to relate these forces to ink formulation, ink rheology, tendency to pick paper, and traditional quantities such as 'Inkometer number'.

(4) It is understood that our use of Equation 6 is confined solely to the mechanical spectrometer in the laboratory. The situation in the nip remains to be measured.

F. B. Jansen, I.G.T. (Communicated) (1) I think part of the conclusions are based on incorrect measurements or on wrong interpretation of measurements or phenomena. First: (in reference to IGT test liquids) it is stated that the silicone fluid is a Newtonian fluid. I think the authors tested this at too low values for the shearing stress. We tested it with shearing stresses which are normal in printing (Tollenaar bar viscosimeter) together with IGT normal pick test oil. The results you see in the graph (Figure 1D). The IGT oil is really Newtonian but Dow Corning 200 absolutely not; it is strongly pseudo plastic. This oil shows no picking tendency not because of its low tack values (as stated by the authors) but because of the low viscosity values involved.

(2) To this I will add a second remark: the statement that a liquid has a high tack because it shows severe picking is the opinion of the authors and not proven, whereas the correlation between viscosity and pick is proven by many authors. Thus the Institute of Paper Chemistry introduced in 1946 the V.V.P., the viscosity velocity product.

(3) My third and last remark is about the supposedly high temperature sensitivity of polyisobutene oils. This supposition is based on Lyne's article in Tappi (1980) but is in my opinion a wrong interpretation of his measurements. What he has measured is the rise in temperature on an IGT inking unit. This heating is caused by the resistance against distortion. Is not this another word for viscosity? So if you want to say something about this rise in temperature you should start with liquids of the same viscosity, but if you want to know something about the temperature sensitivity of this viscosity the simplest way is measuring the viscosities at different temperatures. We did this for the IGT oils and an offset news ink. The results are in Figure 2D. We see practically no difference.

<u>J. Aspler</u> We agree with Mr Jansen that the silicone fluid shows pseudoplastic behaviour at high shear rate. However, Dr Oittinen has suggested that the shear rate in the nip of a press may be as low as 10^3 sec^{-1}, under which conditions the silicone fluid still has a high viscosity. Dr Oittinen has also shown that polyisobutene is non-Newtonian at high shear rate. Dr Oittinen has kindly supplied the accompanying graphed results for IGT oil and for a silicone oil (Figure 3D). Finally, we do not believe that the bar viscometer is particularly suitable for determining viscosity as a function of shear rate, since one does not have a constant shear rate during the experiment.

The silicone fluid has a Churchill Inkometer tack rating of 18.2 at 400 rpm and 23°. Under the same conditions, the IGT medium viscosity polyisobutene fluid has a tack rating of 30, the IPI tack-graded ink No. 3 has a rating of 20, and the IPI tack-graded ink No. 2 has a rating of 18.2.

By the accepted standard, the silicone fluid should be called a high tack fluid because of its Inkometer number. We consider it to be a low tack fluid simply because it will not pick the surface of a sheet of paper. The high Inkometer number for the silicone fluid must result from the effect shown by Professor Mewis[4], that a major contribution to tack on an Inkometer is the transmission of shear through a viscous fluid.

<u>M. B. Lyne</u> It would be desirable to measure directly the tack force experienced by the paper in a surface strength test. We are hoping to do this with a miniature force transducer in the nip of an offset press. Failing direct measurement, picking of the surface of paper in response to the tack forces would seem to be a reasonable indicator of the level of tackiness of a fluid.

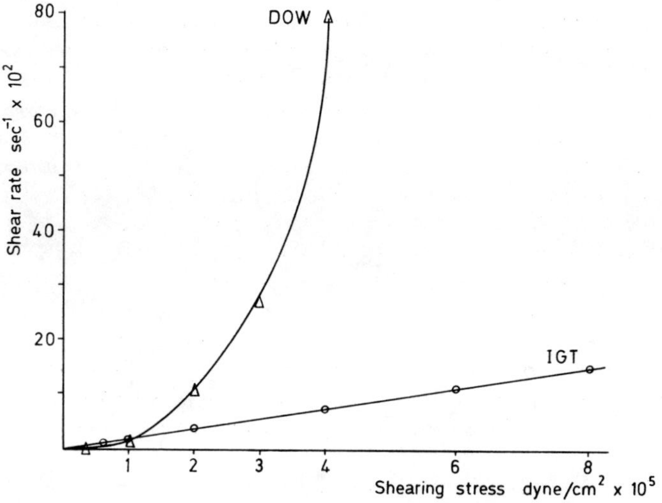

Fig. 1D. Viscosity of Dow Corning 200 and IGT Picktest oil normal.

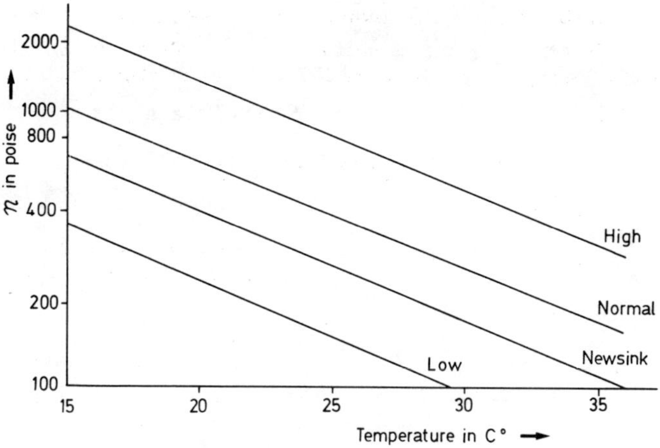

Fig. 2D. Viscosity changes in IGT oils and a news ink.

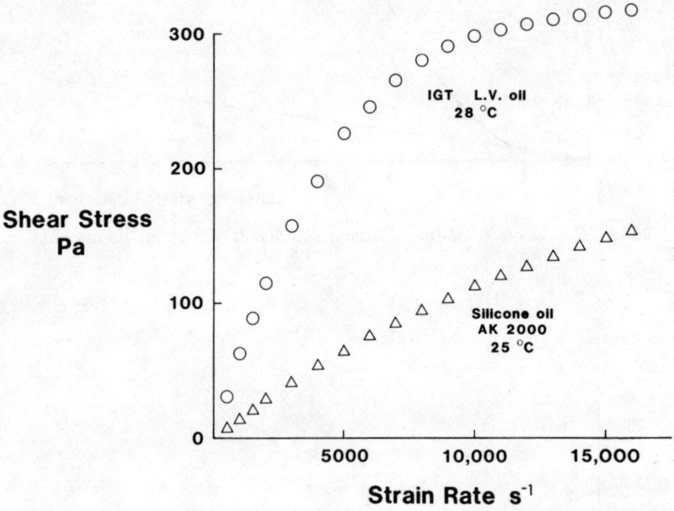
300
200
100
0
Shear Stress
Pa
IGT L.V. oil
28 °C
Silicone oil
AK 2000
25 °C
5000
10,000
15,000
Strain Rate s⁻¹

Fig. 3D.

Attempts to calculate the tack force as a velocity-viscosity product completely neglect other aspects known to affect tack such as elasticity, surface tension, and general molecular mobility. In any case, the relationship between tack force and V.V.P breaks down when examining inks having different chemistrys. Perhaps V.V.P. has been adopted because of its convenience, in lieu of a proper scientific definition for tack.

The TAPPI article mentioned by Mr Jansen was a report on a study of sources of error in the IGT surface strength test when using polyisobutene oils. My infrared investigation showed that the oils heated up during distribution on the IGT inking device (e.g. +2°C for high tack oil). In another part of the study Roy Meret of Woeftellen Bloedel Research compared the change in the amount of fibre picked from the surface of paper in the pick test using PIB, and in a pilot offset press using lithographic ink as a function of temperature. He found that the amount of fibre picked varied between 75 and 100% for different news prints per degree centigrade for the PIB, but varied less than 10% per degree centigrade for lithographic inks. It is apparent from these experiments that variation of tack with temperature is much greater for PIB than for lithographic ink and that since the PIB heats up in the IGT inking device the tack temperature sensitivity of PIB probably contributes significantly to the lack of reproducibility found for the IGT surface strength test in NBS and TAPPI round robin studies. Finally, we agree with Mr Jansen that the rheology of PIB is very different from that for lithographic ink, and it follows that the picking with PIB is not likely to be similar to that with ink.

J. M. Fetsko We agree. Our work showed that oil sometimes gave a 'whiskering' type of picking that had never been seen with inks. The problems of changes with time of test inks and of samples can be minimized by reporting pick results on test papers relative to a reference paper run on the same day.

W. H. Banks Discussions of picking under conditions where the paper is stripped from a forme covered with a tacky fluid, as in web printing, must surely take account of the flexural rigidity of the paper. The fact that picking resistance is anisotropic in the plane of the sheet is a demonstration of this. Treating the test strip as an elastic beam of modulus E and Moment of inertia I, the forces S, acting normal to the surface of the paper must satisfy $S = -EI.d^4y/d^4x$, in which x and y are coordinates of a point on the strip referred to the orme as axis.

In the matter of testing methods, B. Hsu (1960) showed how with measurements made on an inkometer of an ink, and a knowledge of the picking velocity from an IGT tester, it was possible to express picking strength of a paper that was almost independent of the testing liquid, over a wide range of viscosities.

W. Hansen   Rheological properties including tack are normally measured in the laboratory at 25°C (77°F) or 32°C (90°F). On a newspress the temperature will be about 45°C (110-120°F), and the rheological properties will be completely different. Have you any comments on the relationship between laboratory measurements and practical experiences?

J. Aspler   Both the mechanical spectrometer and our research printing press allow temperature control. In the next stage of our work the rheological measurements in the laboratory and force and other measurements on the press will be done at the same temperature.

TACK OF A PRINTED INK FILM AND ITS RELATIONS TO THE PROPERTIES OF INK AND PAPER SURFACE

Poul Halling Sørensen, The Graphic College of Denmark, Copenhagen

ABSTRACT

Tack is an important ink property; it is of importance to both the runability of the ink on the press and how the ink dries in the pile after printing. Through the years much work has been done on tack measurement and many measuring devices have been constructed. A common feature of these studies is largely a concern with the tack of the ink before printing and the rheological relationship with other ink properties, and how the ink behaves during and after printing.

A different basis has been chosen for this study. On a specially constructed device a series of measurements were made of the increasing adhesiveness in the printed film that provides an insight into the tack increase taking place in the ink film after printing. As both the evaporation of solvents and the absorbency of the paper and its affinity for the ink affect this tack increase, a tack measurement of the ink itself before printing gives no information.

The paper is in three parts:

(1) An analysis of the tack of a printed ink film on two papers with different absorbencies printed with eight inks in all, with controlled variations of setting speed, plastic voscosity and shortness.

(2) From actual printings, where different set-off tendencies in the paper pile after printing have been found, an analysis has been made of the tack values of the paper/ink combinations applied.

(3) From well known set-off procedures on testing equipment the connection between these values and the corresponding tack values has been analysed.

1. INTRODUCTION

The relation between the setting speed of a printing ink and the absorbency of the paper surface, i.e. its ability to remove the mineral oil content of the ink after printing, is important for a number of conditions during and after offset printing.

A quick-set ink can deform the paper sheet or the web, it can set on the rubber blanket and cause picking, whereas a slow-set ink can give set-off problems. With very absorbent papers, a sharp decrease in density of the print can occur after printing, often with gloss differences. Chalking also occurs with this type of paper. On the other hand, papers with low absorbencies can cause ink drying problems.

The choice of ink and paper is largely a matter of compromise and a subject of much research. This report is concerned with the tack increase in an ink film after printing and its measurement. Such a measurement implies an investigation of the paper surface and the printing ink together, which it was hoped could be used for estimating the interaction between different combinations of ink and paper.

The paper describes the apparatus used and reports a number of measurements, but it must be regarded as of a preliminary nature and an introduction to the subject.

2. DESCRIPTION OF THE EQUIPMENT

Essentially, the instrument is designed to measure the tensile force needed to separate a rubber-coated wheel from a printed sample, after the wheel has previously been held in contact with the printed sample for a fixed duration, with a constant force (Figure 1).

On the prototype instrument, provision is made for taking five such measurements on each sample, the first measurement being taken approximately 5s after the completion of the printing operation, and the remainder at 12s intervals.

The paper sample (1) to be printed is fixed to a shaped sample wheel (2), and the printing operation is carried out using an IGT printing forme (3), the printing pressure being controlled by an adjustable loading spring, but the printing movement is carried out by hand by turning a cranked handle (4) through about 120°. When the printing operation has been completed, a microswitch actuates an electric drive motor which rotates the shaped sample wheel, so that the rubber-covered measuring forme (5) contacts the freshly printed ink on the paper sample. This measuring wheel is of lightweight alloy construction and mounted freely by a concentric hollow shaft on two knife edges in order to minimise frictional resistance.

Mounted behind the measurement forme on a sliding carriage are the force measurement transducer (6) and a solenoid which, when energised, is used to apply a constant force to the measuring wheel shaft via two parallel side runners of the carriage. A damped spring is used to withdraw the solenoid armature and the side runners away from the measuring wheel shaft during a measurement.

A control circuit energises the solenoid at the same time that

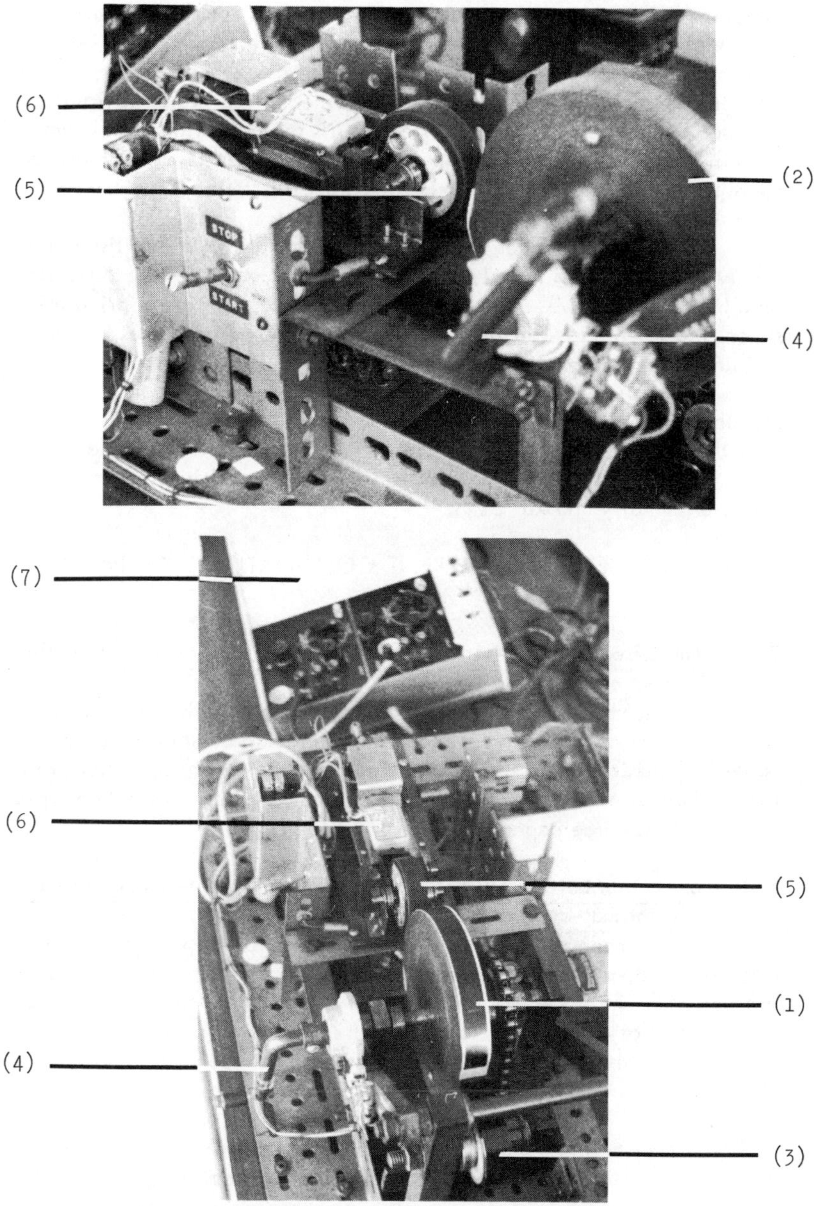

Fig. 1. *The ink-on-paper tack tester seen from the side and from above, (1) paper sample, (2) sample wheel, (3) IGT printing forme, (4) cranked handle, (5) measurement forme, (6) force measurement transducer, (7) Y-T recorder.*

the drive motor operates, so that the wheel is forced against the inked sample which is slowly revolving. Just prior to a tensile force measurement taking place, the drive motor is stopped by a microswitch, and the solenoid armature slowly moves away from the measuring wheel. As this happens, a strain gauge force transducer, the body of which is rigidly mounted between the two side runners, pulls the measuring wheel away from the inked surface via a shaped grab device fixed to the transducer measurement spindle and contacting the wheel shaft in such a manner as to ensure a 'square' and even pull.

The 'bridge-type' force transducer is fed from a stabilised power supply and produces a signal, which is displayed as a vertical deflection on a Y-T recorder (7). After a small time delay, the motor drive restarts, the solenoid is re-energised and the sequence continues for the remaining four measurements.

The diameter of the measuring wheel is large enough to ensure that the surface which comes into contact with the inked sample is clean, i.e. the wheel circumference is greater than the length of sample print.

All measurements have been made in a conditioned room at 50% relative humidity and 23°C.

3. MEASUREMENT OF DIFFERENT COMBINATIONS OF PAPER AND INK

3.1 Comparison of quick-set and slow-set ink of the same make on three different coated papers

A number of experiments were made with a quick-set and slow-set type of ink on a size-press coated, blade coated and air-knife coated paper. Figures 2 and 3 show the variation in tack with time after printing. There is a marked difference in the tack increase of the two types of ink and the three papers.

3.2 Comparison between two slow-set inks of different makes on seven different mat-coated qualities

A number of experiments have been performed with two slow-set inks, A and B, printed on seven different dull-coated paper qualities (marked m 1 - m 7). Figures 4 - 7 show the results of the ink-on-paper tack (in mm) as a function of the time after printing (seconds).

Figures 8 and 9 show the relationship between the tack increase after 10s, read from Figures 4 - 7, and the absorbency of each of the seven papers. The absorbency was measured with a porometry ink from Lorilleux Trykfarver A/S and is expressed as the density of the absorbed ink spot after 10s, measured with a MacBeth densitometer.

It is seen that there is a reasonable connection between increasing tack value after printing and increasing absorbency of the papers m1 - m5 and m7. This is not the case for paper m6, as far as the inks A and B are concerned, which may be due to the fact that this paper is specially glazed and less dull than the others, and does not really belong to the

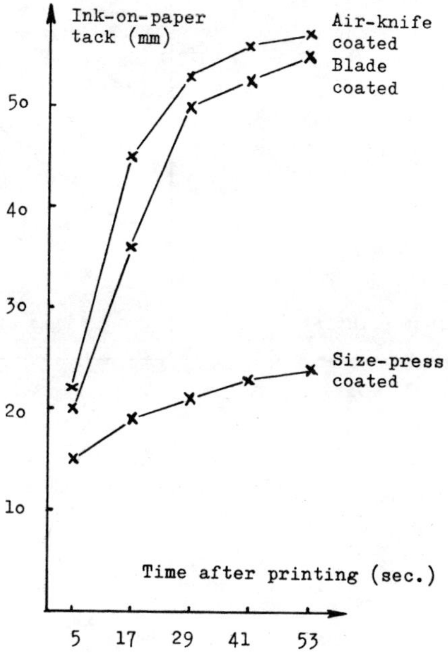

Fig. 2. A quick-set ink printed on three coated papers.

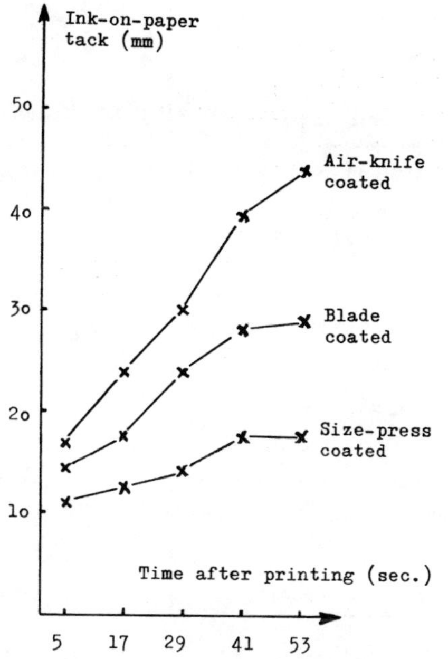

Fig. 3. A slow-set ink printed on three coated papers.

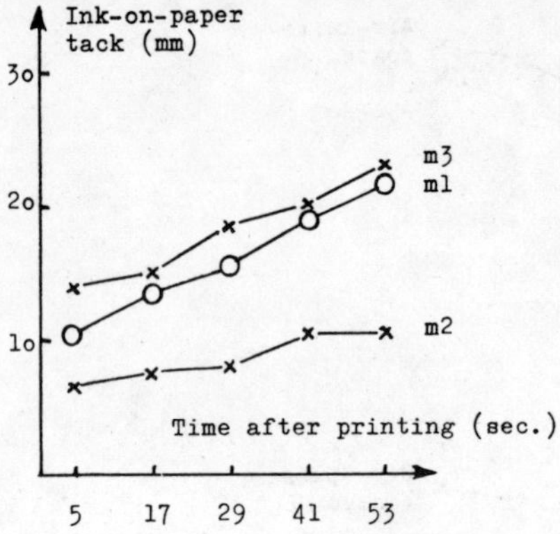

Fig. 4. Ink A/papers m1–m3.

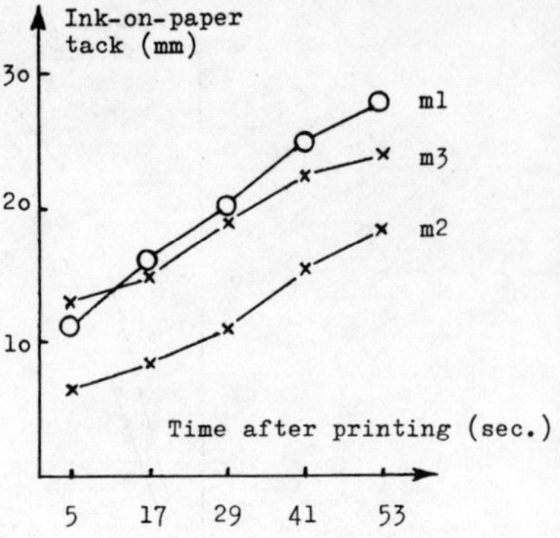

Fig. 5. Ink B/papers m1–m3.

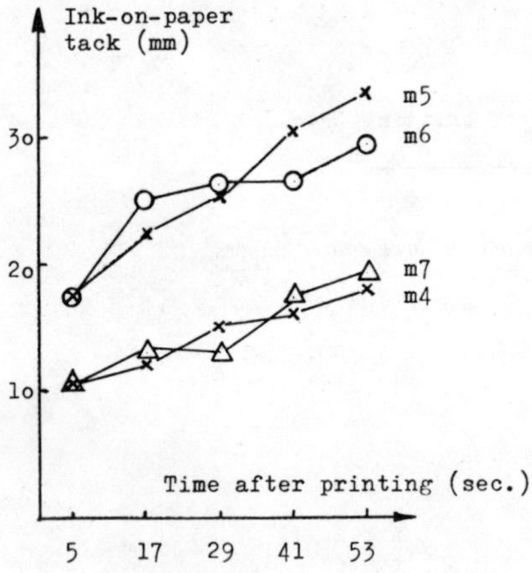

Fig. 6. Ink A/papers m4–m7.

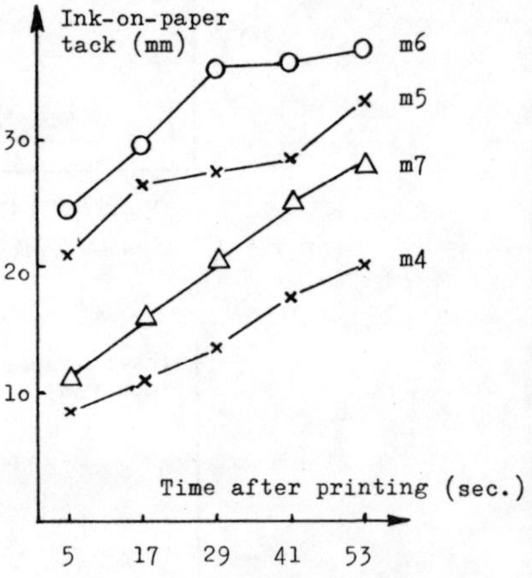

Fig. 7. Ink B/papers m4–m7

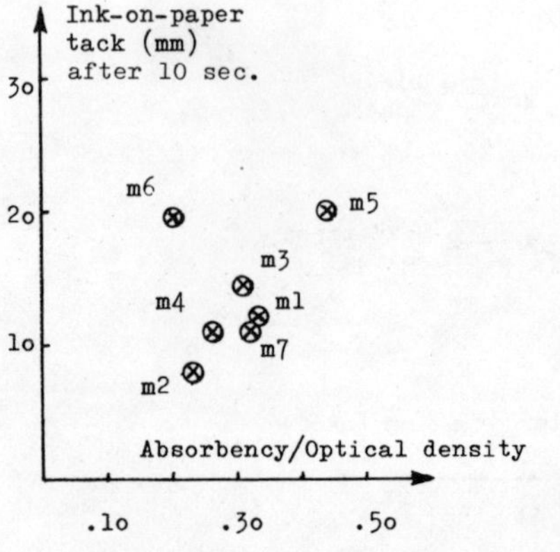

Fig. 8. Ink A on 7 papers m1–m7.

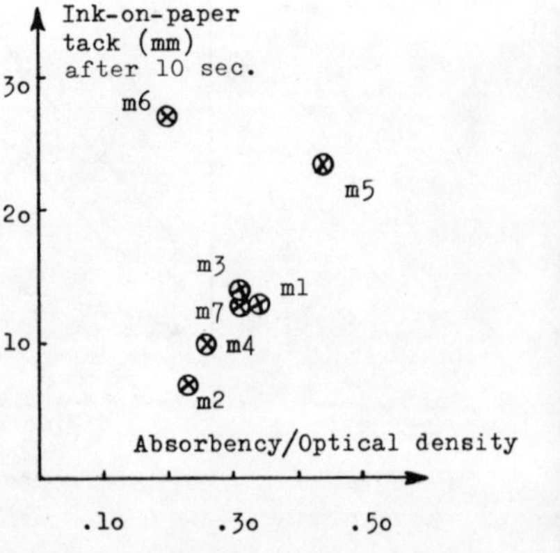

Fig. 9. Ink B on 7 papers m1–m7.

same group.

3.3 Comparison of eight different inks printed on a non-absorbent plastic surface

To estimate the limiting tack value at the time t = 0, a number of experiments were made on a non-absorbent surface, Polyart from Bakelite Xylonite Limited (England), on which the inks do not set after printing.

The eight inks are experimental inks but with a composition which is as close as possible to commercial inks in pigment and drier content. The contents of binder and mineral oil are varied in such a way that viscosity, shortness and setting are each kept at two levels.

Experimental values obtained for the eight inks are shown in Table 1.

TABLE 1 Properties of inks 1 - 8

Ink	Plastic viscosity (1)	Shortness (2)	Setting (3)	Tack (4)
1	136	2,400	Quick	13.3
2	77	1,170	Slow	9.8
3	84	890	Slow	12.2
4	141	1,060	Quick	16.8
5	150	1,600	Slow	17.0
6	77	1,540	Quick	10.7
7	74	880	Quick	12.2
8	152	960	Slow	17.6

(1) Measured with bar viscometer (Fink-o-Meter) at 25°C, plastic viscosity U (poise).
(2) Shortness according to Zettlemoyer, z (s^{-1}).
(3) Determined by set-off experiments on Prüfbau.
(4) Measured with electronic inkometer, Model 101, at 800 r.p.m. and 32°C.

Table 2 gives the values of the ink-on-paper tack in mm for all eight inks, 5, 17, 29, 41 and 53s after printing. The film thickness on the forme before printing corresponds to a rolled-up ink quantity of 0.6 cm^3 on an IGT inking unit. It is seen that there is no significant increase in the ink-on-paper tack after printing on a non-absorbent plastic surface, the small variation being attributed to experimental uncertainty. Table 2 shows the average value for each ink and its standard deviation σ Measurement of the ink-on-paper tack is subject to a standard deviation of $\pm$1.4 - 3.5 mm. Furthermore, the measured value is an expression of the tack of the printing ink before the absorption into the paper surface begins, and is discussed in Section 3.5.

An investigation of the relationship between the properties of inks 1 - 8 in Table 1 and the average values of the ink-on-non-absorbent surface tack in Table 2, gives the correlation coefficients shown in Table 3.

TABLE 2 Values of ink-on-paper tack measured on non-absorbent surface

Ink	5s	17s	29s	41s	53s	Average $\pm\sigma$
1	16.5	21.5	22.0	19.0	20.0	19.8 ± 2.2
2	14.0	14.0	12.0	15.5	15.5	14.2 ± 1.4
3	16.5	15.0	17.5	15.5	14.0	15.7 ± 1.4
4	24.5	23.0	24.5	26.0	26.0	24.8 ± 1.3
5	17.5	15.5	19.5	17.5	19.5	17.9 ± 1.7
6	18.5	15.0	15.5	17.5	15.0	16.3 ± 1.6
7	6.0	14.0	15.0	12.5	12.0	11.9 ± 3.5
8	23.0	24.5	28.5	26.0	26.5	25.7 ± 2.1

TABLE 3 Correlation coefficients - ink properties / ink-on-non-absorbent surface (mm)

	Ink-on-paper tack measured on non-absorbent surface
Plastic viscosity (poise)	0.835
Shortness, z (s^{-1})	0.071
Setting of inks	−0.022
Tack (inkometer)	0.793

3.4 <u>Comparison of eight different inks printed on two papers</u>

A series of measurements were made, with the same inks as used in Section 3.3, on two glossy blade-coated papers 1 and 2 having, respectively, absorbencies of 0.24 and 0.70. The absorbencies were determined according to B.S. 4570:1970, using a porometer ink from Lorilleux Trykfarver A/S. The results are expressed as optical densities as measured with a MacBeth densitometer.

The measurements of the ink-on-paper tack after printing of inks 1 - 8 on papers 1 and 2 are shown in Figures 10 - 13, for inks 1 and 2 with the values at zero time taken from Table 2. The film thickness of the ink-on-forme before printing is given, corresponding to the rolled-up ink quantity on the IGT inking unit, and the rolled-up ink quantity for each curve is indicated by $V = 0.6\ cm^3$; $V = 0.7\ cm^3$, etc.

The film thickness of the prints has, as was expected, a significant influence.

The variation of the ink-on-paper tack with time is supposed by many authors to follow a curve like Figure 14, e.g. Williams[4].

Curves I and II represent quick-set inks; curves III and IV slow-set inks. Within the two groups, I gives less risk of set-off than does II

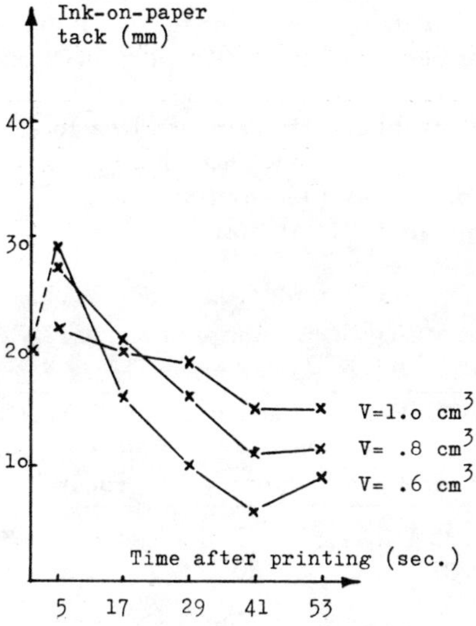

Fig. 10. Ink 1/paper 1.

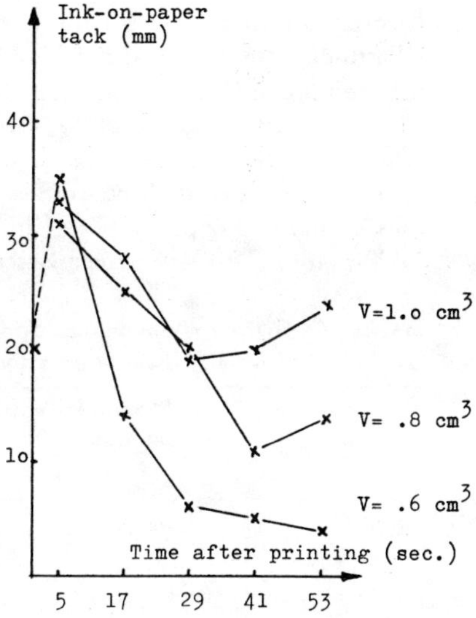

Fig. 11. Ink 1/paper 2.

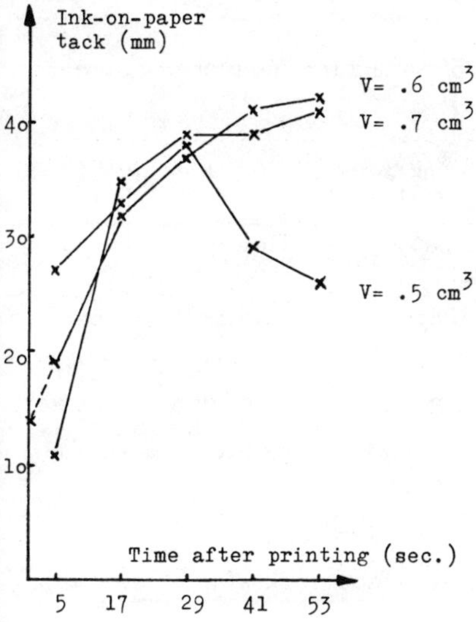

Fig. 12. Ink 2/paper 1.

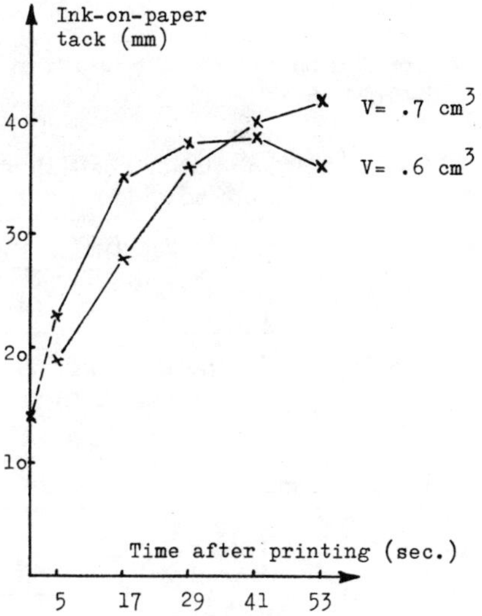

Ink 13. Ink 2/paper 2.

and correspondingly III less than does IV.

From Figures 10 and 11, ink 1 has its maximum tack value within 5 s after printing, whereas ink 2 reaches its maximum, for certain of the film thicknesses, after the device has been stopped, thus setting limits to the measurements.

It has not, therefore, been possible to determine the time for maximum tack of the slow-set inks. However, an estimate has been made of maximum tack, and the corresponding times after printing, for all eight inks on the two papers. The film thickness is in all cases 0.6 cm^3 on the inking unit. Table 4 summarises the results.

TABLE 4 Tack maximum and corresponding times after printing of inks 1 - 8 on papers 1 and 2

	Maximum value of ink-on-paper tack (mm)		Time after printing for maximum value (s)	
	Paper 1	Paper 2	Paper 1	Paper 2
Ink 1 (quick-set)	29	35	5	5
2 (slow-set)	42	39	80	35
3 (slow-set)	30	55	80	100
4 (quick-set)	35	44	80	50
5 (slow-set	32	47	100	60
6 (quick-set)	30	36	17	17
7 (quick-set)	34	29	17	17
8 (slow-set)	32	50	29	100

Correlation between the results in Table 4 and the ink properties are given in Table 5.

TABLE 5 Correlation coefficients - ink properties / tack maximum and corresponding time after printing

	Maximum value of ink-on-paper tack (mm)		Time after printing for maximum value (s)	
	Paper 1	Paper 2	Paper 1	Paper 2
Plastic viscosity (poise)	- 0.27	0.39	0.15	0.30
Shortness, z (s^{-1})	- 0.39	- 0.33	- 0.33	- 0.75
Setting	- 0.26	- 0.73	- 0.61	- 0.75
Tack (inkometer)	- 0.24	0.47	0.23	0.48

3.5 Comparison of values for the ink-on-paper tack with corresponding inkometer values

Figures 10 - 13 show how the tack value of an ink increases after impression, reaches its maximum and decreases again as a function of the time after printing.

As the tack value of a printing ink is normally given in inkometer units (or units from corresponding tackmeters), the values given in Figures 10 - 13 were converted into inkometer units. The values given in Table 2 for the ink-on-paper tack measured on a non-absorbent plastic surface were compared with the values given in Table 1 in inkometer units for the same inks, giving the results shown in Table 6.

TABLE 6 Conversion of ink-on-paper tack into inkometer values

	Ink-on-paper tack measured on non-absorbent surface (mm)	Tack measured on inkometer	1 mm ink-on-paper tack corresponds to the following inkometer units
Ink 1	19.8	13.3	0.67
2	14.2	9.8	0.69
3	15.7	12.2	0.78
4	24.8	16.8	0.68
5	17.9	17.0	0.95
6	16.3	10.7	0.66
7	11.9	12.2	1.03
8	25.7	17.6	0.68

There is a reasonably good correlation between the two values of tack measurement. Figures 15 and 16 correspond to Figures 10 - 13 with inkometer values instead of ink-on-paper tack values for inks 1 and 2 on papers 1 and 2 (film thickness 0.6 cm^3).

It is seen that before printing, ink 2 has a tack corresponding to about 10 inkometer units, and after printing on paper 2 this increases to a maximum of 26 inkometer units at about 30 s, after which the value decreases.

4 INVESTIGATION OF SET-OFF FROM PRINTED CAST-COATED PAPERS

In certain cases, printing on two-sided cast-coated paper can give problems, if attention is not drawn to the connection between printing ink, absorbency of the material in question, pile height, printing speed and the use of spray powder.

As part of an investigation an experimental printing was carried out to study, in practice, the connection between the above parameters. Six different printing inks of different makes were used, I - VI, specially recommended by the manufacturers for printing on cast-coated paper. Two qualities of cast-coated paper were used, one having an absorbency of 0.32 units (paper 3), the other 0.52 (paper 4).

The experiment was performed by printing 500 sheets with each ink/paper combination, without using spray powder, and with a total of 500 sheets in each pile. (Printing press Heidelberg KORD, forme made

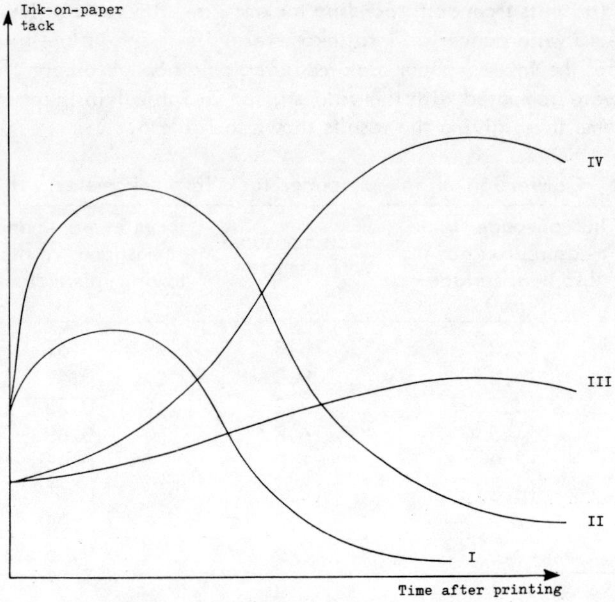

Fig. 14. Tack course fan ink film after printing.

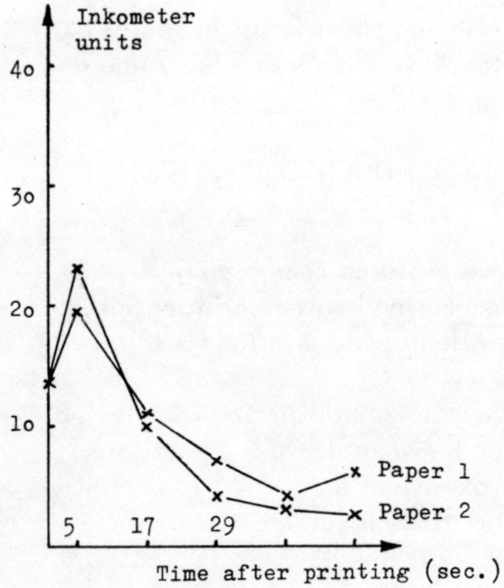

Fig. 15. Tack measurement of ink 1.

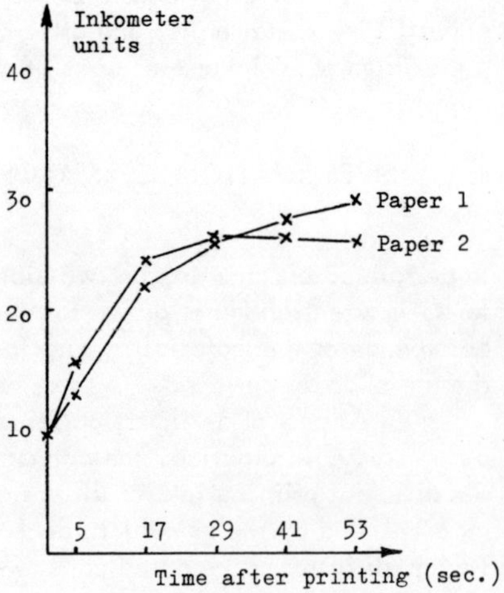

Fig. 16. Tack measurement of ink 2.

from IGT proof film, size A3, basis weight 250 g/m^2, printing speed 3,500 sheets per hour). The result, i.e. an estimate of the set-off tendency for each ink/paper combination, was judged by going through the impression in each case, deciding whether there was set-off or not, and if so, when it has started. The results are shown in Table 7.

TABLE 7 Visual estimate of set-off in pile after printing inks I - VI on papers 3 and 4

	Paper 3 (low absorbency)	Paper 4 (high absorbency)
Ink I	Set-off after 400 sheets	Set-off after 450 sheets
II	Slight set-off through whole impression	No set-off
III	Set-off after 400 sheets	No set-off
IV	Set-off after 400 sheets	No set-off
V	Set-off after 50 sheets	No set-off
VI	Set-off from start	No set-off

It can be seen that there are varying degrees of set-off tendency in the pile among the various ink/paper combinations. To see whether there is a connection between the set-off so demonstrated in practice and the use of the ink-on-paper tack tester, a number of tack measurements were made with inks I - VI on papers 3 and 4, and the values are illustrated in Figures 17 - 22.

A small value of the ink-on-paper tack is essential to avoid set-off in the pile after printing, but no specific values can be given for the ink/paper combinations investigated. However, two conditions require comment:

(1) A maximum value reached in 53 s after printing is probably too short for present purposes, and an extension of the curves might look like those in Figure 23. The maximum value of the ink-on-paper tack and the corresponding time after printing are important in deciding whether the ink/paper combination in question will cause set-off in practice.

(2) If a comparison is made between Figures 8 and 9, where increasing tack values are found with increasing absorbency of the paper surface of dull-coated qualities, and Figures 17 - 22, it is seen that for the two cast-coated qualities the situation is reversed.

The cause of these differences is to be sought in the fact that a proportionally thicker part of the paper surface is used for measuring absorbency on paper with porometry ink than the one used for printing and measuring the ink-on-paper tack, and that the absorbency of the coat film is not uniform throughout the film thickness.

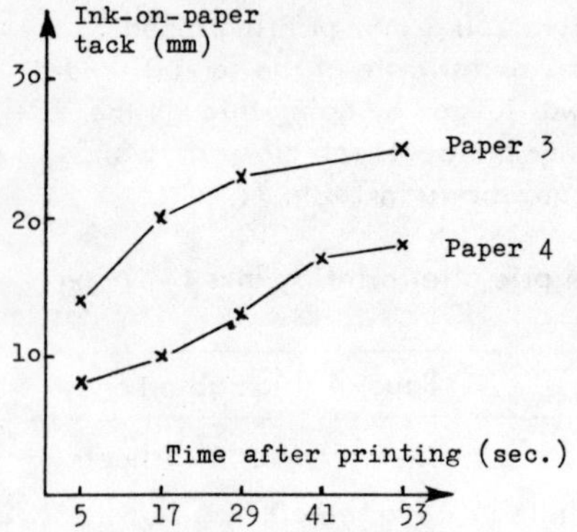

Fig. 17. Ink I.

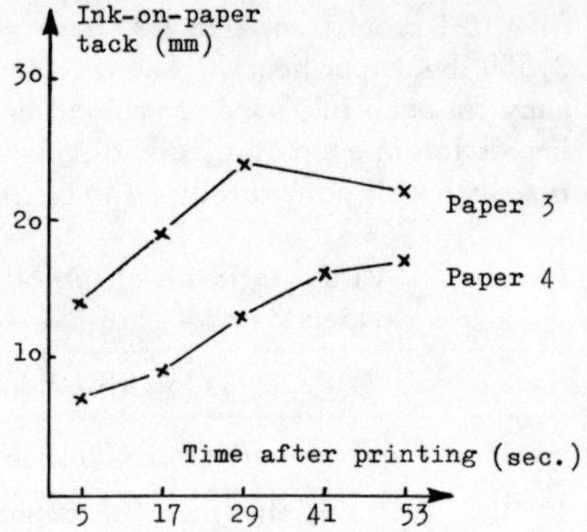

Fig. 18. Ink II.

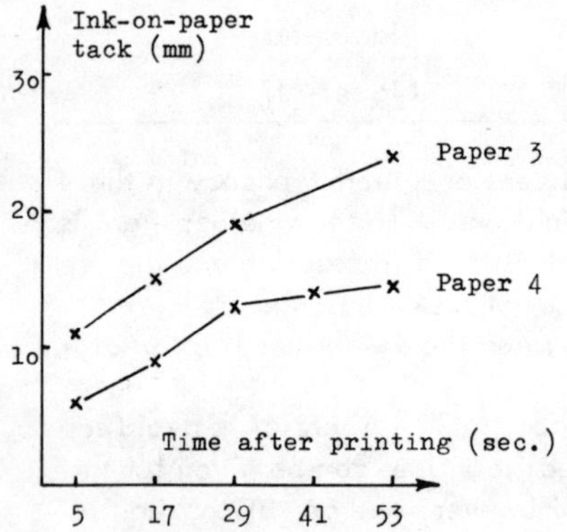

Fig. 19. Ink III.

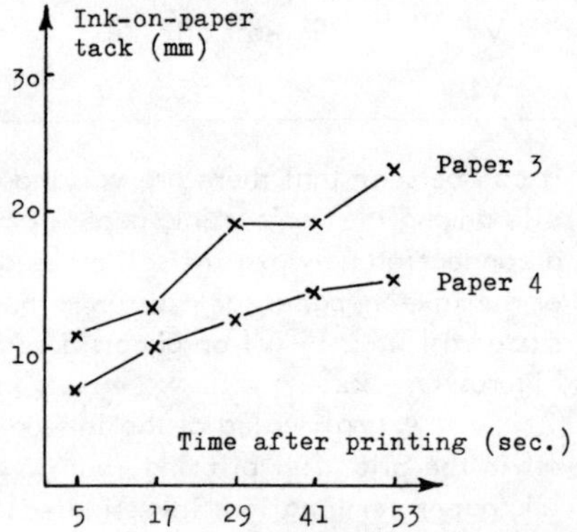

Fig. 20. Ink IV.

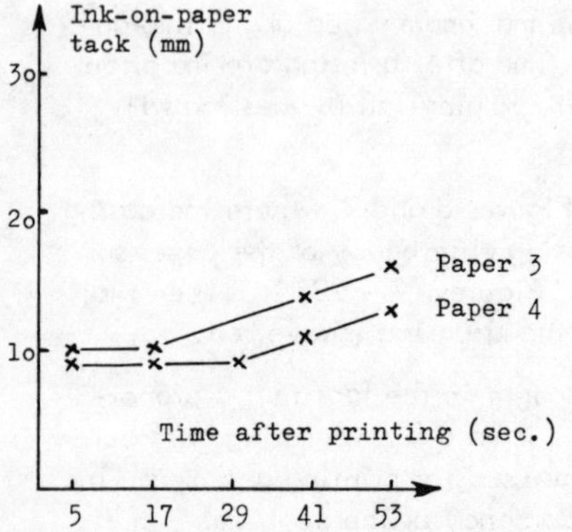

Fig. 21. Ink V.

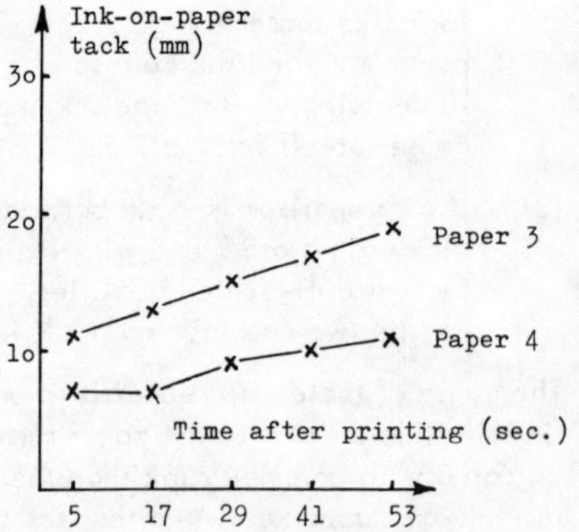

Fig. 22. Ink VI.

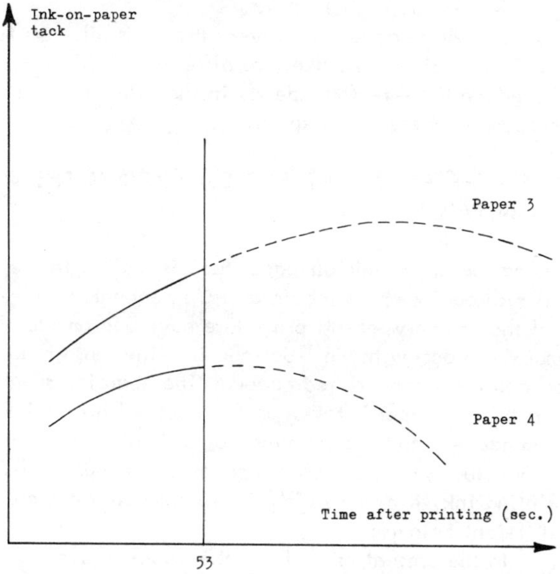

Fig. 23. Tack course extended beyond the working interval of the ink-on-paper tack tester.

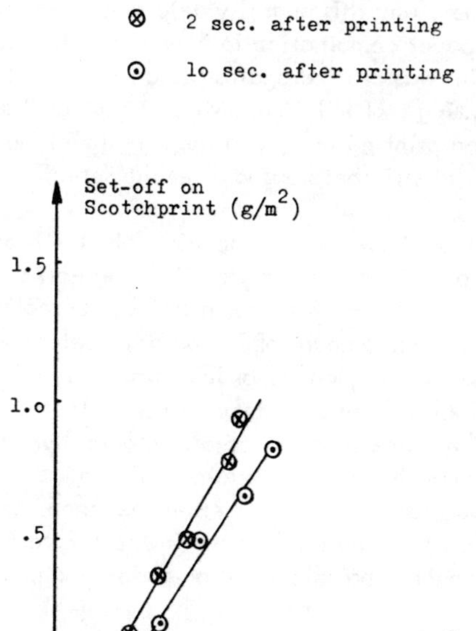

Fig. 24. Set-off of ink 3 on paper 1 (low absorbency).

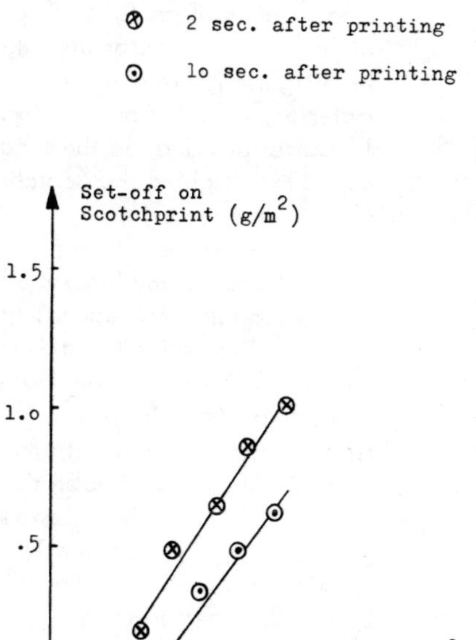

Fig. 25. Set-off of ink 3 on paper 2 (high absorbancy).

It must be added, however, that it is difficult to compare commercial inks of different makes, as different additives may be used in the inks to reduce the set-off tendency in the pile without changing the setting properties of the ink in question.

NEW SET-OFF PROCEDURE FOR ESTIMATING THE SETTING OF INKS ON PAPER

Determination of the ink-on-paper tack is subject to the inconvenience that special equipment is required. Consequently a number of modifications of the ordinary set-off procedure have been made, involving making a number of proofs with well-defined ink films on a proof press, and setting off against a non-printed paper. The quantity of ink transferred to the non-printed paper is taken as a measure of how quickly the ink in question has set, and is determined by a densitometer or a spectrophotometer (or visually). As such a procedure depends on the pigmentation of the printing ink, it may be difficult to compare different inks, especially with different colours.

In the present procedure, the determination of transferred ink quantity in set-off is made by a gravimetric measurement, which is time-consuming but also gives exact results.

An IGT tester, Model AC 2, with two printing units, was used. In the first printing unit a print is made by transferring a predetermined amount of ink from forme to paper, and this quantity determined like a normal ink transfer for the ink/paper combination in question. In this investigation, however, the print was pressed against a non-absorbent material, i.e. Scotchprint (printing foil 121 from 3M) at a time of 2 and 10 s after printing, in the second printing unit, and the set-off ink determined by weighing the Scotchprint strip before and after the set-off process against the print.

A number of investigations have been made with inks 1 - 8 on papers 1 and 2, and Figures 24 and 25 show examples of the set-off conditions for ink 3 on papers 1 and 2, 2 and 10 s after printing, respectively.

The horizontal axis shows the amount of ink on print before set-off (in g/m^2) and the vertical axis, the quantity of ink transferred to Scotchprint (also in g/m^2). Figures 24 and 25 demonstrate a linear relation between ink on print and the corresponding set-off on Scotchprint when the amount of ink on the print is above a certain minimum.

An F-test has been made for estimating whether the linear regression is significant (see Table 8) for the entire experimental material for all the set-off experiments concerning inks 1 - 8 on papers 1 and 2, 2 and 10 s after printing.

It may be assumed that there is a linear connection between the ink quantity on the print and the corresponding set-off on Scotchprint.

In the further analysis of the set-off experiments a number of parameters were chosen for describing the ink/paper combinations investigated. The parameters are marked a – h, cf. Figure 26:

Parameter a: The slope of the straight line at a set-off 2 s after printing ($\Delta t = 2$ s).

Parameter b: Minimum ink quantity on paper (g/m^2), before set-off occurs on Scotchprint ($\Delta t = 2$ s). Results from intersection of the straight line with horizontal axis.

Parameter c: Calculated quantity of set-off on Scotchprint (in g/m^2) at an ink quantity on paper of 1.5 g/m^2 ($\Delta t = 2$ s).

Parameter d: The slope of the straight line at set-off 10 s after printing ($\Delta t = 10$ s).

Parameter e: Minimum ink quantity on paper (g/m^2), before set-off occurs on Scotchprint ($\Delta t = 10$ s).

Parameter f: Calculated quantity of set-off on Scotchprint (in g/m^2) at an ink quantity on paper of 1.5 g/m^2 ($\Delta t = 10$ s).

Parameter g: Difference between the parameters e and b, graphically illustrated as the distance between the intersection of the straight lines with the horizontal axis.

Parameter h: Difference between the parameters c and f, graphically illustrated as the vertical distance between the two straight lines for an ink quantity on paper of 1.5 g/m^2.

TABLE 8 F values for estimating whether the linear regression is significant – set-off on Scotchprint (g/m^2) as function of ink quantity on paper (g/m^2)

	Set-off 2 s after printing		Set-off 10 s after printing	
	Print on paper 1	Print on paper 2	Print on paper 1	Print on paper 2
Ink 1	-	5.4	4,319 ^{xx}	21.7^{x}
2	57.5^{xx}	186.4^{xxx}	56.2^{xx}	34.3^{xx}
3	152.1^{xx}	142.8^{xx}	194.1^{xxx}	17.8
4	36.9^{x}	-	7.2	-
5	14.7	93.6^{xx}	1,410 ^{xxx}	55,689 ^{xx}
6	731.6^{xxx}	431.5^{xxx}	91.1^{xxx}	19.8^{x}
7	188.4^{xx}	20.2^{x}	17.2	7.0
8	165.3^{xx}	72.6^{x}	139.7^{xx}	88.3^{x}

x Value greater than F_{95}, i.e. significance at level 95 %
xx Value greater than F_{99}, i.e. significance at level 99 %
xxx Value greater than F_{999}, i.e. significance at level 99.9%

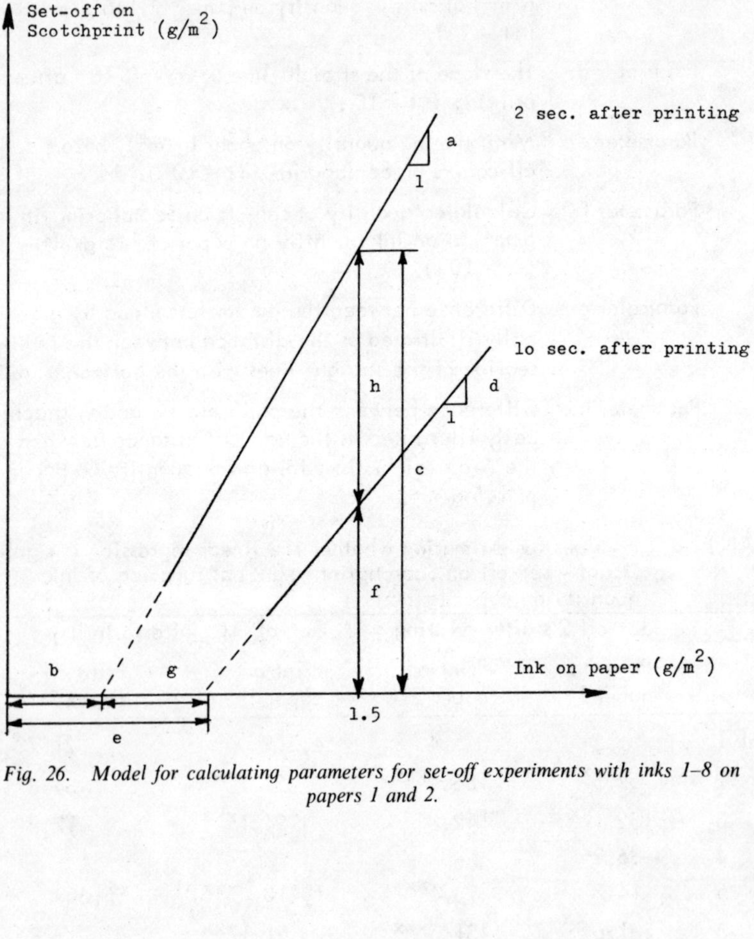

Fig. 26. Model for calculating parameters for set-off experiments with inks 1–8 on papers 1 and 2.

To see which of the above parameters can be used for describing the investigated ink/paper combinations, a correlation analysis has been made of the properties of the printing inks 1 - 8 (see Table 1) and the values of the parameters a - h (see Table 9).

TABLE 9 Correlation coefficients: properties of inks 1 - 8 and parameters a - h, cf. Figure 26

		Set-off 2 s after printing			Set-off 10 s after printing				
		a	b	c	d	e	f	g	h
Printing on paper 1	Plastic viscosity	0.59	0.68	-0.19	-0.17	-0.23	0.21	-0.49	-0.37
	Shortness	-0.27	-0.05	-0.25	-0.52	-0.05	-0.42	-0.02	0.37
	Setting	-0.58	0.05	-0.68	-0.20	0.04	-0.98	0.58	0.82
	Tack	0.72	0.78	-0.14	0.12	-0.10	0.26	-0.40	-0.41
Printing on paper 2	Plastic viscosity	0.23	0.65	-0.35	0.20	0.10	0.02	-0.38	-0.39
	Shortness	0.04	0.36	-0.40	-0.83	-0.32	-0.50	-0.42	-0.08
	Setting	-0.41	0.22	-0.62	-0.57	0.16	-0.81	-0.58	-0.11
	Tack	0.24	0.47	-0.22	0.47	0.38	0.07	-0.11	-0.29

From Table 9 it can be seen that parameter f, i.e. set-off on Scotchprint in g/m^2 for an ink quantity on paper of 1.5 g/m^2 (10 s after printing) is useful for describing the setting speed of the inks 1 - 8 on papers 1 and 2.

To investigate further the connection between the setting speed of the inks, the absorbencies of the paper surfaces and the above set-off on Scotchprint, a two-sided variance analysis has been made. In Table 10 the set-off on Scotchprint after 10 s is shown in g/m^2, corresponding to parameter f in Figure 26. Factor A is the absorbency of the paper (see Section 4) and factor B the setting speed of the printing ink (see Table 1). Further, the confidence intervals of the individual set-off values are given from a t-test of the figure material.

TABLE 10 Set -off on Scotchprint in g/m^2 for an ink quantity on paper of 1.5 g/m^2, 10 s after printing, inks 1 - 8 on papers 1 and 2

		Factor B/ink Slow-set		Quick-set	
Factor A/	Low absorbency	0.52 ±0.13	0.50 ±0.04	0.29 ±0.04	0.28 ±0.08
		0.51 ±0.06	0.56 ±0.06	0.34 ±0.22	0.30 ±0.34
	High absorbency	0.30 ±0.13	0.33 ±0.16	0.17 ±0.08	0.18
		0.18 ±0.02	0.33 ±0.21	0.12 ±0.16	0.16 ±1.41

It can be seen from the table that there is some variation in the size of the confidence intervals, but there is a distinct dependence between the size of the set-off, the absorbency of the paper and the setting speed of the ink. Further, there is an interaction between the absorbency of the paper and the setting speed of the ink, as is also shown by the analysis of variance in Table 11.

TABLE 11 Variance analysis of the values given in Table 10

Source of variation	Sum of squares	Degrees of freedom	Mean square	F
Paper absorbency/A	0.147	1	0.147	
Ink setting/B	0.122	1	0.122	
Interaction/AB	0.030	1	0.030	18.75^xxx
Residual	0.020	12	0.0016	
Total	0.299	15		

xxx Value greater than F_{999}, i.e. significance at level 99.9%.

There is a distinct interaction between the two factors A and B, and for a further analysis of the relation between the setting speed of the inks and the absorbency of the paper other methods must be used.

DISCUSSION AND CONCLUSIONS

On a specially constructed piece of equipment, essentially based on elements from the IGT tester, a number of investigations have been made of the ink-on-paper tack the first minute after printing. Investigations of slow-set and quick-set inks on a number of coated papers of different qualities gave characteristic differences in the change in tack.

An analysis of the tack change for a non-absorbent material formed the basis of an investigation of the uncertainty to which these results are subject. Further, it was possible to make a conversion from ink-on-paper tack into inkometer values for the printing ink, so that the tack increase in the ink film on the paper after printing could be compared with the original value before printing.

An experiment with eight special inks on two papers showed that it was possible to quantify the connection between the setting speed of the ink when printed on different papers and the duration from printing until the ink film reached its maximum value for the ink-on-paper tack. However, some uncertainty is attached to these results, as the device was not specifically designed for such investigations.

An investigation of set-off when printing on cast-coated paper gave information about the connection between the settings of different inks and the absorbencies of the paper surfaces.

Parallel with the investigations of the ink-on-paper tack tester was work with a set-off procedure on a gravimetric basis, which gave exact results for the set-off of a number of ink/paper combinations. As the ink-on-paper tack measurements for the same ink/paper combinations are subject to a rather great uncertainty, it has not been possible to make any actual comparisons between the two methods.

ACKNOWLEDGMENTS

The author thanks the following for help and support during the work: Mr. S. F. Loveday and Mr. A. E. Kingsland, C. Townsend Hook & Company Limited; Mr. A. T. Franklin, Reed Paper & Board (UK) Limited; Mr. E. Stenfeldt-Madsen, Sadolins Trykfarver A/S.

REFERENCES

1. Franklin, A. E., 'Paper/ink/press relationships', Professional Printer, 24, 2: 2 - 5 (1980).

2. Swan, A., 'Carry-over ink piling on litho presses', Printing Equipment and Materials, 8: 32 - 33 (1973).

3. Terry, M. P., 'Brief note on principle of operation and construction of the ink-on-paper tack tester', Internal paper for Reed Paper & Board (UK) Limited (1974).

4. Williams, C., 'The chemistry of quick setting inks', Lithoprinter, 9: 49 - 51 (1970).

GRANULOMETRIC ANALYSIS OF RESIDUAL INK IN DE-INKED PULP

G. Calabro and D. Pugliesi, Centro di Sperimentazione Grafica,
Siva S.p.A., Rome

ABSTRACT

Because of the low contrast it is not possible to measure directly the granulometric distribution of residual ink particles on a sample of de-inked pulp by means of a video-analyzer. A method has been worked to increase the contrast of the sample by means of an intermediate photographic step. The paper describes conditions required to obtain reproducible measurements that agree with visual countings. The method can also be applied to the analysis of prints and, if necessary, to sharpen the contrast.

METHOD

In the development of a process for de-inking and related de-inking agents it should be possible to make a granulometric estimation of residual ink particles in paper samples obtained from a sheet former through the use of de-inked pulp. This is required both for the study of the type of action of the de-inking agents, and for the choice of steps to be applied in the elimination of the particles. Clearly the most effective of these steps would be flotation or washing, according to the size range of the particles.

Counting the particles visually by means of a microscope may be rather tiring, and therefore, an attempt has been made to develop an appropriate and reliable technique capable of providing a sufficiently detailed classification. We have, therefore, attempted to utilize an image analyzer, the Zeiss 'Microvideomat' (Figure 1). This apparatus comprises an optical system which produces the image of the object being examined and, depending upon the magnification desired, can be a photographic objective lens or a microscope; a black and white TV camera which projects the image produced onto a monitor. An electronic device scans the image and differentiating the areas of diverse optical density on the basis of a level of discrimination determined by the technician, allows for an adequate counting of the particles, an estimation of their diameter, area, etc.

Because this type of discrimination is based on a difference in the optical density between the various areas of the image, it is necessary to measure areas having densities included in a given range or, at least, a density level above or below a determined limit, and having a density level different from that of all the other areas of the image.

This generally implies the need to obtain hard contrast images with clearly marked outlines. Furthermore, since there will inevitably be a wide gradient of density at the border between two areas of different density, the technician may be able to alter the apparent size of the two areas by varying the density of discrimination within the gradient. To a certain extent, this may alter the actual size of the areas and, so as to guarantee greater reliability of the calculations, it is necessary to use hard contrast images, particularly keeping in mind that the original contrast of the object being examined will inevitably be attenuated by light diffusion from the optical system and by the inherent characteristics of the televising system.

With specific regard to granulometric calculation of the ink particles, it must be possible to measure the particles present throughout the thickness of the sample under study. The sample is therefore made transparent through the use of a liquid having the appropriate refractive index. However, the sample will inevitably become greyish, the ink contrast will no longer be as sharp, and shadows not pertinent to ink particles will appear on the inside of the sample. In these circumstances, the difficulties linked with the use of the 'Microvideomat' lead to two distinct phenomena:

(1) The density gradient inevitably present at the margin of the ink particles on the image hinders an exact measurement of the area, since it is not possible to distinguish clearly the actual outline of the particles when varying the density discrimination level. The apparent size may vary gradually within a wide range, from the complete inclusion to the complete exclusion of the particle itself. This uncertainty with regard to the discrimination level makes for a rather vague area measurement. Also, the measurement may vary within a relatively wide range in repeat observations or between observers.

(2) The uncertainty in the discrimination levels in (1) above entails a further inconvenience, i.e. when varying the level itself, non-ink particles and therefore areas of discrimination with lower density levels appear or disappear. These zones cause uncertainty both in the evaluation of the total area and in the calculation of the number of particles involved. This uncertainty is not easily resolved in the rather imprecise image obtainable on the screen. This is due to the fact that the image on the screen shows areas which may either consist of tiny ink particles having a density gradient which

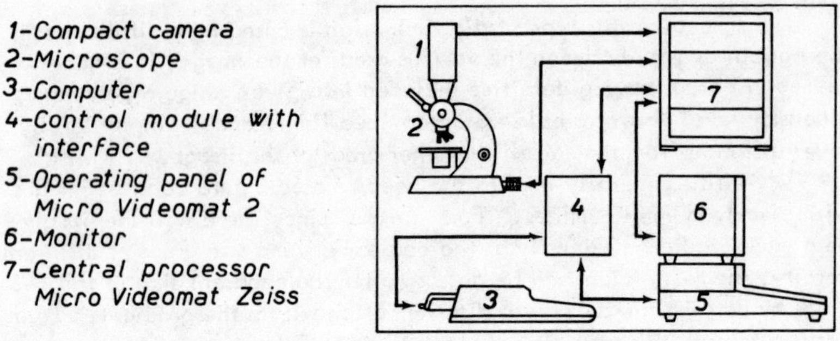

Fig. 1. Block pattern of Microvideomat-Zeiss

(1) Compact camera
(2) Microscope
(3) Computer
(4) Control module with interface
(5) Operating panel
(6)Monitor
(7) Central processor

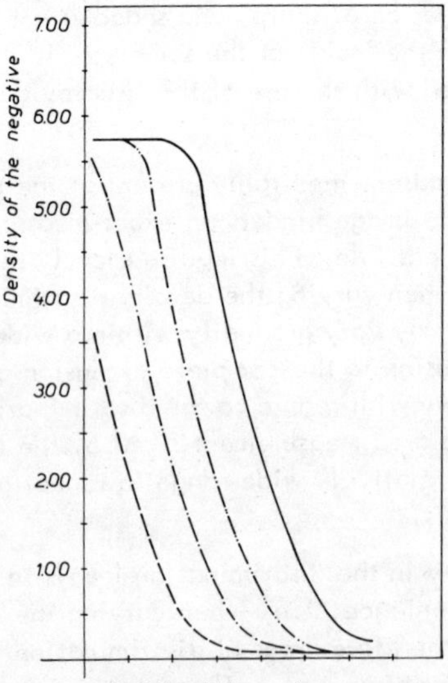

Fig. 2. Series of sensitometric curves obtained with different exposures for the Agfaortho 25 negative film.

reaches the central area of the image, or in denser parts of the shadows accidentally caused by the intertwining of fibers within the sheet.

Due to these conditions of uncertainty in calculating both the number and the areas of the particles, a precise analysis is not possible. This situation stems from selected observation conditions since, as mentioned earlier, the sample is examined in transmitted light and immersed in a liquid with a refractive index near to that of the fibers. This situation cannot be modified since it is indispensible to examine a transparent sample so as to count all particles present throughout the thickness of the sheet. It is therefore impossible to sharpen the contrast of the image, even having decreased the thickness of the sample to the extent necessary for a simultaneous focus on the total thickness. Also, a further reduction of the thickness is not possible since the layer of fibers would no longer retain the ink particles.

Thus, since the conditions of direct observation cannot be altered, it was thought opportune to overcome the difficulty by producing a photograph with the sharpest contrast and use it for the calculations. Although this calls for a more complex method, in that an additional stage would be introduced, it would eliminate the shadows below a certain density level and give a clearer margin of the areas actually reproduced, once the correct exposure and the appropriate contrast factors of the photographic material had been determined. Of course, for reproducibility, a standardization of operating conditions is necessary.

By exposing hard contrast negative material to a step wedge in a sensitometer it is possible to trace curves such as those shown in Figure 2, where for each of the exposures the density of the wedge is compared with the density of the respective steps obtained on the negative. If the conditions of treatment remain constant the shape of the curves are the same, the different positions being due to the steps in the original wedge.

By plotting equal density levels of the negative versus the logarithm of the exposures, it is possible to obtain a single characteristic curve of the material used. Curves representing the different photographic material being tested are shown in the diagrams in Figure 3. Of course, the logarithmic scale of the exposures is arbitrary, since it involves uniquely the exposure times and the densities of the step wedge independently of the illumination given off by the wedge itself which is kept constant throughout the series of tests.

It was considered that the most appropriate photographic material was the N. 1 film (orthochromatic 15° DIN film used for the photographing of documents) which has the highest contrast factor and is capable of reaching a very high level of density. In view of the fact that it is a material used for the reproduction of documents, it was assumed that both its resolving power and its sharpness were quite high.

This material was used in the form of sheet film for the photomicrograph of the transparent sample. An objective with a relatively low

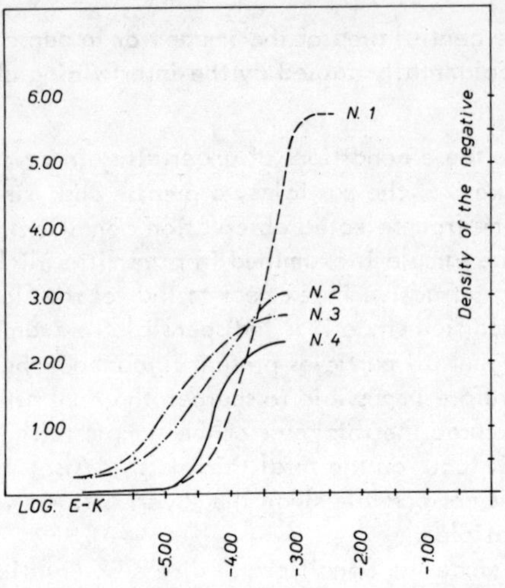

Fig. 3. Comparison of characteristic curves of the negative materials tested

(1) Agfaortho 25 (25 ASA)
(2) Kodak Panatomic-X (32 ASA)
(3) Ilford Pan-F (50 ASA)
(4) Kodak microfilm 5786 (20–50 ASA)

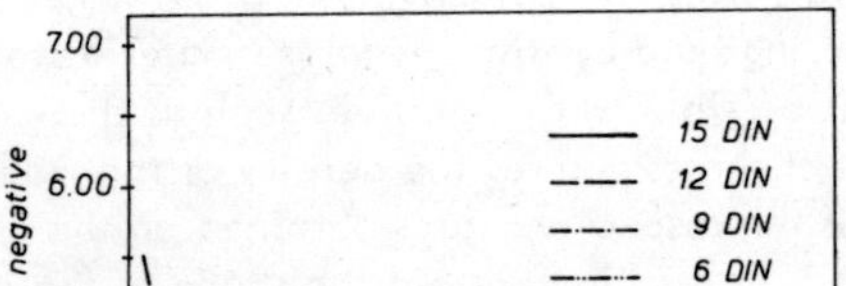

Fig. 4. Series of sensitometric curves obtained through the microscope by means of test objects and with different exposures (Agfaortho 25).

magnifying power was employed so that the whole thickness of the sheet was in focus. The photograph was taken on a 9x12 cm format sheet so that a contact print could be made, thus avoiding the attenuation of sharpness and contrast resulting from the use of an enlarger. The 9x12 format cannot be tested through a single scanning by means of the 'Microvideomat' apparatus with the optical system used. It can, however, be done by making use of an automatic scanning stage which exposes adjacent areas to the lens in succession.

With the same arrangement of the microscope, a series of sensitometric test objects, each of which had four different known density levels, were photographed. Considering the fact that the microscope's optical system inevitably attenuates the contrast of the image, the presence of a set of test objects at a known density level would allow for a quantitative evaluation of this attenuation.

The conditions of treatment of the sensitive material were controlled; however, the various exposures were modified within a wide range by setting the exposure-meter of the microscope's photographing system on all the possible DIN degrees within a wide space around those established by the manufacturer, since the different exposure times could not be set directly. The sensitometric photographs have made it possible to trace a series of curves conceptually analogous to those shown in Figure 2. Each of these curves represents the exposure obtained by setting the microscope's exposure-meter at different ratings (Figure 4). Here too, it is possible to trace a characteristic curve by replacing the original density levels with the logarithms of the exposures. However, since the exposure-meter was an automatic instrument, and the exact exposures were not known, their ratios were taken into account by assuming the particular exposure relative to the exposure-meter set at 15° DIN as equal to one, and by considering the other exposures as inversely proportional to the sensitivity used. This has made it possible to trace the diagram in Figure 5, where the different curves overlap. Naturally, the scale of the exposure logarithms is once again arbitrary and does not correspond to the one shown in Figure 3, due to the different way in which the exposure is evaluated.

In any case, scale differences, which depend on an arbitrary constant, may lead to a different localization of the curves, but not to a difference in their slope.

The difference in slope obtained experimentally should be considered therefore as an effect resulting from the attenuated contrast of the image of the test object due to the above-mentioned phenomena of light diffusion. Precisely, the ratio between the slopes of the rectilinear sections of the two curves of Figures 5 and 3 (N.1) will give us the attenuation factor of the apparent densities seen through the microscope. Actually, once the density of the sensitometric test object has been modified according to this ratio it is possible to retrace the curve shown in Figure 3 (N.1). In Figure 6, the Figure 5 curve is compared with the curve traced using the attenuated density levels.

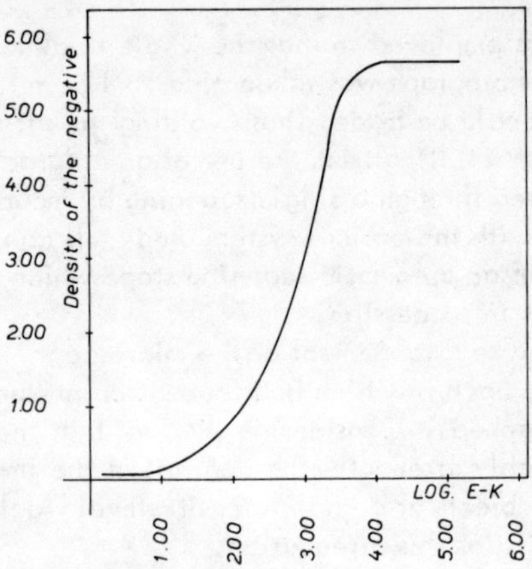

Fig. 5. Characteristic curve derived from the curves of Fig. 4.

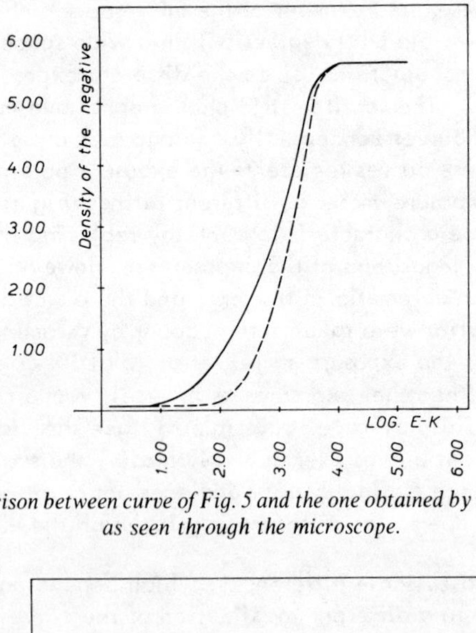

Fig. 6. Comparison between curve of Fig. 5 and the one obtained by using density values as seen through the microscope.

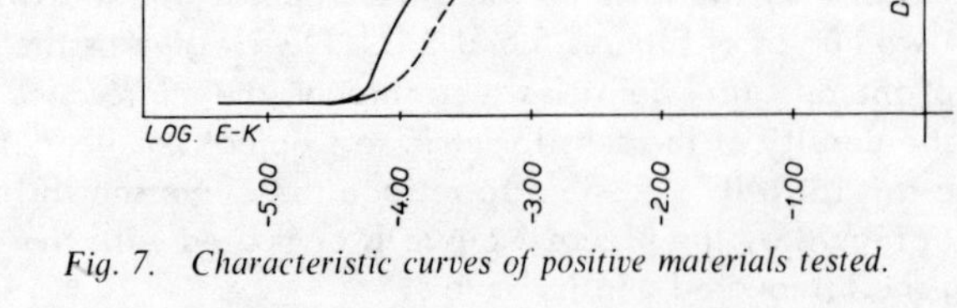

Fig. 7. Characteristic curves of positive materials tested.

Through a visual analysis of the negative images photographed, we immediately rejected exposures which had an insufficiently high maximum density level or vice versa, i.e. those in which the smallest particles had been completely erased by light diffusion due to excessive exposure. Although it would be possible to analyze the negatives directly through the analyzer, it was preferable to make positive prints of the images themselves. This is due both to the fact that an automatic scanning stage is best suited for the images lit by reflection, particularly when dealing with relatively large formats, and that although a positive print is yet another step in the process, it has certain advantages: it increases markedly the contrast factor of the photographing system and makes the process more flexible because it can offset a lower exposure in the negative with a greater one in the positive, or vice versa.

Two different papers whose characteristic curves are shown in Figure 7 were tested through the sensitometer. The one represented by curve 1 was selected. This paper showed a higher contrast factor and good sharpness features, since it is a type of material used for documents.

Figure 8 shows the curves relative to the reproduced densities as against the original densities, with regard to the material involved, for a series of exposures.

Having thus chosen the positive material to employ, a negative image of the field concerned was selected, which was obtained through an intermediate exposure with respect to the range of useful exposures visually determined beforehand. This negative was printed with a series of exposures and the prints produced were then analyzed through the Microvideomat. The visually most satisfying level of discrimination was selected; then the results were compared to the calculations made visually through the microscope on the paper sample. Thus, it was possible to determine a particular exposure which enabled us to obtain satisfactory values. The overall image is shown in Figure 9, while Figure 10 shows a soft contrast photograph of the sample as seen through a microscope.

It is then possible to trace a curve representing the entire photographic process, from the curves of the negative stage (see Figure 4) to those of the positive stage (Figure 8).

With reference to the diagrams in Figure 11, going from each density level in the original to a curve of the negative, and then passing to the appropriate curve of the positive, experimentally selected, it is possible to determine the density levels of the positive corresponding to each of the originals. Thus, the curve of the reproduction, as shown in quadrant IV (Figure 11), can be traced. Obviously, it is possible to trace back to the curve of the reproduction by using a different curve of the negative, provided that the appropriate positive curve is used.

Thus, a series of conditions based on the appropriate coupling of negative and positive exposures can be determined. These conditions are equivalent, provided that the curve zones used are those nearer to the approximately rectilinear section. In a test made to obtain repro-

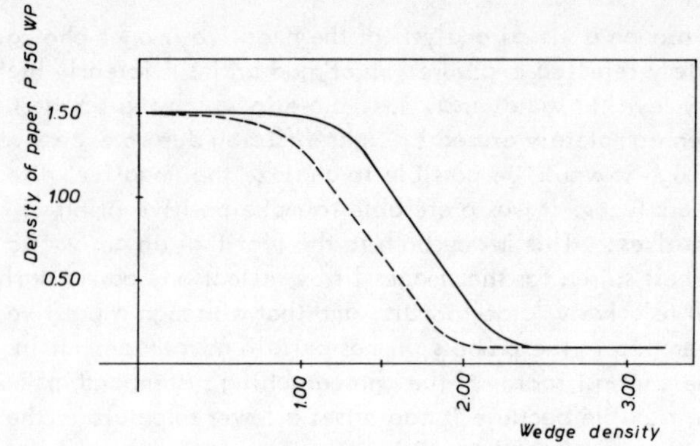

Fig. 8. Series of sensitometric curves obtained with different exposure for the paper Agfa-Gevaert P 150 WP.

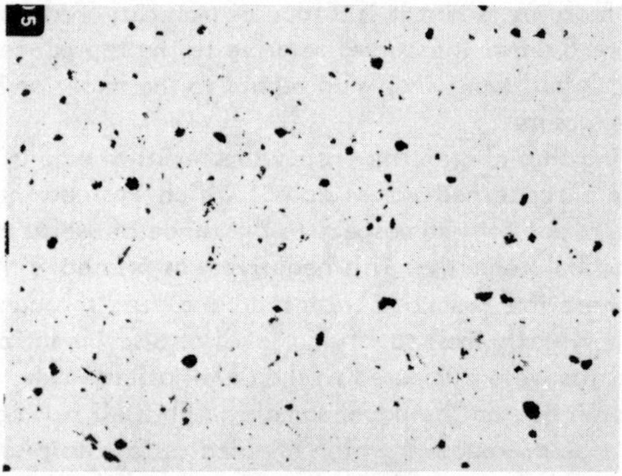

Fig. 9. Photograph of the sample using hard contrast materials under the operative conditions described in the paper.

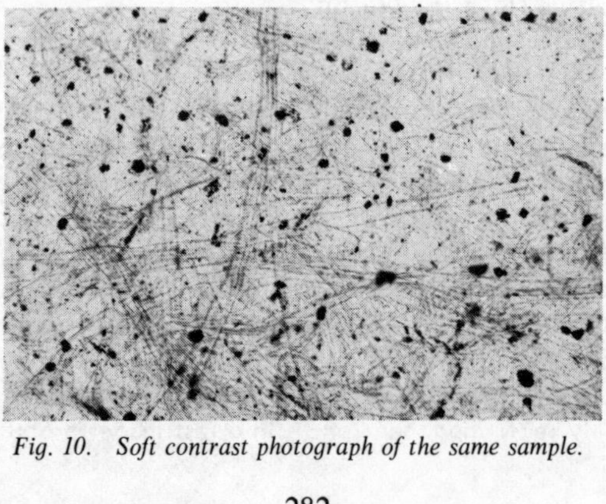

Fig. 10. Soft contrast photograph of the same sample.

ductions of the same field by exposing the negative material on 6°, 9°, 10°, 11° DIN, the results agreed satisfactorily.

Figures 12, 13 and 14 show distributions according to size of ink particles in four classes, whose width is 10μ m each, for an overall scale from 0 to 40 μm. The first classification was determined visually, while the remaining two were made by using the negatives obtained through the microscope with the exposure-meter set at 6° and 11° DIN, respectively.

Despite the different conditions of exposure, the correspondence between the two instrumental calculations seems rather good; whereas, the correspondence between the instrumental calculations and the visual calculation is excellent with regard to the total number of particles, but less so with regard to the distribution of the particles into the various size-classes, since the smallest-size class contains a low number of particles, while the following one consequently contains more. This is a result of the inevitable magnification of the image of the particles due to light diffusion in the photographic material. The magnification stems from a phenomenon which takes place around each of the particles and, as the percentages show, it is more sensitive to the smaller particles, as a consequence of the higher perimeter solidus area ratio, and gradually decreases as the particles become bigger, until it becomes inappreciable for the last size class.

Due to this phenomenon, the particles in the smallest size class become less numerous and, consequently, the subsequent class contains a greater number, whereas the third and fourth classes remain practically unaffected. However, it is believed that this phenomenon exclusively involves classes with the smallest particles. Furthermore, if the number of particles in each class is almost equal, the intermediate classes will automatically be offset, and thus, a correct calculation is possible. In our case, this cannot be verified since the field selected for the study contains a very large group of particles in the first size-class and a rather scanty one in the last class.

Without considering the particular sensitive material used, reference may be made to the overall reproduction curve (Figure 11, IV) for a synthesis of operating conditions necessary for an exact calculation. This curve must have a high slope for the rectilinear section ($\gamma \geqslant 4 \cdot 8$) and its central point must correspond to the density of the original which is 1·3.

The meaning of this density limit is that all the density levels beyond that value in the original will most likely be black or, at least, sharper when reproduced. On the other hand, the areas whose density level is below that value in the original will tend to be white or, at least, lighter when reproduced. This means that on the sheet being analyzed, once it has been made transparent, the 1·3 density value is the dividing line between the ink's density level and the possible shadows present on the sample itself.

Of course this value is valid for black inks only; in the case of coloured inks it would have to be modified experimentally.

The method suggested may be employed not only for the calculation of residue ink particles through transparent samples, but also for the analysis of prints on other surfaces whenever it is necessary to sharpen the contrast of the image.

EXPERIMENTAL INVESTIGATION

Considering that the method described was employed in the estimation of residue ink particles on de-inked pulp, it was essential to make a simultaneous evaluation throughout the entire thickness of the sheet which was obtained from the sheet former and then tested. In the case studies an overall 100X enlargement, given by a 10X objective lens, was used for the photograph, giving a shallow depth of field suitable only for sheets of less than 15 - 20 g/m^2.

Although the manufacture of sheets so thin is not feasible, it may be possible to resort to thicker sheets which are then split into layers (e.g. with the use of a Beloit 'Sheet-Splitter'). However, it must be kept in mind that the distribution of particles in the original thickness is not homogeneous.

Having obtained a sheet of the proper grammage, it is compressed at high pressure so as to further reduce its thickness. It is then immersed into benzyl alcohol, whose refractive index is near to that of the cellulose, and put into vacuum to eliminate air bubbles. The sample is then placed on an object-holder and covered with a slide preventing the formation of air bubbles and allowing the benzyl alcohol to fill up completely the space below the slide.

The slide is subsequently observed through the (AXIOMAT Zeiss) microscope in transmitted light and photographed using a 9 x 12 cm format sheet film (Agfaortho 25). After exposure, the film was developed under standard conditions with a developer having a methol-hydroquinone base. A contact positive was made using a bromide and a point source lamp and hard contrast positive material (Agfa-Gevaert P150 WP). This material was also developed under standard conditions using a methol-hydroquinone developer.

The sensitometric evaluation of both the negative film and the positive were made by means of the Kodak 7A sensitometer fitted with a 21-step wedge. For the sensitometric evaluation through the microscope's optical system, use was made of test objects obtained in the following way: 2 or 3 grey gelatine filters with different densities were overlapped in such a way as to obtain in the cross-over area four different-density quadrants, all of which could be simultaneously photographed through the microscope. The overlapping arrangement is shown in Figure 15.

The analysis of the photographs was made with the Zeiss-Microvideomat image-analyzer fitted with an optical system made up of an objective lens (the Zeiss 50 mm Orthoplanar) and interfaced with a computer.

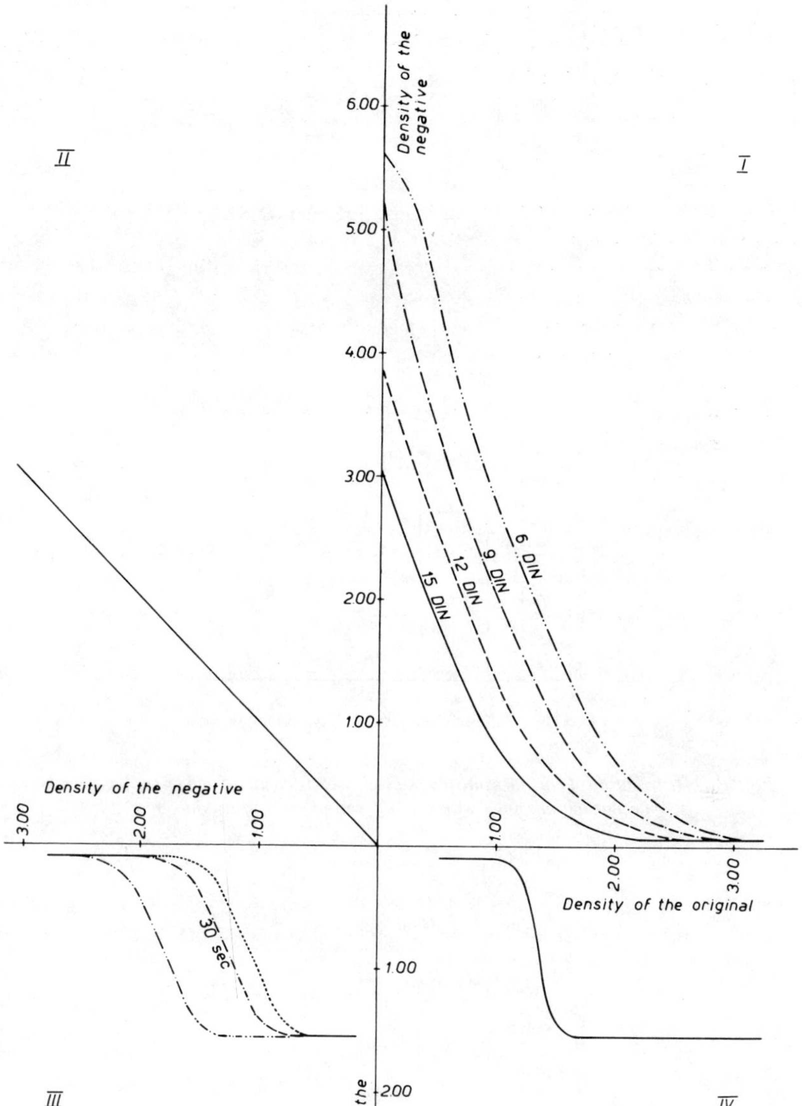

Fig. 11. Reproduction curve of the complete photographic process.

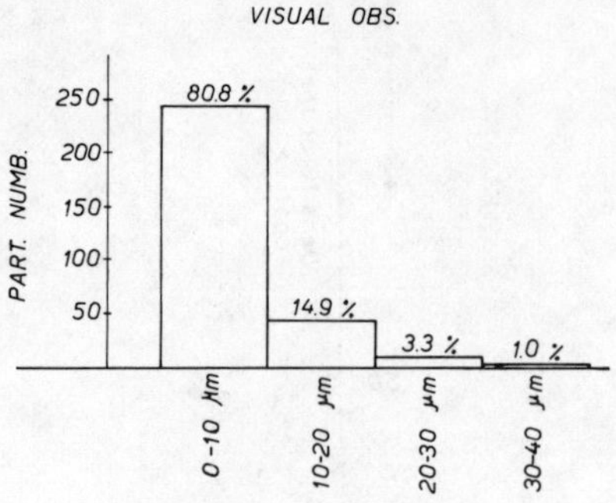

Fig. 12. Distribution of particles into four size-classes by means of visual count.

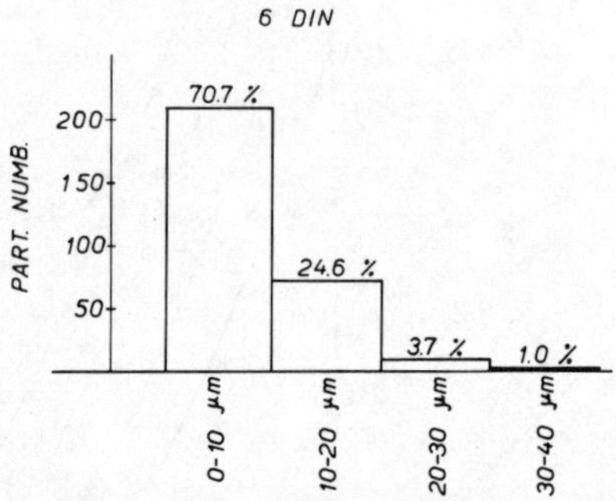

Fig. 13. Distribution of particles into four size-classes by means of image-analyzer on the photograph obtained with 6° DIN rating of exposure-meter.

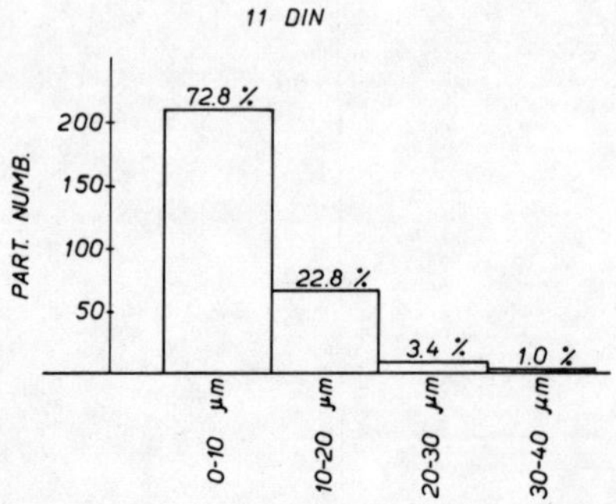

Fig. 14. Distribution of particles into four size-classes by means of image-analyzer on the photograph obtained with 11° DIN rating of exposure-meter.

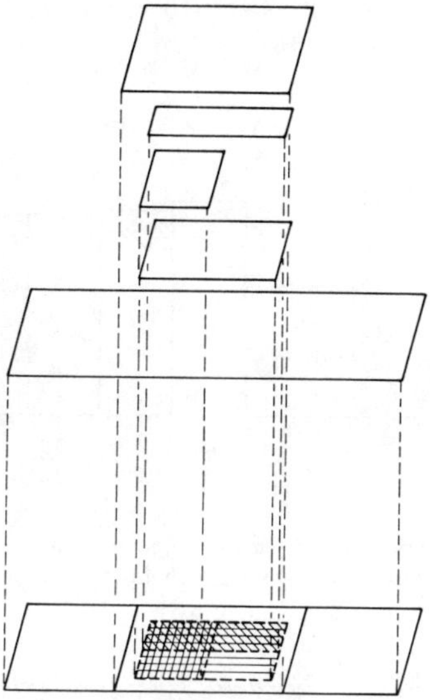

Fig. 15. Exploded and assembled views of the sensitometric test object.

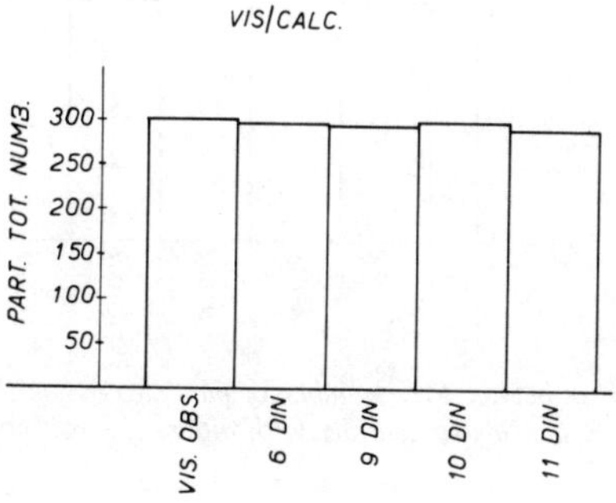

Fig. 16. Comparison between the total number of particles obtained visually and instrumentally with different sensitivity ratings.

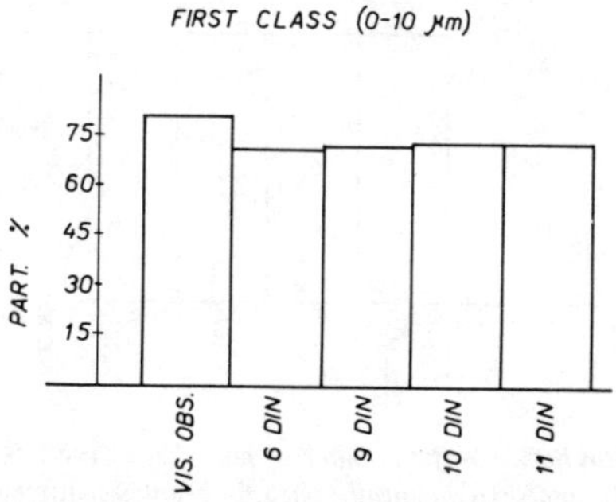

Fig. 17. Comparison between the number of particles in the first size-class obtained visually and instrumentally with different sensitivity ratings.

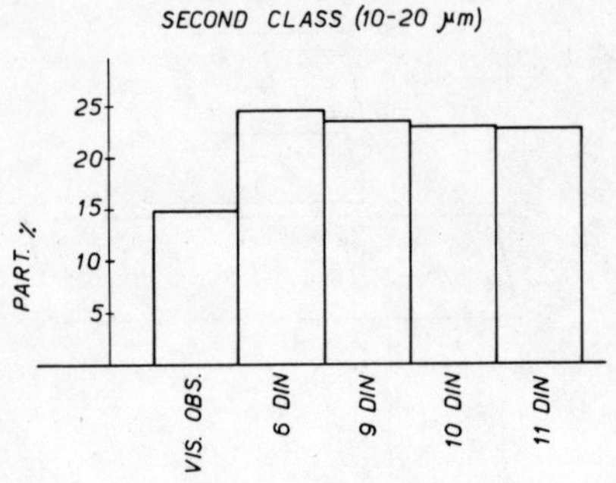

Fig. 18. *Comparison between the number of particles in the second size-class obtained visually and instrumentally with different sensitivity ratings.*

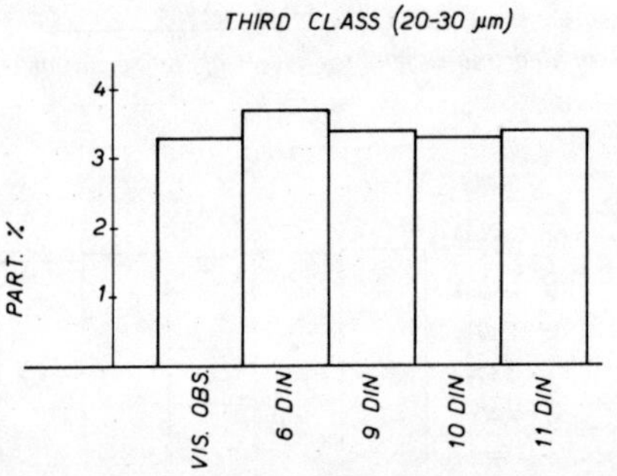

Fig. 19. *Comparison between the number of particles in the third size-class obtained visually and instrumentally with different sensitivity ratings.*

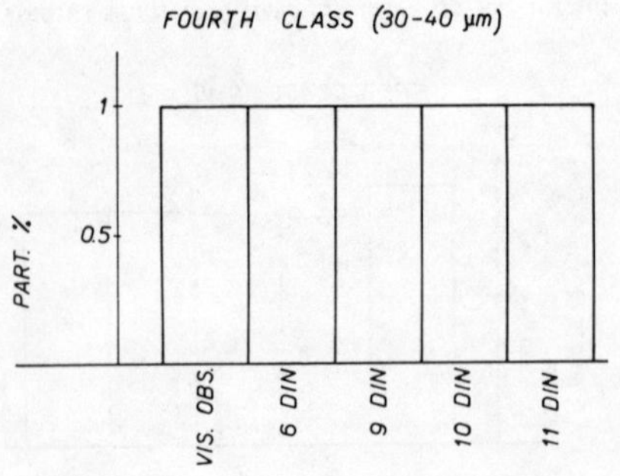

Fig. 20. *Comparison between the number of particles in the fourth size-class obtained visually and instrumentally with different sensitivity ratings.*

The discrimination of the limit density level is done visually, and requires an experienced operator. Once the level of discrimination has been determined, the calculations are made by the computer which uses the appropriate program and provides for field-scanning as well as collection and processing of data. The estimations made involve a number of fields, each of which corresponds to a surface of 1.1 mm^2 on the sample. The number of fields depends on the number of particles in each field, so that the total number obtained may be statistically significant.

The photograph of each of the fields is analyzed by means of a scanning stage, programmed accordingly, in eight adjacent areas. The computer then supplies the total number of particles, their subdivision according to diameter into size-classes selected by the operator, and the percentages, in relation to all eight fields in each figure. The histograms representing the calculations made are shown in Figures 16 to 20.

The first of these histograms compares the total number of the particles obtained visually and the number resulting from photographs obtained with the various exposures of the negative and of the positive.

The following histograms, Figures 17 - 20, compare the percentages of the particles belonging to each class. As already stated, the series of histograms clearly show that the total number of particles is calculated satisfactorily. However, the subdivision into size-classes is rather poor for the first class, and favours the following class due to light diffusion. The diameter variations in the larger particles belonging to the last classes are practically insignificant.

Under these operating conditions (enlargement of the photographic image and magnification by the Microvideomat's optic system), the smallest particle which can be measured is about 1 μm diameter.

The size-classes which we have selected in view of the type of sample used amounted to four with 10 μm spaces. Of course, different magnifying conditions and size-classes may be chosen according to the sample to be analyzed.

This research has been carried out under Research Contract no. RPC 026 I (S) sponsored by The European Economic Community.

THE IMPORTANCE OF PAPER FORMATION FOR GRAVURE PRINT QUALITY

B. J. Corsham, European Rotogravure Association e.V., Munich, West Germany

ABSTRACT

Countless articles have been written in support of one or other paper surface tests as a guide to how the paper will behave in the gravure press. Many others have pointed out that there is little general correlation between surface properties, as such, and the printability of the paper.

The author argues that formation, a very important characteristic of the paper sheet which invalidates many of the conventional tests, has been largely ignored. This is probably due to the fact that it is difficult to make a sensible measurement of formation and even then to do anything about it.

Some measurements have been made to show the degree of variation of fibre mass and how these variations affect the conventional paper properties and the performance of the paper in the press.

The author suggests that more attention should be paid to the effect of formation on paper for gravure printing and that raises the question 'whether a paper with lower standards in terms of traditional tests but with greater uniformity would not be a better economic proposition'?

INTRODUCTION

It has always been recognised that paper is a variable product. By its very nature it cannot be uniform. Even if a sheet is formed from fibres in a theoretically random manner, Kallmes and Corte [1] showed that it will have an irregular mass distribution. Although modern designs of headbox attempt to induce flow characteristics which break up large fibre flocs there are still considerable variations in web formation which result in an irregular or wild appearance on looking through the sheet.

In order to minimise the effect of short-term variations in the sheet and, indeed, variations from sheet to sheet it is an established principle that results of tests on paper should be reported as the average of several readings over as large a number of samples as is practicable. The whole philosophy of testing paper is to describe each property by a number together with some indication of its uncertainty (variance or range) and to use that number as a measure of the difference between papers. The conventional surface tests (apart from the surface profilers) measure over a large probe area and are not able to resolve the differences in physical properties which occur from spot to spot. This reduces the number of readings and the distribution of results but it does not help to discribe the paper accurately.

There has been a considerable improvement in the quality of certain papers over the last few years. Generally speaking the printing performance of s/c mechanical printing papers for gravure is quite good. The control of the combination of surface properties, strength and substance has been much improved with the increasing use of scanning, nuclear mass-detectors and moisture meters and improved head boxes have helped the papermaker to provide a more regular sheet over the whole width of the deckle.

Indeed, the papermakers are approaching the best that can be achieved in smoothness as measured by air-leak tests given the standards of substance and price (machine speed) which they have to achieve. Better methods of testing paper are needed to assess papers sensibly in order to maintain standards and to modify the paper to give acceptable performance with physical properties that are easier and cheaper to attain. Although the new papermaking equipment has improved the macro-formation of the sheet the micro-formation is still unacceptably variable. The measurement and assessment of the effects of uneven micro-formation (mass changes of wave length less than 10 mm) is difficult, time consuming and therefore costly. But further improvements in the quality and the efficiency of papermaking depend on looking more closely at the variability of paper and its effect on important properties.

PAPER TESTING FOR GRAVURE

For many years it has been recognised that the air-leak methods of testing paper are unsatisfactory for gravure papers. Blokhuis[2] reported this in 1965 and so did Pritchard[3]. More evidence was given by Falter[4] in 1970 when he was reporting on the relevance of the Kontaktanteilmessung (KAM) in testing papers for gravure. Pritchard had found that Chapman test equipment could produce contact patterns which were similar to the formation pattern of the sheet. Notwithstanding their lack of correlation with print performance the Bekk and Bendtsen test instruments are widespread and easy to use and there are many laboratories employing these air-leak methods to assess paper for gravure printing. The effective use of these tests must surely be limited to either:

(1) the routine examination of a single paper product in a mill to monitor some change in the surface of the paper (not necessarily relevant

to print quality), or

(2) the routine examination by the printer of a single type of paper from a single mill (the results used only as a possible supporting factor if the paper performs badly on the press).

These tests fail lamentably to differentiate between similar papers of high quality which nevertheless give observable differences in print performance. When Bristow[5] introduced the concept of Printing Pressure Requirement he chose a very wide range of papers to establish worthwhile differences and to achieve a satisfactory correlation. Even then there was some overlapping in results between what one would have expected to be very different papers. This test was developed in acknowledgment of the fact that the paper is under pressure when printed. The poor agreement between print quality and the results of tests on the standard Parker-Print-Surf is, Bristow stated, 'usually related to a suspected variation in the surface compressibility of the material'.

The main point of this paper is to suggest that the basic reason for the lack of correlation between air-leak tests and print performance is the uneven formation of the sheet. This factor invalidates these tests because they cannot detect the differences occurring within the test probe area, differences which nevertheless affect the quality of the print.

THE CONSEQUENCES OF POOR FORMATION

Consider the nature of paper. The uneven distribution of fibres not only produces a variable mass over the area of the sheet but it also affects other properties relevant to the transfer and distribution of ink and the runnability of the paper web in a rotary press.

Figure 1 showing a cross-sectional view of the paper indicates positions A of high fibre mass and B of low mass resulting from flocculation of the stock and uneven distribution of suspended fibres on drainage. (The question of possible uneven distribution of filler and fines will not be discussed here nor the possible effects of wire mark and press felts). The uneven web is dried and machine calendered and then supercalendered In the supercalender the heavy areas A will develop a higher nip pressure than the areas B and will consequently be compressed and glazed more than B. When the supercalendered paper arrives at the gravure nip the areas A will not only have more fibre but be thicker, denser and more glazed than areas B. The gravure nip is to all intents and purposes a fixed gap and the areas A will develop a greater printing pressure which, with the other factors, will lead to a better transfer of ink at areas A than at areas B. If there are missing dots they will form preferentially in areas B. Missing dots occurring in clusters (in low mass areas) will be more offensive to the observer and will give a more uneven print than if they are evenly distributed. Morton[6] reported finding most missing dots in areas of low fibre mass measured by β-radiography. Furthermore, the more densely packed fibres in areas A and the glazed surface will affect

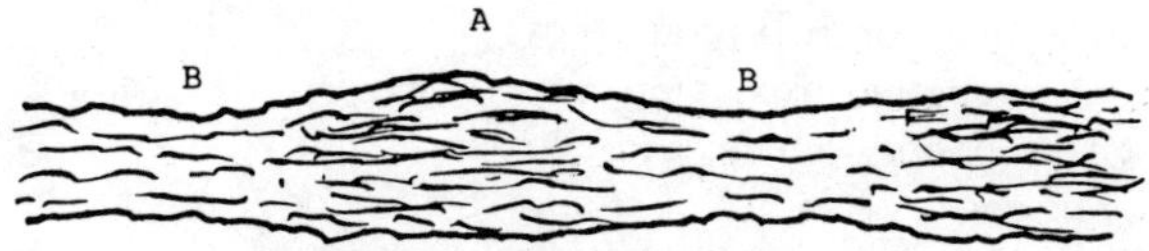

Fig. 1.

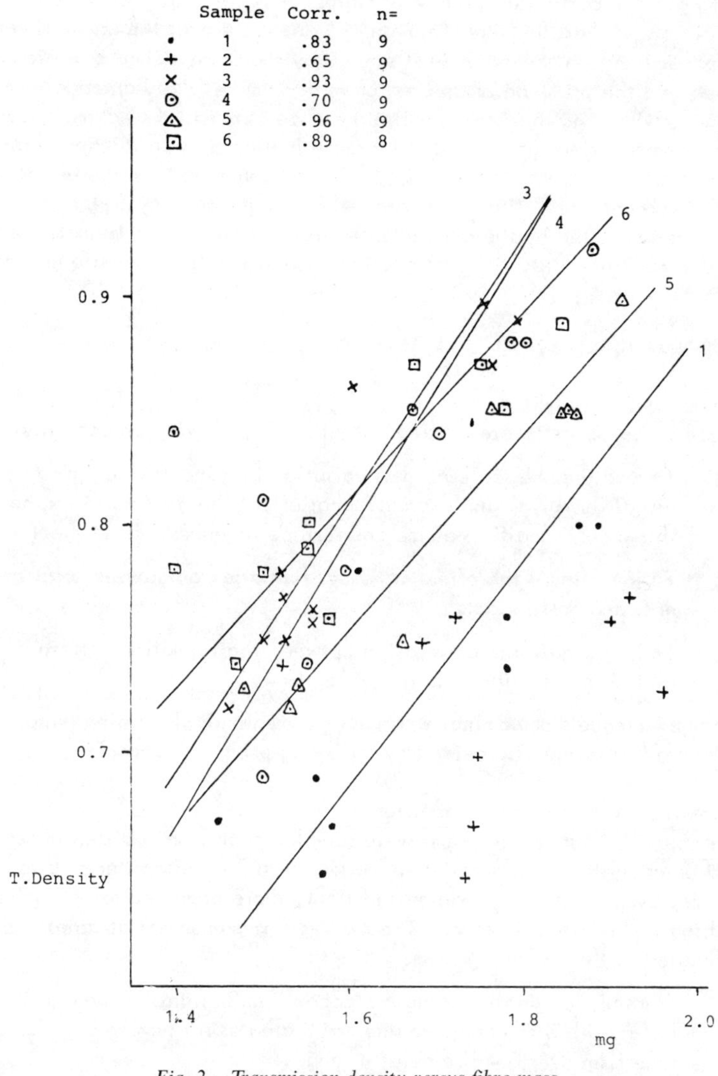

Fig. 2. Transmission density versus fibre mass.

the distribution of the constituents of the ink and give rise to possible mottling effects in the full tone areas.

Yet another aspect of formation is its effect on the physical strength of the paper. It is held by some that a wild sheet is stronger than an even one. This belief stems from the days of rag paper when a wild sheet was synonymous with strength and character. It is true that strong papers made for packaging are made from fibres which, due to their nature, tend to flocculate and give a wild sheet (uneven formation) But, given a particular mixture of fibres, it is better to distribute them uniformly so that the fibres form an even network over the entire sheet. A paper break in the machine is caused by the coincidence of a high stress and the presence of an area of web (or more likely an edge) not able to withstand that stress. The presence of areas of weakness (low mass and few fibre bonds) might increase its tendency to fail in relatively low stress conditions. Göttsching[7] has reported work on the effect of sheet-forming speed (including uneven fibre distribution) on the strength of paper as tested by the conventional instruments. Poor formation tends to decrease the tensile strength but not necessarily the tear strength of paper.

EXPERIMENTAL WORK

Objectives

Three series of tests were carried out with the following objects in view:

(1) To measure the scale of mass variation in a sheet which has uneven visual formation and examine what effect the mass variation has on thickness, specific volume and surface roughness of the sheet.

(2) To investigate the effect of visual formation on gravure print evenness and missing dots.

(3) To investigate the correlation between the formation pattern given by transmitted light and by β-radiation.

A series of sheets of nominally 60 gsm s/c mechanical printing paper were selected from samples provided by mills and printing companies.

Formation and Physical Properties

Methods Six sheets of paper were sampled each from a different source and on each sheet ten areas were marked with 7 mm diameter circles where, by inspection by transmitted light, there appeared to be an area of high or low fibre density. The marked and numbered test areas were subjected to the following tests:

(1) Transmission density using a Macbeth TR527 transmission densitometer with 3 mm aperture and calibrated as for testing continuous tone film over 0 - 1.6 density range.

(2) Surface roughness using a Perthometer C50 surface profile recorder

and analyser. A single measurement was made on each area with a 1.5 mm stroke.
(From the centre of each test circle a disc of paper was cut from the sheet using an ordinary desk paper punch. Disc diameter approximately 5.6 mm.)

(3) The thickness of each disc was measured on a standard Frank caliper gauge with micrometer clock.

(4) Each disc was weighed on an analytical balance.

Results

(1) Transmission density and fibre mass (Figure 2) shows there to be a correlation between the two methods of assessing mass distribution for all papers except number 2. There was no apparent reason for the irregular performance of this sample.

(2) Surface roughness and fibre mass (Figure 3) shows no correlation. However, Figure 4, showing roughness plotted against transmission density, does suggest a relationship and statistical treatment of the results gave a correlation with 99% confidence.

(3) Caliper and fibre mass (Figure 5) shows expected correlations although no relative reduction in caliper due to compression in high mass areas was detectable.

(4) Apparent specific volume and fibre mass (Figure 6) shows some uncertainty in relationship. However, two paper samples gave correlations with 90% and 95% confidence.

Formation and Print Evenness

Method Ten sheets of paper were sampled and on each sheet ten areas were marked as for the previous series. The sample sheets were then printed using black ink and the standard FOGRA test cylinder on a Haindl/Einlehner gravure test press.

(1) It was intended to set the press to print with a light impression load and to exaggerate the production of missing dots. However, the press printed too well and no missing dots appeared in the areas marked for tests. Therefore, reflection density measurements were made on the print in the specified sample areas.

(2) At a later stage some light tone areas on the printing forme which had exhibited some missing dots were examined. Areas were chosen which were observed, by transmitted light, to be printed on high or low mass areas. These areas were examined for the appearance of missing dots within a 3 x 3 mm^2 square.

Results

(1) Figure 7, showing reflection density vs transmission density, gives

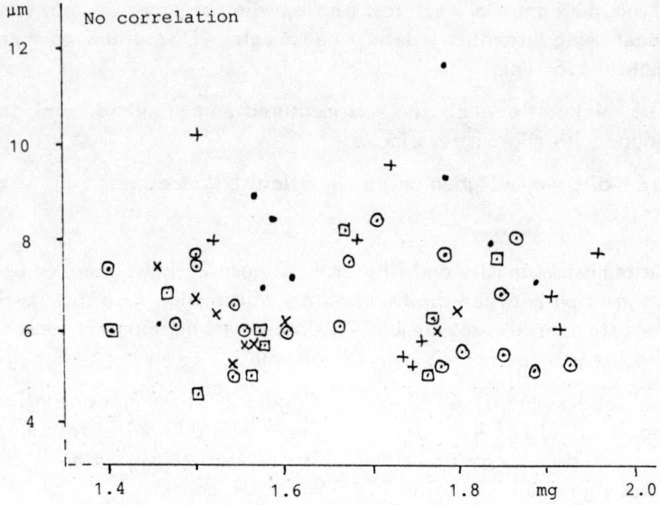

Fig. 3. Surface roughness versus fibre mass.

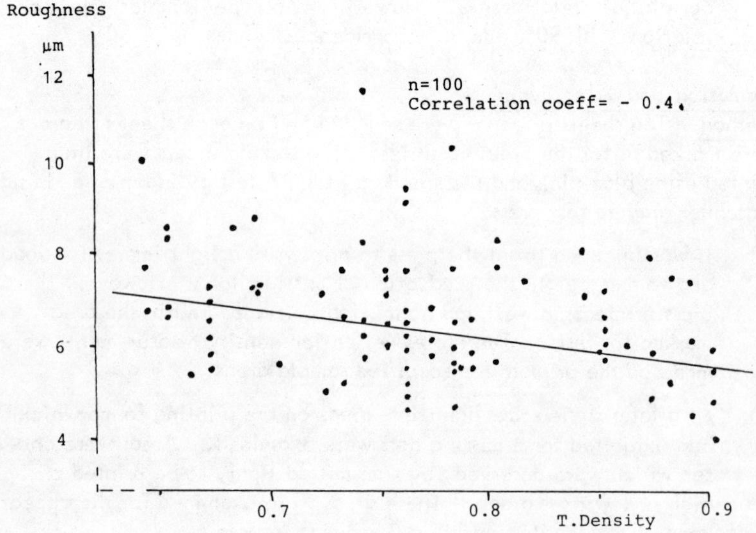

Fig. 4. Surface roughness versus transmission density.

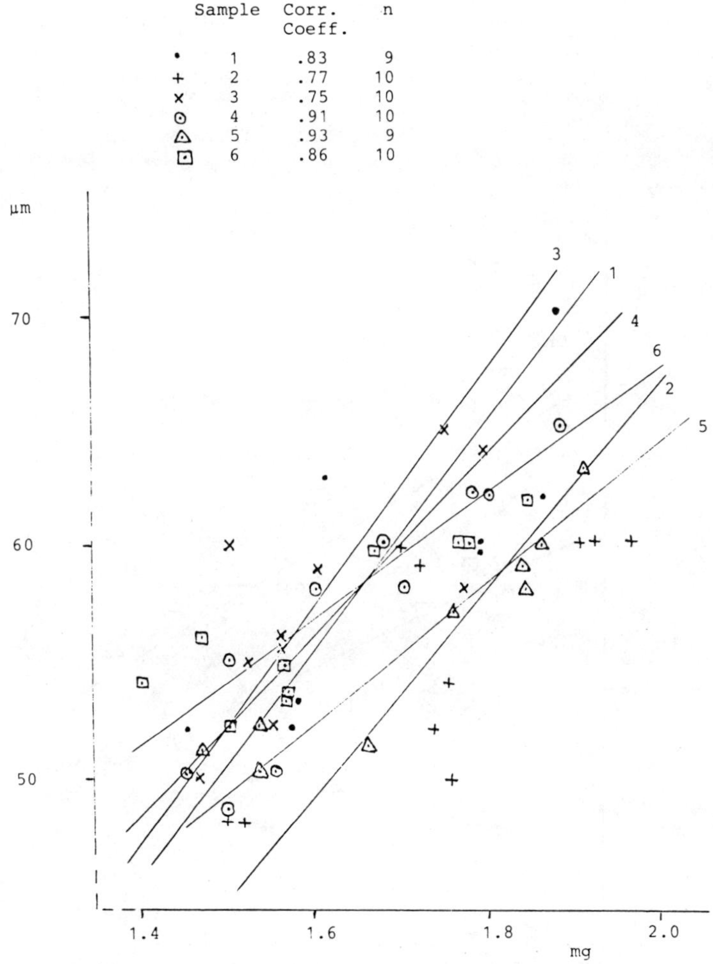

Fig. 5. Caliper versus fibre mass.

	Sample	Corr. coeff.	n
•	1	--	9
+	2	--	10
×	3	.2	10
⊙	4	--	10
△	5	.51	9
⊡	6	.7	10

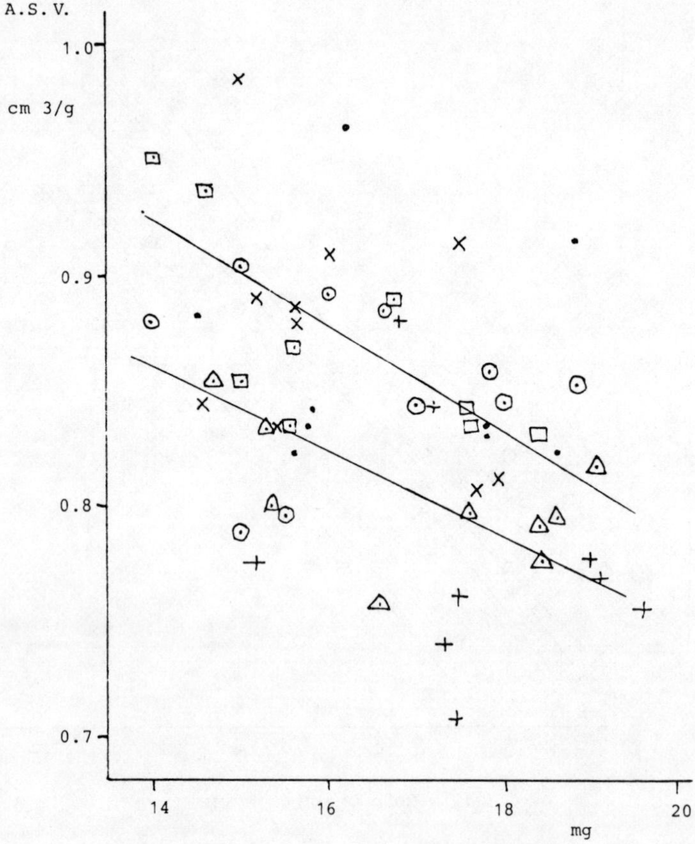

Fig. 6. Apparent specific volume versus fibre mass.

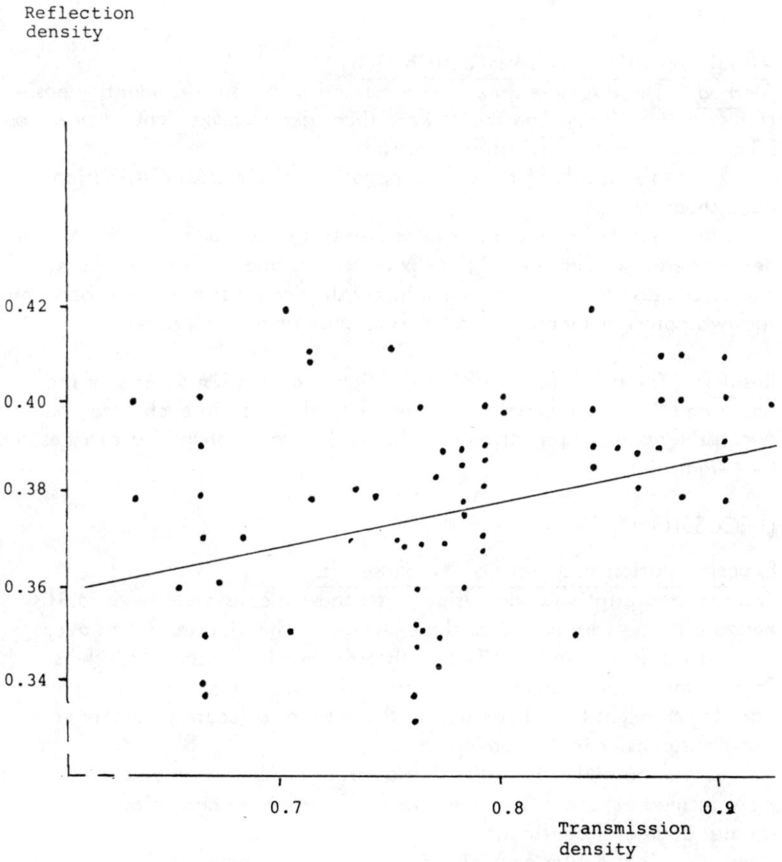

Fig. 7. Reflection density of print versus transmission density. Correlation coefficient = 0.34, n = 71.

a rather wild distribution of points. However, there appeared to be some suggestion of shape and, with the large number of results, a correlation was shown to exist.
No determination of fibre mass was possible in this instance.

(2) The count of missing dots showed there to be a significantly greater number (2.07) in the low mass areas than (1.13) in the high mass areas. These counts are the averages of estimates made in ten areas.

Visual Formation and Mass Distribution

Method Three sample sheets were marked with points to identify position and direction. The sheets were then used to make contact prints on film exposed to a β-radiation source.

The resulting films were negatives of the mass distribution in each sheet.

Transmission density measurements were made with the Macbeth densitometer on both the original paper sheets and the β radiographs. Measurements were made at 4 mm intervals along a straight line between the two points of identification at 10 cms apart on each sheet.

Results The graph (Figure 8) of the light transmission density of the paper and the β radiograph show clear correlations, in each case, between the apparent formation and the actual mass distribution as measured by β-radiation.

DISCUSSION

Experimentation and Variability of Results

The reported printing and physical tests showed considerable variability because of the very nature of the exercise. The aim was not to average out test results but to identify the differences which occur over the surface of the sheet. Some of the variability and the lack of anticipated correlation might have been due to the failure to locate precisely each test measurement in the same place.

The author believes that a larger number of observations and more refined experimental techniques will produce correlations with a greater degree of confidence.

The few tests which were carried out supported the proposal that variations in the fibre mass over the area of the sheet are relevant to the printing performance of paper in the gravure press.

SURFACE SMOOTHNESS

Consider cross-sectional views of paper sheets under a smoothness test probe (Figure 9). The straight line represents the surface of the contact probe of an air-leak smoothness tester (Bekk, Bendtsen or Parker-Print-Surf). The smoothness of the surface is an inverse function of the total

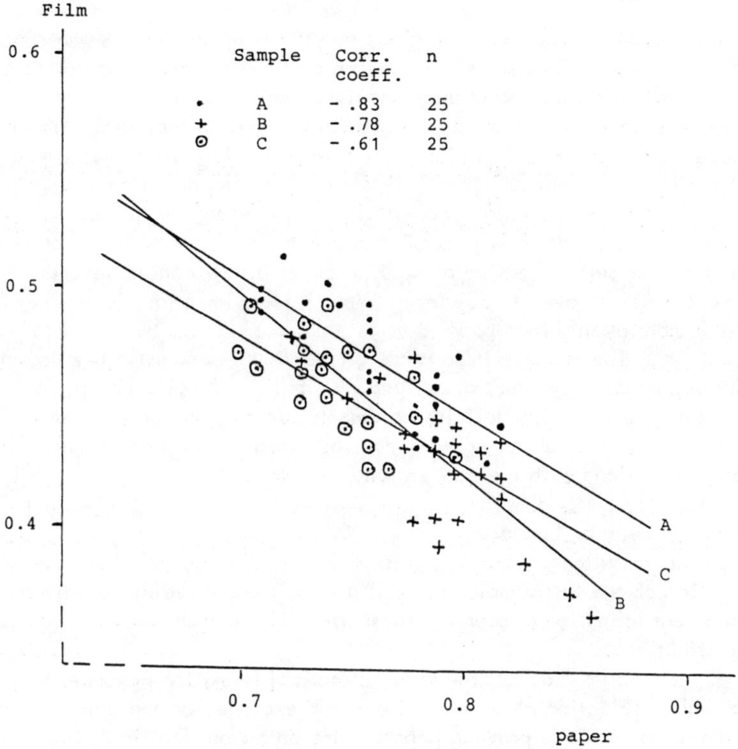

Fig. 8. Transmission density β radiograph versus paper.

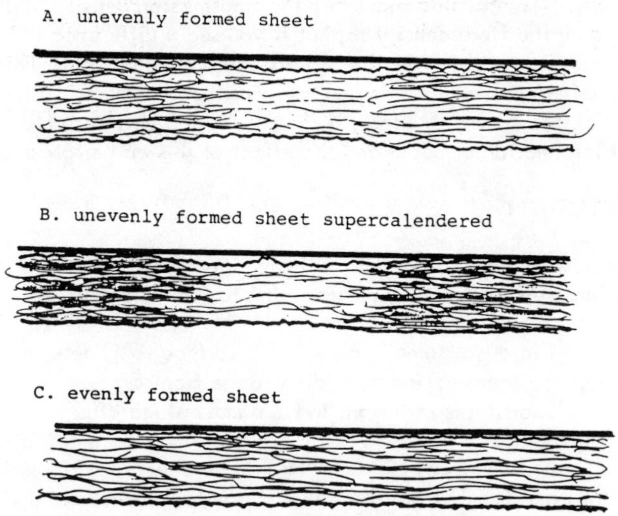

Fig. 9.

effective air gap between the probe and the paper. The calendering of paper B will cause it to be very smooth at the high mass spots and to give test readings indicating better smoothness than paper A. However, due to the scale and depth of the air gaps in the low fibre areas and the higher developed nip pressure at high fibre areas the print will not be even. It is not unknown for a printer to ask for lightly calendered paper in order to achieve a more even result notwithstanding the rougher surface.

Consider now paper C and assume it has the same smoothness, that is, the same effective air gaps as paper B. It can achieve this smoothness without calendering. What is more important is that it will print more evenly than paper B.

The relative importance of evenness, compressibility and micro surface structure in ink transfer and distribution can give rise to considerable discussion. No definite statements can be made until a study of the surface and structure of paper and its performance has been made. This cannot be done with the conventional test procedures.

Measurement of Formation

Both the β radiographs and the test weighings showed that, with the exception of one test sample, there is a very strong correlation between apparent formation as seen by transmitted light and the actual fibre mass distribution.

Sara[8] doubts the value of optical formation meters except for a limited selection of papers. He would exclude, among others, supercalendered gravure printing papers. He points out that in supercalendered papers the sheet is preferentially consolidated in the high fibre mass areas and this changes the light transmission properties in an uncontrolled way making optical formation measurement unrepresentative of the mass variation. True as this might be, the transmission density will probably undervalue the fibre density so that if you see a difference it is there.

The use of formation testers based on light transmission should not be rejected even if their readings must be considered with caution. However, they are only useful if the results can be interpreted to give useful information of paper and the effect of this on performance.

CONCLUSIONS

The density of light tones and the incidence of missing dots in gravure test prints were shown to vary according to the fibre mass distribution in the sheet.

There was some evidence that surface roughness and the fibre density of the sheet varied according to the fibre mass.

Short term variations in fibre mass which affect the gravure print quality cannot be resolved by air-leak smoothness test methods.

New methods of testing should be developed to assess the effect of mass distribution in the sheet in order to maintain and improve the performance of papers within the economic constraints which prevail.

Formation meters and analysers based on light transmission should not be rejected.

ACKNOWLEDGMENTS

The author wishes to thank Mr.K. A. Falter of FOGRA for permission to use laboratory facilities for the paper and printing tests and Ms. Ingrid Fineman of STFI for obtaining the β radiographs.

REFERENCES

1. Kallmes, O. and Corte, H., 'The Structure of Paper', TAPPI (43) 9 (1960).
2. Blokhuis, G., 'Paper for Photogravure Printing', Printing Technology, Vol. 10, No. 1 (1966).
3. Pritchard, E. S., 'The Effect of Roughness Distributions on Speckle in Gravure Printing', 8th Int. Conf. Printing Res. Inst. 1965, Pergamon Press, London (1966).
4. Falter, K. A., 'Bedruckbarkeitsuntersuchungen von ungestrichenen und gestrichenen Tiefdruck-Illustrationspapieren', FOGRA-Mitteilungen, No. 64, May (1970).
5. Bristow, J. A., 'The Surface Compressibility of Paper - a Printability Property', 15th Int. Conf. Printing Res. Inst. 1979, Pentech Press, London (1980).
6. Marten, J., 'Novel Approaches for Printability Testing of Gravure Papers', Gravure Research Inst., 31st Annual Meeting, Key Biscayne, (1978).
7. Göttsching, L., 'Einfluss der Formation (Durchsicht) von Papier auf verschiedene Festigkeitseigenschaften', Wochenblatt für Papierfabrikation, 7 (1979).
8. Sara, H., 'Microscale Uniformity of Paper', European Rotogravure Association, Report PA 7 (1980).

DISCUSSION

M. B. Lyne   In the literature you will find that those that have compared the R.M.S. or coefficient of variation of mass in paper with the R.M.S. or coefficient of variation of print density on the same paper have found no significant correlation. However, those that have superimposed beta radiographic negatives of the paper over the print have found the result was a uniform grey, i.e. a highly significant relationship between mass distribution and print density. Dr Jordan of our institute has confirmed via image analysis that first order statistics do not correlate, but that second order statistics (describing the spatial distribution of flocs) do correlate for mass and print density on the same printed sheet of paper.

J. A. Bristow It is important that the influence of paper formation be documented. I think I can encourage you by confirming that the Swedish industry is aware of this. We have a project concerned with coated boards for gravure and it is quite clear that the sins of the paper or board maker cannot be covered by a coating. A study of board machines confirms that boardmakers are aware of the importance of good formation. This is also true of light-weight coated paper manufacture.

Your paper implies that you have no fillers in your papers. Most uncoated gravure papers contain about 20% of a clay filler. How do you define formation and what sort of formation do you hope for in a paper containing about 20% of a clay filler or other inorganic mineral component?

B. J. Corsham I am heartened by Mr. Bristow's assurance that the papermaking industry is aware of the influence of the paper formation. It is, of course, in the papermaker's interest to produce a uniform sheet. Yes, the papers taken for my tests had a significant amount of clay filler but I do not know how much nor, indeed, the distribution.

A precise definition of formation and the question of whether it should relate only to the fibre or to the total mass of the sheet could be debated at some length. I cannot say what the formation should be for that is my question. My plea is that, assuming the paper cannot be entirely uniform, we should measure differences by the best practical means we have available and investigate the magnitude and nature of the differences which lead to acceptably uniform printing results. That being done, the problem of controlling the property on the paper machine can be tackled.

J. L. Poujade Two comments - First, I doubt that optical transmission evaluation can in any case replace fibre mass evaluation in describing paper formation, because it does not work at least in one case, when flocks are calendered to such an extent that blackening occurs and that their measured opacity becomes lower than that of the rest of the sheet.

On the other hand, I would like to support your conclusions on the influence of formation on printability using the example of direct lithography which similar to gravure, in the sense that the non-resilient forme is in direct and continuous contact with the paper and that printing pressure in the nip is low: in newspaper printing, we have achieved much better results with well formed double-wire newsprint supplies than with traditional Fourdrinier nipped newsprint.

B. J. Corsham Indeed, hard calendering will change the optical properties of the sheet in such a way that will make optical formation measurements after calendering an uncertain indicator of mass variations.

However, formation should be measured before calendering and ideally on the paper machine itself. Can we not look ahead to the day when a signal representing the micro variations in mass can be monitored continuously, then analysed and fed back to the head box in order to modulate the necessary variables to keep the variations within acceptable limits of period and amplitude.

The blackening of high mass areas in the calender does lend support to my suggestion that calendering affects the density of the sheet and will result in areas of high pressure in the printing nip.

M. Ginman You criticize the air-leak roughness testers but I would like to put a good word for the Printsurf tester. In a number of studies we have found correlation coefficients of 0.8 - 0.9 between missing dots and Printsurf roughness for both uncoated and coated gravure papers.

Missing dots are not considered to be a main problem for uncoated supercalendered paper today. A larger problem is the mottle in dark tones caused by uneven absorbency which depends on uneven mass distribution.

B. J. Corsham The Printsurf is probably the best instrument for measuring the surface of paper by the air-leak principle and correlations may be found between roughness results and missing dots on samples of paper which are significantly different. However, with standards of paper being what they now are, improvements can be achieved only in small steps and air-leak methods have great difficulty in discriminating sensibly between papers of the same grade and specification which nevertheless perform differently on the press.

I agree with the second comment regarding the effect of formation on gloss mottle in the full tones. I was not able to support this phenomenon with any evidence and thank Mr. Ginman for his observation.

PRINTING BEHAVIOUR OF OFFSET NEWS INKS

P. Oittinen, The Finnish Pulp and Paper Research Institute, Helsinki,
U. Lindqvist, Technical Research Centre of Finland Graphic Arts Laboratory, Espoo.

ABSTRACT

Samples of European commercial black offset news inks have been characterized by measurements of the ink properties, the quality of solid printing as tested in the laboratory, and by pilot scale printing behaviour. The purpose has been to document the quality level of news inks, and elucidate the role of ink in print quality.

It was found that the optical print quality of solids arose mainly from ink and paper interactions. On the other hand, it became evident that the halftone printing behaviour, set-off and rub-off were governed by the ink properties to a more marked extent.

INTRODUCTION

This paper reports a study of black offset news inks (Lindqvist and Oittinen[4]). The objectives of the study were:

(1) Documentation of the quality level of black offset news inks currently in use in Europe.

(2) Exploration of the variations in print quality which arise from variations in ink properties.

(3) The correlation of laboratory-scale and pilot-scale printing behaviour with respect to the interactions of ink with dampening water and with paper.

Those ink properties of greatest significance to ink and paper interactions, i.e. the rheological properties, are summarised. Also, the ink and paper interactions are discussed on the basis of laboratory and pilot-scale tests. The approach adopted is orientated towards the user of printing ink.

Efforts to improve the quality level are likely to prove more feasible when they are directed by a knowledge of the mechanism of print-quality formation. It is generally accepted that in newspaper printing, the quality level largely results from ink and paper interaction. This finds its origin in the porous, compressible character of the paper and in the fluid-like character of the ink. It has not been shown just how important the rheological properties of the paper are in comparison with the rheological properties of the ink. A characteristic of an interaction is that it can occur only on the condition that it is favoured by both the interactive components. In other words, the final result is obtained in a way which is multiplicative rather than additive. The multiplicative nature implies a limitation of the possibilities for compensating for deficiencies of the properties of one of the contributing components by means of the other components. Furthermore, a change of a given magnitude in one of the components becomes apparent in the end result to an extent which depends upon the other components.

Dimensionless numbers are expressions of the multiplicity of interactions. Some dimensionless numbers for the characterization of ink behaviour in a nip have been listed in Table 1. The variables represented by these numbers have the dimensions of length, time or force.

TABLE 1 Dimensionless numbers characterizing the nip behaviour of ink. Purely viscous ink behaviour is assumed

	Length scale	Force scale	Time scale
Contact formation with paper	x/σ	pt/η_1	
Halftone dot spreading	x/d	pt/η_1	
Pressure penetration	x/r x/r'	pt/η	
Capillary absorption	x/r	$\frac{\gamma t \cos\theta}{r\eta_1}$	
Splitting resistance	R/x	$\eta_h U/\gamma$	
Deformation of paper relative to ink flow	x/x_o		Et/η

Symbols:

d	halftone dot diameter (μm)	p	nip pressure (Pa)
r	capillary radius (μm)	r'	one-dimensional pore volume (μm)
t	nip contact time (s)	x	ink film thickness (μm)
x_o	paper thickness (μm)	E	elastic modulus of paper in compression (Pa)
R	roller radius (μm)	U	rotational speed (m/s)
γ	surface tension of ink (N/m)	θ	contact angle between ink and paper

Symbols (continued)

η	ink viscosity (shear rate unspecified) (Pas)	η_l	ink viscosity measured at a low shear rate (Pas)
η_h	ink viscosity measured at a high shear rate (Pas)	σ	root mean square value of profile roughness of paper under compression (m)

The length-scale numbers express the ratio of the ink film thickness and other length variables, such as paper roughness or pore dimensions, half-tone dot size or roller geometry. As the ink film thickness is the only length scale variable directly related to the ink, it is certain to assume an extreme importance in all nip phenomena.

In the case of a purely viscous ink, the force-scale dimensionless numbers concern the magnitude of the viscous forces in relation to the other forces which operate in the nip, and attempt to give rise to ink flow. Nip pressure is the primary cause of flow inside the nip, whereas capillary suction plays an increasingly important part outside the nip. At the stage of splitting the ink, a force is generated in the ink layer, since the ink opposes the deformation to which it is subjected with increase in the distance between the roller surfaces. In purely viscous splitting, the surface tension ultimately determines the height of rise of the tension in the ink layer prior to rupture. In dispersive systems, such as inks, other mechanisms operate as well, such as cavitation or elastic deformation.

A purely viscous ink has no inherent time scale, but thixotropy and elasticity constitute sources of such a scale (Oittinen [6]). However, this topic lies outside the scope of this paper. A phenomenon which is related to the time scale is the relative deformation under nip pressure of the ink, and of the paper. An indication of this is provided by a dimensionless time variable (cf. Table 1). According to this, when a diminution occurs in the contact time, a decrease becomes apparent in the significance of ink viscosity as a source of ink and paper interactions, whereas that of paper elasticity increases.

In simple laboratory tests, it can be expected that ink and paper interactions are comparable with those occurring in offset printing only if the laboratory tests are scaled-down versions of practical printing. If possible geometrical differences are ignored, this can be the case only if the dampening water of offset printing does not influence the interactions. Another factor to be considered is the discontinuous nature of laboratory printing as compared with full-scale printing, a factor which inevitably limits the ability of laboratory printing to forecast cumulative phenomena. This applies in particular if the cumulative phenomenon tends towards a saturation level with the number of revolutions. Linting (Lindqvist and Meinander [2]) falls within this category of phenomena.

EXPERIMENTAL

Black industrial offset news inks, about 30 in number, were collected from printers in ten European countries. The samples were taken from regular consignments, and represented processes in which printing was effected under widely different conditions. The mean values of the properties measured may be regarded as characteristic of a typical - or average - European offset news ink.

TABLE 2 Results obtained in laboratory-scale tests

Property	Lowest value	Highest value	Mean value	Variation coefficient %
Ink properties				
Viscosity (Ferranti) (Pas)				
23°C 100 s^{-1}	3.2	37.7	13.8	54.3
500 s^{-1}	1.9	14.5	7.8	41.0
1000 s^{-1}	1.5	10.6	6.2	38.7
10°C 500 s^{-1}	2.6	24.9*	13.1	40.5
40°C 500 s^{-1}	0.9	6.1	2.9	44.8
Tack 23°C, 3 μm, 2 m/s	43	163	102	24.8
5 μm, 2 m/s	49	156	96	27.1
5 μm, 4 m/s	52	197	120	26.7
Surface energy (mN/m)	33.7	49.6	39.5	10.9
Polarity (%)	0.6	11.4	6.8	83.8
Particle size (m)				
3 largest particles	14	43	28	39.3
group of particles	6	27	16	37.5
Interaction ink/paper				
Ink transfer (%), 3 g/m^2 on plate	61.7	75.0	69.3	3.3
Ink requirement (g/m^2)				
D 0.85	1.05	1.70	1.38	10.1
D 0.95	1.50	2.65	1.97	10.7
Print-through (D)				
D 0.85	0.044	0.061	0.051	9.8
D 0.95	0.049	0.079	0.060	11.7
2 g/m^2 on paper	0.049	0.070	0.061	8.9
Set-off (D)				
D 0.85	0.123	0.176	0.146	10.3
D 0.95	0.185	0.283	0.222	9.0
2 g/m^2 on paper	0.143	0.297	0.209	13.3
Rub-off (D)				
D 0.85	0.017	0.102	0.048	52.1
D 0.95	0.019	0.119	0.058	51.7
2 g/m^2 on paper	0.017	0.124	0.059	54.0

*The most viscous inks could not be measured.

The laboratory tests made can be divided into three main categories: pure ink properties, properties which depict the ink and water interactions, and those which depict the interactions between the ink and the paper. Mean values, minimum and maximum values, and the variation coefficients of the test results, have been listed in Table 2. The ink and water interactions are discussed separately (Lindqvist [3], this conference).

Pilot-scale printing tests were made for determining the printability properties of six inks - selected to represent different levels of ink requirement, set-off and print-through. For the pilot-scale linting test, four inks were chosen characterized by different viscosity and tack levels.

Table 3 presents a brief description of the laboratory test methods and the pilot-scale tests. A more detailed description is to be found elsewhere (Lindqvist and Oittinen [4]).

TABLE 3 Methods of Measurement

Type of method	Property	Apparatus	Measuring conditions
Ink rheology	Viscosity (Pas)	Ferranti-Shirley	Plane/cone geometry; 23, 10, 40°C; rising curve; 100, 500, 1000 s^{-1}; acceleration 2.7 s^{-1}/s.
	Tack (relative units)	Tack-0-Scope	23°C; ink thickness 3, 5 µm; 2, 4 m/s; test time 30 s/level.
Colloidal properties of ink	Degree of dispersion (m)	Grindometer	Groove depth 0-50 µm; width 40 mm; (a) three biggest particles, (b) first group of particles.
Surface energetic properties of ink	Surface energy and its components	Contact angle projector	Thin ink layer on glass. Contact angle measurement from droplets of 50 µl of model liquid. Calculation.
	Ink pH	Potentiometer	Dilution of 3 g of ink in 27 ml xylene/metanol; dilution with 30 ml water; titration with NaOH and HCl
Ink/water interaction (Lithographic)	Interfacial tension	du-Noüy-tensiometer	Dilution of 20% ink in toluene. Centrifugation; addition of water phase and tensiometer measurement. (Laraignou [1]).
	Spreading pressure	du Noüy-tensiometer	Application of ink droplet to a water surface; tensiometer measurement.
	Water content	ULTRA TURREX mixer	Ink sample and 50% water mixed for 4 min at 25000 rpm. Removal of water excess. Drying at 105 ± 2°C for 15 h. Weighing.

TABLE 3 (continued)

Type of method	Property	Apparatus	Measuring conditions
	Water content of ink as a function of time	Sunbeam Mixmaster 235W mixer	Ink sample mixed with excess of fountain solution at 90 rpm for 10 min; determination of increase in weight once a minute. (Surland [10]).
	Change in tack	Tack-0-Scope	Injection of 0.5 ml of water in the drum nip at constant ink amount 10 m.
Ink/paper inter-actions	Laboratory test printing	IGT AC 2-press	Rubber disc; line pressure 20 kN/m; speed 4 m/s.
	Ink trans-fer (%)	Balance	Determination of ink transfer at diff-erent amounts on disc. Calculation.
	Density and print-through	Elrepho	Reflection measurement; SCAN-P36: 76; density level 0.95 and 0.85.
	Ink requirement (g/m^2)	Elrepho	Density as a function of ink film thickness; calculation of required ink film for D, 0.85 and 0.95.
	Set-off (D)	IGT AC 2	Determination in second nip after 0.017 s; same paper used. 2 g/m^2. (SCAN-P36:76)
	Rub-off (D)	PATRA	Pressure 2 kPa; 100 rev.; 90 s; 2 g/m^2.
Pilot scale printing test		Goss-Community	Speed 2.8 m/s; compressible rubber Ultra 303; QUADRI plate.
	NCI-level	Densito-meter	10-12 inking levels; density measure-ment; graphical determination of maximum relative contrast level. (Schirmer and Tollenaar [9]).
	Print-through	Elrepho	As in the laboratory tests.
	Evenness and general appearance	Subjective assessment	Assessed by 5-8 experts using the ranking order method. Quality factor P (0...100).
	Set-off	Analytical balance	Brass plates (9 x 13 cm) fastened on first bend pole in press.
Linting test	Lint amount (g/m^2)	Analytical balance	Lint collected on tape (200 cm^2) after 2000 rev. Non-image and 50% dot area. (Monni [5]).

RESULTS AND DISCUSSION

Printing ink properties

Pure ink properties are discussed here with the aim of characterizing offset news inks, and to provide a background for the subsequent recommendation of methods of measurement for these inks.

Viscosity is the primary rheological parameter of offset news inks. The inks are non-Nowtonian dispersions; on average, the viscosity was found to decrease from about 50 Pas at the shear rate of 1 s^{-1}, to about 4 Pas at 10^4 s^{-1}. The range throughout which a viscosity measurement on a rotational viscometer proved to be best controlled is 10^2 to 10^3 s^{-1}. The dependence of viscosity on the shear rate is more marked at low temperatures. It follows from this that the agglomeration of the pigment particles is the primary cause of the shear rate dependence of the viscosity. European offset news inks vary markedly in their viscosity level (Table 2). The differences are most pronounced at low shear rates.

Within the group of news inks, viscosity and tack are separate properties, despite their strong correlation. The correlation, taken together with the influence of the ink film thickness and the speed on tack (Table 2), bear out the assumption (cf. Table 1) that tack is primarily controlled by hydrodynamic flow in the nip, i.e. the product of viscosity and velocity. The closest agreement was achieved when viscosity was expressed at a shear rate of about 10^3 s^{-1}. The differences in the level of the surface energy of the inks were relatively small (Table 2), and it is unlikely that these represent a significant variable as far as variations in tack are concerned.

The readings of tack decreased with an increase in the ink film thickness, and increased with a rise in the speed; however, the order of ranking of the inks remained unchanged. From this, it can be inferred that the choice of measuring conditions for tack is not a critical factor. As regards the dependence of tack upon the time of running on the tackmeter, considerable differences were apparent between the inks. However, in absolute tack units, the increase in tack as a function of time is small; during the 10 minute-long period on the Tack-0-Scope, the increase in tack averaged about 30%.

Internations between printing ink and paper

A characteristic of ink transfer to the paper in the laboratory test was a high percentage of transfer and small differences between the inks, despite the pronounced variations in the viscosity. The form of the ink transfer curves suggests that a considerable fraction of the quantity of ink transferred becomes immobilized in the pores. This implies that the ink transfer is principally controlled by the ratio of the ink film thickness and a one-dimensional measure of the pore volume. It remains to be investigated whether this situation prevails with other qualities of newsprint, and whether variations in the compressibility of the paper surface enter as a significant variable. It may be noted that in the present case, deficient contact formation in the nip between the paper surface and the ink layer - assessed from the ink transfer curves - did not restrict the ink transfer at practical ink-film thicknesses.

It was found that variations in the measured optical print properties, ink requirement and print-through, were of the order of 10%. It is unlikely that the differences in the optical properties of the inks are as small as this. It is more probable that, in addition to the optical parameters of the inks, it is the physical interaction with the paper structure which significantly controls the optical density. The penetration of ink into the paper structure - about 4 g/m^2 would be needed to fill the available surface pore volume of the paper - and the inevitable variation in the ink film thickness, make the ink less efficient in the production of density. The increase in density above the level of 0.95, for which about 2 g/m^2 of ink was required, occurred only slowly. The often-quoted density level of 1.4 for a matt surface was not attained at reasonable ink film thicknesses.

The show-through level of the paper imposes a limit on the lowest print-through which is attainable. On the grade of newsprint used, the print-through value resulting from the show-through component was about 0.03. This is about one half of the average print-through value at 0.95. The variations between the inks were largely attributable to the variations in the quantity of ink on the paper at a fixed level of density.

Set-off measured at a short time interval in one impression is largely an ink-transfer phenomenon. In the present study, it was mainly controlled by the quantity of the ink on the paper. Some influence of the viscosity was also discerned in that a higher ink viscosity implied more set-off. This arises from slower absorption into the paper of a more viscous ink (cf. Table 1). Nevertheless, other factors, for which no explicit explanation can be made, also played a part. It is extremely probable that one such factor is the rate at which the oil becomes separated from the ink.

As regards the interactive quality factors of print, rub-off displayed the greatest variation (cf. Table 2). As physical ink properties cannot explain this, it must be concluded that it is the differences in ink composition which constitute the underlying reason. This is understandable, as long term changes in the ink layer have time to occur between printing and the rub-off measurement. These are inevitably controlled by the chemical composition of an ink.

Printing behaviour on a pilot scale

Success in the prediction of ink behaviour in offset printing by means of simple laboratory tests is a question of great practical importance. Many open questions remain as a result of the limited data on the subject.

It has become common practice to assess the result of test printings on the level of the solid density at which the relative contrast between a dark halftone area and a solid is highest, the so-called NCI level (Schirmer and Tollenaar[9]). A maximum in the curve of contrast as

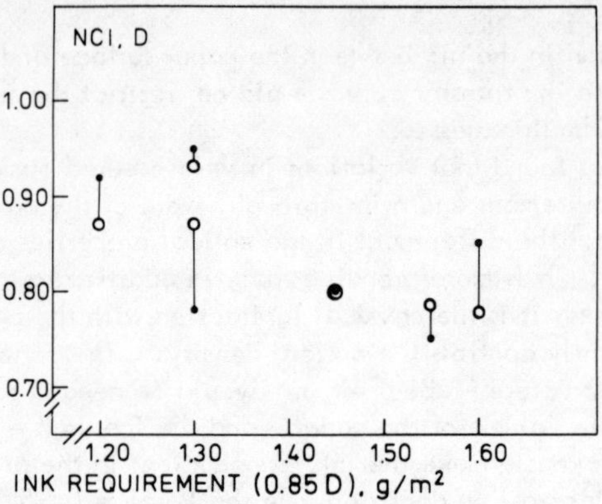

Fig. 1. The relation between ink requirement and the NCl-level. Circles represent a linear regression model which accounts for the influences of ink requirement and viscosity.

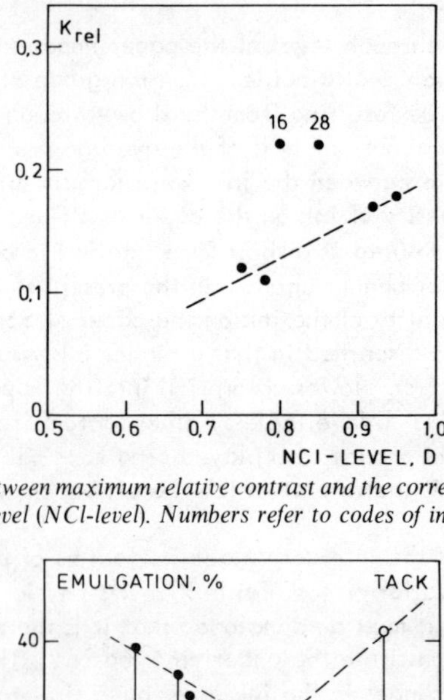

Fig. 2. Relation between maximum relative contrast and the corresponding solid density level (NCl-level). Numbers refer to codes of inks.

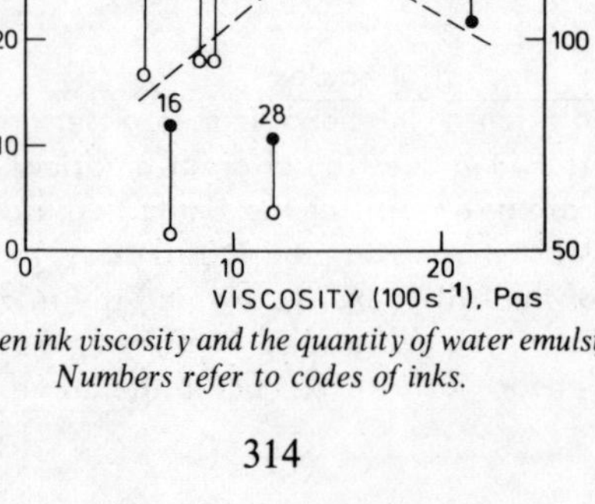

Fig. 3. Relation between ink viscosity and the quantity of water emulsified (●) or tack (○). Numbers refer to codes of inks.

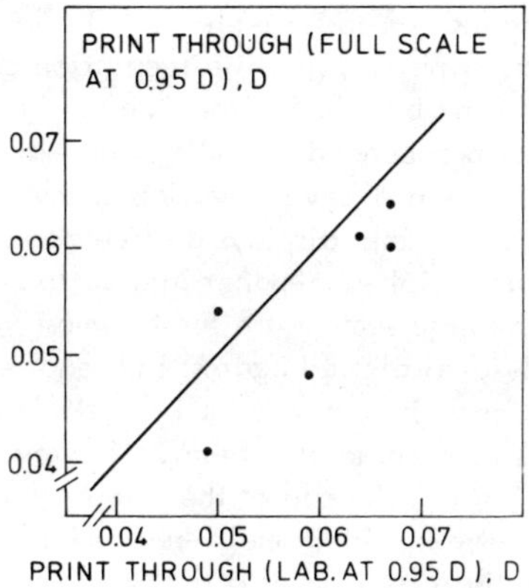

Fig. 4. *A comparison of print-through, measured from laboratory prints and pilot scale prints.*

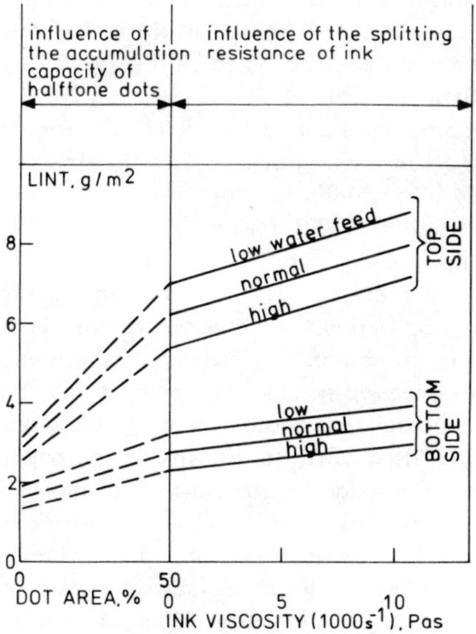

Fig. 5. *A schematic presentation of the contributions to linting of different factors.*

plotted against solid density arises as an interplay of two opposite influences, an increase in contrast predictably by means of the classical models for halftone area density, such as the Murray-Davies and the Yule-Neilsen model, and that of dot spreading. The latter can be of two different types, optical or physical, while separation of the oil from the ink and spreading of the ink both fall within the category of physical spreading. In the case of pronounced spreading, no maximum contrast is obtained. The involvement of several variables makes it unlikely that the NCI-level for a group of inks displays a close correlation with a single ink property or a simple ink-and-paper interaction effect. Nevertheless ink viscosity and ink requirement are the foremost variables. Figure 1 illustrates the NCI-level plotted against ink requirement for six inks. The expected relation, viz: a lower NCI-level with an increase in the ink requirement, is observable as a trend. However, other factors certainly contribute. The influence of the viscosity also become manifest as a trend - as is shown by linear multi-regression calculations (Figure 1). An increase in the viscosity implies a rise in the NCI-level. The mechanism is less ink-spreading (cf. Table 1).

It is instructive to examine the relation between the NCI-level and the respective relative contrast. The values of the relative contrast are far lower than are those predicted by the above-mentioned models for halftones. The models predict relative contrast values in excess of 0.3, whereas those measured are around 0.2. This carries the implication that dot spreading occurs well before attainment of the NCI-level. In fact, dot spreading has been found to be a systematic and continuous phenomenon (Oittinen et al.[7]).

In regard to the behaviour of the different inks, Figure 2 illustrates that they fall into two groups. Examination of the ink properties discloses that the inks which are characterized by a relatively higher contrast are such that they did not emulsify much water in the laboratory test, but had a low tack (Figure 3). It may perhaps be that the surface water in the case of these inks is large in magnitude and when transferred to the paper, acts by filling the surface pore volume, and reducing the spreading of the halftone dots. However, the role of tack is less obvious. Further studies are needed to elucidate physical halftone dot formation.

Print-through was evaluated at the NCI-level, and at a constant density. As the quantity of the ink on the paper is the major factor which governs print-through, a correlation between print-through measured in laboratory conditions at a constant density cannot be expected to be indicative of NCI-level print-through. When compared at a constant density, the order in which the laboratory test and the pilot test ranked the inks was roughly the same (Figure 4). Nevertheless, the values of the pilot scale print-through tended to be lower. Apparently the presence of water exerts a reducing effect upon ink penetration.

Pilot scale set-off is a property of runability. By reason of the cumulative nature of set-off on the press, the quantity of ink on the paper is of minor significance, as opposed to its role in the one-impression set-off test in the laboratory. The influence of the viscosity was similar; an increase in viscosity implied more set-off.

All the materials involved, viz: the printing ink, the paper and the dampening water, contributed significantly to the amount of lint which collected on the blanket. In magnitude, variations in ink viscosity and those in the water-feed, within practical limits, seemed to exercise almost equal influence upon the quantity of lint. The situation is presented schematically in Figure 5.

CONCLUSIONS

According to the measurements made, black offset news inks of widely different properties are used in commercial production. In view of the optical printing result achieved on solid areas, the properties of the inks are not, however, excessively critical with regard to variations in the final results. Halftone dot formation, on the other hand, proved to be more sensitive to ink properties; the mechanisms require further study. With respect to the factors of mechanical print quality, considerable variations between the inks occurred in rub-off - indicative of variation in the composition of the inks. Furthermore, set-off appears to be closely related to ink properties.

In regard to the relations between laboratory-scale and pilot-scale printing behaviour of the inks, the results obtained gave evidence of a role being played by the dampening water as a modifying factor. The hypothesis was made that the presence of water reduces the magnitude of the interactions between the ink and the pore structure of the paper.

The quality control of news inks can be justified not only on quality aspects, but also on economic grounds. The results obtained have provided a basis for the recommendation of some methods for the routine control of black offset news inks. The methods concern measurement of the rheological properties of the inks - and additionally, their lithographic behaviour - together with laboratory-scale printing behaviour in the form of ink requirement, print-through and rub-off. A compilation of the methods is given in Table 4.

TABLE 4 Methods recommended for the routine control of black offset news inks

Property	Method	Significance
Rheological		
Viscosity (Pas)	Cone and plate geometry; 23°C; accelerating curve; 100 and 1000 s^{-1}	Controls flow behaviour on the press
Influence of temperature	10°C and/or 40°C	As required
Tack	Rotational tackmeter (Tack-0-Scope); 23°C; 5 m; 2 and 4 m/s	Controls splitting resistance
Lithographic		
Emulsification (%)	Emulsification of water in the ink in the presence of excess water; time dependence	Controls emulsification and water balance on the press
Interfacial tension (mN/m)	A 20% solution of ink in xylene. Pigment removed by means of centrifuging. Tensiometric determination	Correlation with water transfer. A low interfacial tension favours emulsification of water in the ink
Printability		
Ink requirement (g/m^2)	IGT; rubber-covered disc; 4 m/s; 20 kN/m; D, 0.95; SCAN-P36:76	Ink mileage
Print-through (D)	D, 0.95; SCAN-P36:76	Penetration of ink and its components into paper
Rub-off (D)	D, 0.95; SCAN-P36:76; PATRA tester; 2 kPa; 100 rev.	Long-term ink setting

REFERENCES

1. Laraignou, R., 'Interfacial Tension Measurements in Offset Inks Related to Halftone Printing in Offset', Advances in Printing Science and Technology, Vol. 3 (Ed. W. H. Banks), Pergamon Press, Oxford (1964).

2. Lindqvist, U., Mainander, S.,'A Pilot Test for Linting Propensity of Papers', TAPPI Printing and Reprography Conference, Chicago, Proceedings (1980).

3. Lindqvist, U., Karttunen, S., Virtanen, J., 'New Models on Offset Lithography'. This conference (1981).

4. Lindqvist, U., Oittinen, P., Study of Black Offset News Inks, STK-IFRA-NATS Project, Otaniemi, (1980) (In Swedish).

5. Monni, O., 'Measurement of Paper Linting on an Offset Press', Graphic Arts Research Institute, Report 5/1973 (1973) (In Finnish).

6. Oittinen, P., 'Fundamental Rheological Properties and Tack of Printing Inks and their Influence on Ink Behaviour in a Printing Nip', Acta Polytechnica Scandinavica, Chemistry Including Metallurgy Series No. 131, Helsinki (1976).

7. Oittinen, P., Saarelma, H., Kuosmanen, R., 'A Laboratory Scale Procedure for Halftone Printing', Graphic Arts Finland 9. 1:11-15. (1980).

8. Oittinen, P., 'Limits of Microscopic Print Quality'. This conference (1981).

9. Schirmer, K.-H., Tollenaar, D., 'Optical Densities of Halftone Prints', Proceedings of the 11th International Conference of Printing Research Institutes (Ed. W. H. Banks), Keppler Verlag, Frankfurt am Main, (1973).

10. Surland, A., 'A Laboratory Test Method for Prediction of Lithographic Ink Performance', TAGA-Proceedings, (1980).

DISCUSSION

M. H. Bruno Did you consider the pH of the fountain solution in your studies? Alkaline fountain solutions are used in the U.S. to reduce linting tendency. Also emulsified inks are used which produce sharper halftone reproduction.

P. Oittinen The pilot scale printing tests were made using the type of fountain solution for each ink as recommended by the ink manufacturer.

J. L. Poujade (1) What ink/water emulsification test did you use?

(2) What was the geographical coverage of your survey and how even was the response?

P. Oittinen The emulsification test used is a high speed (25000 rpm) mixing test. The water pick-up of the ink is determined gravimetrically against a blank. Tests were also made according to the method developed at Sun Chemicals.

The inks tested originate in western European countries and are used in editions of newspapers of different size. We consider the collection to be representative of European offset news inks.

F. Andresen The newspapers are today very concerned about quality and costs.

(1) How do you transmit the knowledge from your project to the newspapers?

(2) Has your report led to any changes in the newspapers choice of ink, or in the ink manufacturers formulations?

(3) Have you determined which of the 30 inks are the most economical for the newspapers?

P. Oittinen

(1) The final report of the project has been circulated to newspapers in Scandinavia via the Scandinavian Newspaper Technical Cooperation Council and IFRA - one of the commissioners of the project - is aware of the results.

(2) The project has been completed only recently, for which reason we may have to wait for its impact.

(3) The price of an ink seems to be related to a marked extent to commercial factors besides aspects of quality.

Y. Bery How can you get a decrease in transfer and an increase in density with increasing ink viscosity?

J. M. Fetsko The relationship between viscosity, ink transfer and print density depends on how viscosity was developed. If you increase ink viscosity by pigment addition, print density may well increase even though transfer decreases. If you increase ink viscosity with a high viscosity liquid component, decrease in ink transfer should be reflected in a corresponding decrease in print density.

P. Oittinen I indicated an increase in the NCI-level with an increase in viscosity; not in ink transfer.

INKS FOR THE EUROPEAN SCALE PROPERTIES AND APPLICATION

W. Hansen, Scandinavian Paint and Printing Ink Research Institute, Denmark

ABSTRACT

Properties of various sets of European Scale inks have been investigated in connection with a project carried out for the IARIGAI Standardization Group. The results are presented and discussed. It concerns rheological properties of the inks, determination of optimum ink thickness, inking levels, density at optimum ink film thickness, gloss, and transparency. Results of full scale printing are discussed in relation to the laboratory tests.

Many factors influence printing ink properties, including the properties of the pigment and the binder. The influence of various binders on the behaviour of pigments of the same type from various manufacturers is demonstrated and discussed.

INTRODUCTION

Inks for the European Scale are generally considered, i.e. the so called 'standard inks'. The only thing standardized is, however, their colour when printed on a selected 'standard paper'. Indirectly this means a certain standardization of the colour strength of the inks, since the standard colours must be obtainable within a defined range of ink film thickness.

Looking for the possibility of standardizing the offset process, from original to final print, the ink naturally plays an important role. A working group among IARIGAI institutes has for some years discussed this problem. A part of these discussions has dealt with the properties of European Scale inks. In order to obtain more information about these inks a project was carried out by SPPIRI, investigating properties of four Scandinavian sets of European Scale inks (yellow, magenta and cyan inks). The laboratory investigations were, with the assistance of FOGRA, supplemented by printing on a commercial press.

This report briefly summarizes the results of these investigations. Furthermore, some results from investigations concerning the influence of the binder on ink properties are presented. These results explain some of the variations which can be expected among European scale inks of different origins.

LABORATORY INVESTIGATIONS AND RESULTS

The investigation of commercial European Scale inks included four sets of inks manufactured by different major ink companies in Scandinavia. The measurements included:

(1) Rheological properties (viscosity and yield value)
(2) 'Optimal ink film thickness'
(3) Density of 'optimal ink film thickness'
(4) Gloss
(5) Transparency.

In what follows, the results of these investigations are indicated and discussed.

Viscosity and Yield Value

Viscosity and yield value were measured on an adjusted falling bar viscometer (Laray) at shear stresses from 10,000 to 110,000 dynes/cm^2. The results are presented in Table 1.

TABLE 1 Viscosity and yield value for four sets of European Scale inks

Ink set / Ink colour	Viscosity (Poise)*				Yield value (dynes/cm^2)**			
	A	B	C	D	A	B	C	D
Yellow	204	382	224	346	7,600	2,400	9,600	9,100
Magenta	455	474	248	331	11,000	7,100	11,300	7,300
Cyan	306	358	168	510	6,000	3,200	5,000	1,200
Black	268	633	166	244	4,600	4,200	6,000	3,400

* 10 Poise = 1 Pa·s; **10 dynes/cm^2 = 1 N/m^2

The results of viscosity measurements indicate a relatively large variation in ink viscosity, not only from one ink set to another, but also within the same set of inks. This could lead to conclusions or guesses concerning the ink sequence used or the type of press used (one-, two- or four-colour press). However, the ink companies delivering the inks stated that no ink sequence was recommended for the inks, and the same inks were used on various types of sheet-fed presses.

It is not possible from the data in Table 1 to draw conclusions about ink sequence, etc. However, the viscosity values could indicate that for ink set A, magenta could be used as the first colour, for ink set B, black could be printed first and for ink set D, cyan could be selected as the first colour.

Optimum Ink Film Thickness

The optimum ink film thickness in this investigation is considered as the ink film thickness where the print is in the best possible accordance with the colour values indicated in the European standard.

Optimum ink film thickness was determined for prints made on 5 different papers:

- N. Standard paper (APCO II/II)
- T. Top-Cote 106, Coated art paper, 135 g/m^2
- U. Uno-Cote 110, Double blade-coated, 115 g/m^2
- S. Semicote 130, Machine coated paper, 100 g/m^2
- E. Tryk E 270, Surface limed, calandered paper, 100 g/m^2

In Table 2, the results of measurements for the standard paper are indicated, demonstrating the accordance of the inks with the European standard.

TABLE 2 Optimum ink film thickness on standard paper and accordance with European standard

Ink set / Ink colour	Optimum ink film thickness (g/m^2)			Colour difference (ΔE_{CEI})		
	Yellow	Magenta	Cyan	Yellow	Magenta	Cyan
A	0.70	1.08	1.33	2.3	3.0	6.2
B	0.85	1.08	0.96	3.0	6.5	2.0
C	0.70	1.19	1.27	3.0	5.8	3.8
D	0.83	1.04	1.31	1.6	4.9	2.8
European standard				2.3	5.0	3.0

The results demonstrate that the optimum ink film thicknesses vary considerably. The results also illustrate that all ink sets are in accordance with the European standard, considering that an error in the measurements corresponding to $\Delta E \pm 2–3$ units can be expected.

Various reasons for using different ink film thicknesses could be presented. One reason is that the ink sequence used influences the choice of inking level. Another reason could be use of the colour strength of the ink to adjust the print colour. Weakening the colour strength of, e.g. cyan, and still printing with the same ink film thickness, leads to 'warmer' colours of the print. The wish for a 'warmer' European Scale has often been put forward by some Scandinavian printers.

For prints on types of paper other than the standard paper, agreement with the European standard is not expected, and the standard does not refer to the demand for agreement for papers other than the standard.

However, the optimum ink film thickness, indicating the best possible agreement with the standard, was determined, and these values and the colour differences are indicated in Appendix 1.

The optimum ink film thickness of various inks printed on various papers are indicated in Table 3.

TABLE 3 Optimum ink film thickness (g/m^2) for prints on various papers

Ink colour	Manufacturer	Paper N	T	U	S	E
Yellow	A	0.75	0.77	0.92	1.38	1.33
	B	0.85	0.92	0.96	1.43	1.38
	C	0.70	0.70	0.88	1.31	1.31
	D	0.83	0.65	0.88	1.42	1.42
Magenta	A	1.08	1.08	1.23	1.38	1.65
	B	1.08	1.12	1.38	1.54	1.54
	C	1.19	1.08	1.27	1.50	1.38
	D	1.04	1.08	1.19	1.54	1.54
Cyan	A	1.33	1.16	1.33	1.33	1.33
	B	0.96	0.92	1.00	1.42	1.31
	C	1.27	1.20	1.23	1.46	1.38
	D	1.31	1.31	1.31	1.50	1.70

An overall overage of the results indicates that the inking levels on the various types of paper should be approximately as in Table 4.

TABLE 4

Ink colour	Inking level on paper: N	T	U	S	E
Yellow	X	X $\pm$ 0	X + 20%	X + 80%	X + 80%
Magenta	X	X $\pm$ 0	X + 15%	X + 40%	X + 40%
Cyan	X	X − 5%	X $\pm$ 0	X + 20%	X + 20%

It must be remembered that indication of these 'optimum ink film thicknesses' and inking levels only refer to the colour of the print, i.e. the possibility of obtaining colours close to the European scale. Other factors, such as printing conditions and paper properties, will normally mean that the inking level used in practice does not lead to the 'optimum ink film thickness', but to an optimum print in other respects.

Gloss

The gloss of prints at optimum ink film thickness was measured at a 60° angle using a Multigloss (Mallinkrodt) glossmeter. The results are given in Table 5.

TABLE 5 Gloss of prints at optimum ink film thickness (%)

Ink colour	Manufac- turer	Paper N	T	U	S	E
Yellow	A	59	77	48	18	11
	B	71	80	50	22	14
	C	70	70	43	19	12
	D	65	77	43	24	11
Magenta	A	65	79	53	19	9
	B	68	70	42	18	9
	C	69	78	54	18	11
	D	65	70	46	18	13
Cyan	A	49	63	35	14	9
	B	59	67	38	20	8
	C	56	65	37	15	7
	D	54	59	42	16	8
Average -	Yellow	64	76	46	21	12
	Magenta	67	74	49	18	10
	Cyan	55	64	48	16	8

While the variation of gloss between inks of different manufacture is relatively small and hardly significant (less than ± 10%), the results indicate the highest gloss on commercial coated art paper (T), a little lower gloss on standard paper (N), and a significantly reduced gloss on blade-coated paper (U). As expected, the gloss of prints on machine-coated paper (S) and calandered paper (E) are very low.

Transparency

Transparency of the inks as determined according to a method developed at SPPIRI (see Appendix 4), indicated the following results:

TABLE 6 Transparency of inks investigated

Ink manufacturer	Colour Yellow	Magenta	Cyan
A	76%	91%	89%
B	85%	85%	93%
C	86%	88%	88%
D	84%	90%	91%
Average ± std. dev.	83% ± 5%	89% ± 3%	90% ± 2%

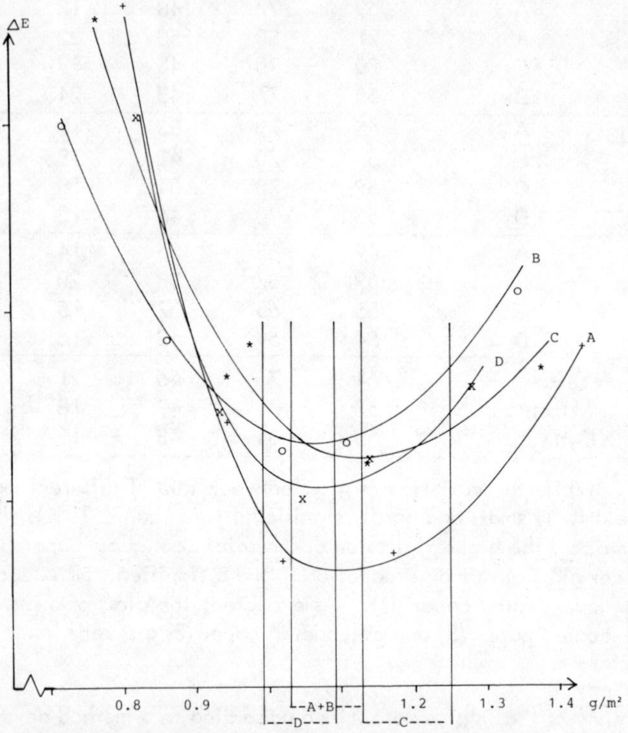

Fig. 1. Colour difference (ΔE_{CIE}) as a function of ink film thickness (g/m^2) for 4 magenta inks indicating corresponding density ranges ($\pm 5\%$) for the individual inks, when printed on standard coated art paper (APCO II/II).

The results illustrate a transparency for yellow, which is a little lower than for magenta and cyan (as expected). Variations of transparency between inks of different manufacture are not significant.

Density at Optimum Ink Film Thickness

The density of all prints with optimum ink film thicknesses was measured using a Macbeth RD-100R densitometer. The results are given in Table 7.

TABLE 7 Density of prints with optimum ink film thickness

Ink colour	Manufacturer	Paper N	T	U	S	E
Yellow	A	0.81	0.91	0.93	0.90	0.82
	B	0.82	0.90	0.88	0.93	0.82
	C	0.79	0.84	0.88	0.83	0.77
	D	0.92	0.88	0.85	1.05	0.87
Magenta	A	1.30	1.32	1.34	1.18	0.94
	B	1.31	1.32	1.34	1.18	0.94
	C	1.40	1.36	1.38	0.98	0.90
	D	1.30	1.33	1.36	1.12	1.08
Cyan	A	1.46	1.47	1.46	1.05	0.93
	B	1.70	1.65	1.38	1.36	1.08
	C	1.56	1.58	1.44	1.30	0.94
	D	1.55	1.64	1.58	1.24	1.13

The results show that a variation of density levels of at least $\pm$10% between inks of the same colour printed at optimum ink film thickness must be expected. The differences are caused by variations in the raw materials used in inks from different manufacturers.

Relationship between ink film thickness, minimum colour difference and density

Density measurements are frequently used in the press room to control inking level during printing. When printing with European Scale inks, the inking level should normally correspond to the ink film thickness representing the smallest possible difference from the European standard.

From curves of colour difference (ΔE_{CEI}) as a function of ink film thickness the inking level representing minimum ΔE_{CEI} can be found. Table 8 shows minimum colour differences and corresponding ink film thickness and density for 4 commercial European Scale ink sets printed on standard paper. Furthermore, variations in ink film thickness at various density tolerances are indicated. These values are calculated from the measurements indicated in Appendices 2A and 2B. From these measurements the curves of Figure 1 are made for the magenta inks.

When printing at minimum ΔE, different inking levels must be used for the different inks, depending upon colour strength and colour hue of the inks. As seen from Figure 1, different density levels must also be used. For inks A and B the inking level is from 1.03 to 1.13 g/m^2, corresponding to a density range of 1.29 $\pm$ 5%. The inking level for ink D is from 0.99 to 1.09 g/m^2, corresponding to a density range of 1.25 $\pm$ 5%. This means that if a density level of 1.29 $\pm$ 5% is standardized, ink D can be used, just as well as inks A and B.

TABLE 8 Minimum colour difference and corresponding ink film thickness and density for 4 commercial European Scale ink sets (A – D) printed on standard paper (APCO II/II)

Ink		$\Delta E_{min.}$	g/m^2	Density	Variation in ink film thickness for density tolerances of $\pm$ 1%	$\pm$ 2%	$\pm$ 5%
Yellow	A	2.3	0.70	0.80	0.68-0.72	0.66-0.74	0.61-0.79
	B	3.0	0.85	0.83	0.83-0.88	0.80-0.90	0.71-0.95
	C	3.0	0.70	0.80	0.68-0.72	0.66-0.74	0.61-0.79
	D	1.6	0.83	0.93	0.82-0.84	0.80-0.85	0.76-0.89
Magenta	A	3.0	1.08	1.29	1.06-1.10	1.05-1.11	1.03-1.13
	B	6.5	1.08	1.29	1.06-1.10	1.05-1.11	1.03-1.13
	C	5.8	1.19	1.40	1.17-1.21	1.16-1.22	1.13-1.25
	D	4.9	1.04	1.25	1.02-1.06	1.01-1.07	0.99-1.09
Cyan	A	6.2	1.33	1.46	1.31-1.35	1.30-1.36	1.26-1.40
	B	2.0	0.96	1.65	0.94-0.98	0.93-0.99	0.87-1.05
	C	3.8	1.27	1.53	1.23-1.31	1.21-1.33	1.13-1.41
	D	2.8	1.31	1.53	1.19- ??	1.15- ??	1.08- ??

Ink C, however, demonstrates an inking level of 1.13 to 1.25 at minimum colour difference for a density range of 1.40 $\pm$ 5%. This means that in spite of the good agreement with the European Scale, the inking level needed for ink C is not suitable for a standard density range of 1.29 $\pm$ 5%, unless a larger colour difference is accepted. The conclusion could be the following: The density level shall be standardized and ink C must be reformulated or rejected.

Making the same calculation for the same magenta inks when printed on a double-coated paper (Table, Appendix 2), the curves indicated in Figure 2 are obtained.

From these curves and the density levels indicated it is seen that ink A when printed in ink film thicknesses from 1.17 to 1.29 g/m^2 has a density range of 1.36 $\pm$ 5%. Inks C and D printed in approximately the same inking level (1.21 to 1.33 g/m^2 for ink C and 1.13 to 1.25 g/m^2 for ink D) will have density values within the same range as ink A (1.38 $\pm$ 5% and 1.36 $\pm$ 5%). For ink B, however, the inking level is considerably higher (1.31 to 1.45 g/m^2) if minimum colour difference is to be established. The corresponding density range is 1.32 $\pm$ 5%, and ink B is, in principle, not suitable for printing on double blade-coated

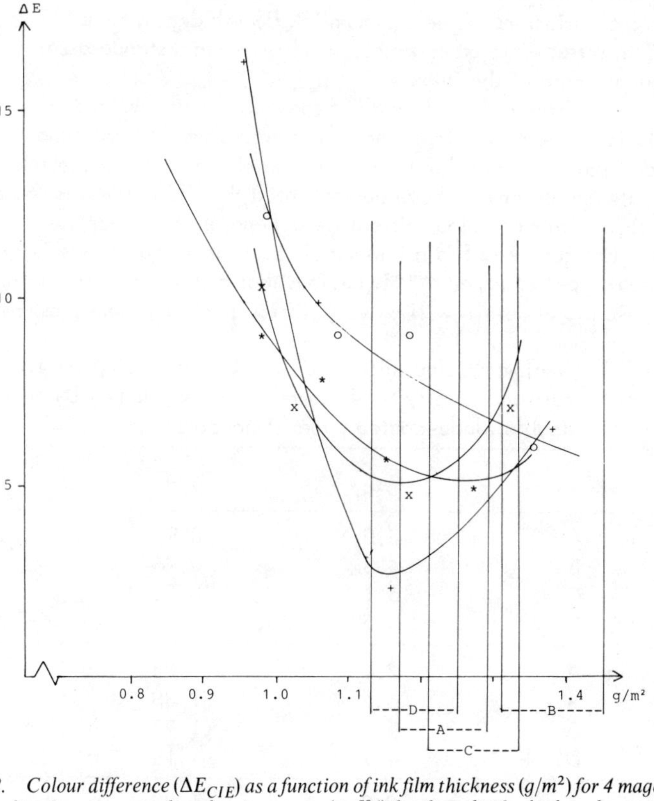

Fig. 2. Colour difference (ΔE_{CIE}) as a function of ink film thickness (g/m^2) for 4 magenta inks indicating corresponding density ranges ($\pm 5\%$) for the individual inks, when printed on double blade-coated paper (Uno-cote).

paper, if minimum colour differences are to be established.

Corresponding calculations can be made for yellow and cyan inks, indicating that minimum colour differences for various inks are obtained not only at different inking levels but also at different density levels for various inks.

If density ranges are standardized for European Scale inks, this would also mean a standardization of the optimum ink film thickness, meaning the ink film thickness at minimum colour differences from the European standard. This would mean a change in the European standard allowing variations of the optimum ink film thickness from 0.7 to 1.1 g/m^2 in offset. In other words, it would mean a standardization of the colour strength of the inks.

If the colour strength of the inks is standardized, a number of limitations would be introduced in the press room. Furthermore, this standardization would limit the selection of pigments for the inks. Finally, the measurements have demonstrated that inking levels and density ranges at minimum colour difference depend on the paper used. This means that European Scale inks should be formulated individually for individual types of paper. It is obvious that a standard introducing many limitations and details will be very difficult to use in the press room.

TABLE 9 Minimum colour difference and corresponding ink film thickness and density for 4 commercial ink sets (A - D) printed on double blade-coated paper (Uno-cote)

Ink		ΔE_{min}	g/m^2	Density
Yellow	A	3.0	0.92	0.87
	B	4.3	0.96	0.82
	C	5.1	0.88	0.86
	D	4.5	0.88	0.82
Magenta	A	2.4	1.23	1.36
	B	6.2	1.38	1.32
	C	5.0	1.27	1.38
	D	4.9	1.19	1.36
Cyan	A	8.0	1.33	1.49
	B	5.3	1.00	1.50
	C	5.6	1.23	1.45
	D	5.0	1.31	1.36

PRINTING ON A COMMERCIAL PRESS

With the assistance of FOGRA, test printing was carried out under the following conditions:

Press: Heidelberg Speedmaster 102 Z, two-colour, 71 x 102 cm

Density control: CPC II, Print Control Device

Papers: APCO II/II, 150 g/m^2
Coated art paper, Ikonolux, 120 g/m^2
Calandered paper (Naturpapier), 100 g/m^2

Printing speed: 4000 impressions per hour

Adjustment of ink film thickness (density) was carried out using a matt offset paper. Deviations in paper properties between this paper and the paper used for the tests may have caused smaller deviations in print density between the first prints and the last prints of each impression. Prints on APCO II/II were made by inserting a number of sheets in the stack of art paper.

The inking unit of the press was prepared for printing two inks simultaneously in the same inking unit. The printing conditions are given in Table 10.

TABLE 10

Ink colour	Ink code left	right	Test No:	Density full tone	Inking unit	Damping solution
Cyan	C	B	01	1.70	59	5.1
			02	1.80	64	5.5
			03	1.90	70	5.7
Magenta	A	B	01	1.60	62	6.3
			02	1.70	71	7.0
			03	1.80	87	7.8
Yellow	D	B	01	0.80	27	4.0
			02	0.85	29	4.1
			03	0.90	31	4.2

Comments The damping solution used contained approximately 10% isopropanol. All inks, except ink B, performed badly with the alcohol solution. The printer has made the following comments concerning each set of inks tested:

Cyan: Ink C differs considerably from ink B in its demand for damping solution, making it impossible to obtain prints of good quality in simultaneous printing. Ink C emulsified in the alcoholic damping solution. Adjustment of damping solution is made relative to the printing of ink B.

Magenta: Both inks A and B demonstrate a high consumption of damping solution, especially ink A. By increasing the print density, a relatively large increase in damping solution supply was needed.

Yellow: At the relatively low ink film thickness used for these prints, no problems appeared.

Originally magenta ink D was selected for printing together with ink B. However, magenta D emulsified in the damping solution to such a high degree that printing was impossible.

Water sensitivity

Sensitivity of the inks towards the damping solution was tested for magenta inks A and B. Tests included adjustment of the damping solution supply in steps of 0.25 from 5.75 to 7.25, and further adjustments to 8.0, 9.25 and 10.0.

The results can be evaluated visually from the test prints. Magenta ink B, as distinct from ink A, was not very sensitive to the damping solution adjustment. Ink B only demonstrated higher covering at low adjustment of damping solution. Ink A demonstrated smearing at low adjustment (5.75) and water marks at high adjustment (10.0) of the damping solution supply.

Accordance with the European standard

For yellow inks, minor increases in the deviation from the European standard has been found for the full scale printing compared with laboratory tests. For magenta and cyan inks, the deviations from the European standard in full scale printing are not significantly different from the results found in laboratory testing.

Minimum colour difference and density level

From the commercial prints on standard paper (APCO II/II) colour measurements and density measurements were carried out and compared with the laboratory measurements. The results are given in Table 11 (also Appendix 3).

TABLE 11 Minimum colour difference and density for European Scale inks printed on a laboratory press and on a commercial press

		Laboratory press			Commercial press	
Ink		ΔE_{min}	Density	Ink film thickness g/m^2	ΔE_{min}	Density
Yellow	B	3.0	0.83	0.85	4.8	0.84
	D	1.6	0.93	0.83	3.7	0.81
Magenta	B	6.5	1.29	1.08	6.5	1.34
	A	3.0	1.29	1.08	5.6	1.38
Cyan	B	2.0	1.65	0.96	1.6	1.69
	C	3.8	1.53	1.27	4.5	1.61

From the results shown in Table 11 it is seen that commercial prints, with the exception of Yellow D, are printed at a higher density level than the laboratory prints. In two cases (Magenta A and Cyan C) the density level for commercial printing would be outside the density range of ±5% relative to a density level based on laboratory measurement.

It is also noted that the minimum colour difference obtained in commercial printing generally is higher than the colour difference found for laboratory prints. One probable reason for this is that the laboratory prints are made without the use of damping solution contrary to the commercial prints.

Dot gain

Ont of the factors used in evaluating print quality is the dot gain. Dot gain in the commercial prints was measured for one sheet of each impression according to the method indicated by FOGRA. The results are presented in Appendix 3.

The results were rather confusing, and are commented on as shown in Table 12. The results of these dot gain determinations do not permit the possibility of establishing inking levels on the basis of dot gain measurements.

TABLE 12

Ink	Code	Dot gain 80% area	Dot gain 50% area
Yellow	B and D	Not significant (1-6%) Independent of ink type, inking level and paper	Negative (2-20%) As for the 80% area
Magenta	B	Dot gain 13-17% Increase proportional to inking level Independent of paper	Dot gain 3-12% As for the 80% area
	A	Dot gain 10-13% Independent of inking level and paper	Not significant dot gain from +5 to +3% Independent of inking level and paper
		Ink A was found very sensitive to the damping solution	
Cyan	B	Dot gain 9-14% Increase with inking level Independent of paper	Not significant dot gain from +2 to +5% Independent of inking level and paper
	C	Dot gain 13-20% Increase with inking level Independent of paper	Dot gain 5-17% As for the 80% area

Discussion

Printing on a commercial press has primarily given an excellent illustration of the difficulties in using laboratory measurements to predict press performance of printing inks. Most interesting is probably the deviation of print density between laboratory made proof prints and commercial prints, and the fact that colour differences relative to the European standard are higher in commercial printing than in laboratory printing, when standard paper is used.

Whether dot gain is a suitable measurement for inking level is questionable on the basis of the results obtained. However, it must be admitted that the number of tests made are far from sufficient for making final conclusions.

The use of damping solution on the commercial press contrary to the laboratory press undoubtedly influences the optical properties of the print to such a degree that parallel conclusions from laboratory measurements to commercial printing are very difficult to make.

BINDER INFLUENCE ON INK PROPERTIES

The formulation of printing inks involves a number of considerations in the selection of pigments and binders. One of the factors influencing the final result is the influence of the binder on the pigment properties. This investigation has primarily concerned the influence of the binder on the optical properties of the printing ink. In the following, some of the results obtained for three magenta pigments, each dispersed in five different binders are presented.

Optical properties, binder influence

The investigation of 3 pigments dispersed in 5 different binders included determination of relative colour strength, transparency and gloss. The results are given in Table 13.

From the results it is seen that different binders lead to different properties of the inks, as could be expected. It is a little more difficult to find a trend for the binder influence, e.g. optimum colour strength for the same binder using different pigments. In other words, the binders seem to influence the pigments in different ways. Comments on the individual properties are as follows:

Colour strength: normally high for pigments dispersed in binders 3 and 5, and relatively low for pigments dispersed in binder 1. The deviations are, however, hardly significant to any higher degree.

Transparency: normally high in binder 3, but primarily dependent upon the pigment itself. The pigments G and I demonstrate a high transparency and the pigment H a relatively low transparency. The binder has only a minor influence on this property.

Gloss: measured in two ink film thicknesses, and binder 5 clearly demonstrates the highest gloss for all three pigments, while binder 4 generally leads to the lower gloss. It is evident that the pigment also influences the gloss of the ink, since all inks based on pigment H demonstrate a significantly lower gloss than do the rest of the inks.

TABLE 13 Optical properties of 3 magenta pigments dispersed in 5 different binder systems

Pigment	Binder	Colour strength relative	Transparency %	Gloss % 1 g/m^2	Gloss % 1.5 g/m^2
	1	0.88	91	70	67
	2	0.97	92	75	72
G	3	1.00	100	82	80
	4	0.89	88	55	54
	5	0.98	100	81	79
	1	0.95	78	41	43
	2	1.02	79	47	42
H	3	1.00	89	48	42
	4	0.96	77	44	37
	5	1.09	90	65	37
	1	0.89	87	62	55
	2	0.89	99	71	70
I	3	1.00	100	71	72
	4	0.98	90	45	47
	5	1.01	89	80	86

Optimum ink film thickness and colour strength

The pigments selected should all be useful for the manufacture of European Scale inks, which was confirmed by the measurements. Table 14 shows the optimum ink film thickness for the individual inks, the minimum colour difference from the European Scale (ΔE_{min}), and the colour strength of the inks.

TABLE 14 Optimum ink film thickness, ΔE_{min}, and relative colour strength for 15 inks (3 pigments each dispersed in 5 binders)

Binder	Pigment G			Pigment H			Pigment I		
	g/m^2	ΔE_{min}	Colour strength	g/m^2	ΔE_{min}	Colour strength	g/m^2	ΔE_{min}	Colour strength
1	0.91	0.7	0.88	1.15	4.7	0.95	0.95	4.4	0.89
2	0.92	3.8	0.97	1.12	5.1	1.02	0.88	5.6	0.98
3	0.92	3.8	1.00	1.26	4.8	1.00	0.80	4.0	1.00
4	0.91	2.7	0.89	1.25	4.5	0.96	0.89	4.4	0.98
5	0.94	3.1	0.98	1.05	4.6	1.09	0.85	2.7	1.01

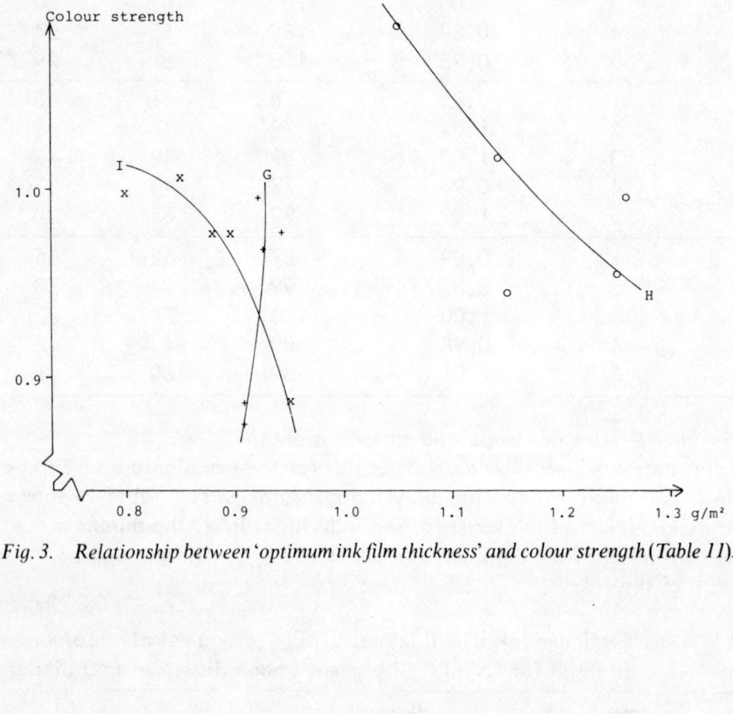

Fig. 3. Relationship between 'optimum ink film thickness' and colour strength (Table 11).

From the results given in Table 14 and illustrated in Figure 3, it is seen that different pigments behave differently in the same binder system. All the pigments in all the binder systems demonstrate agreement with the European Scale. For pigment G, the optimum ink film thickness is independent of the binder used and of the colour strength of the ink system. Pigment I demonstrates a small dependence of colour strength and optimum ink film thickness, meaning that increased ink film thickness is required, when colour strength is decreasing. For pigment H this dependence between ink film thickness and colour strength is more significant and, furthermore, a higher inking level is generally required, corresponding to a generally lower colour strength of this pigment than for pigments G and I.

Discussion

Colour strength of an ink or a pigment is determined by the optical properties of the ink or the pigment: the colour hue and saturation on one side, and the content of black on the other side. When ink film thickness is determined on the basis of densitometer measurements, the black content or the 'dirtiness' of the ink will have a dominating influence.

For pigment G it was found that the optimum ink film thickness is independent of the colour strength of the ink (or pigment). This means that establishment of a density value would lead to different printing results depending upon the colour strength of the ink. Furthermore, it could be expected that for some of the inks, the printing results would not be in agreement with the European Scale, simply because of variations in the colour strength of the inks caused by differences in the binders used.

In general, it seems clear that the binder-pigment relationship influences the optical properties of inks to such a degree and so differently, depending upon the pigment, that general standards can only be established if very narrow specifications for the ink composition are established.

CONCLUSION

The properties of various commercial European Scale inks have been investigated. The results demonstrate that significantly different inking levels must be used for the different inks in order to obtain a minimum colour difference relative to the European Scale. This inking level, furthermore, depends upon the type of paper printed.

The results also demonstrate that density levels for the inks printed in optimum ink film thickness, vary significantly. Printing at fixed density levels can only be done if a number of ink properties are standardized: optimum ink film thickness, colour strength, limited number of pigments, etc. Furthermore, it would be necessary to manufacture different ink sets for different types of paper.

Printing on a commercial press, with the assistance of FOGRA, demonstrated minor differences between laboratory measurements of the optical properties of inks and the result obtained in practice. Furthermore, different inks demonstrated significantly different results in dot gain.

Investigations of the influence of the binder on the optical properties of printing inks have demonstrated that the binder, as expected, influences the colour strength, transparency and gloss of printing inks. On the other hand, the binder system seems in several cases not to influence the optimum ink film thickness significantly.

This means that no correlation can be expected between the colour strength of an ink, the optimum ink film thickness, and the density of the print.

APPENDIX 1 Properties of European Scale inks

Code	Colour	Viscosity (Poise)	Yield Value (dynes per cm^2)	Paper	Optimum ink film thickness (g/m^2)	ΔE CEI	Density	Gloss (%)	Transparency (%)
A	Yellow	204	7,600	N	0.74	2.3	0.81	59	76
				T	0.77	3.6	0.91	77	
				U	0.92	3.2	0.93	48	
				S	1.38	6.1	0.90	18	
				E	1.33	8.1	0.82	11	
	Magenta	455	11,000	N	1.08	3.0	1.30	65	91
				T	1.08	3.8	1.33	79	
				U	1.23	4.0	1.41	53	
				S	1.38	16.0	1.15	19	
				E	1.65	20.0	1.04	9	
	Cyan	306	6,000	N	1.33	6.2	1.46	49	89
				T	1.16	7.6	1.47	63	
				U	1.33	8.2	1.46	35	
				S	1.33	14.5	1.05	14	
				E	1.33	21.0	0.93	9	
B	Yellow	382	2,400	N	0.85	3.0	0.82	71	85
				T	0.92	4.7	0.90	80	
				U	0.96	4.3	0.88	50	
				S	1.42	7.2	0.93	22	
				E	1.38	9.1	0.82	14	
	Magenta	474	7,100	N	1.08	6.5	1.31	68	85
				T	1.12	5.8	1.32	70	
				U	1.38	6.2	1.34	42	
				S	1.54	15.5	1.18	18	
				E	1.54	21.0	0.94	9	
	Cyan	358	3,200	N	0.96	2.0	1.70	59	93
				T	0,92	2.8	1.65	67	
				U	1.00	5.4	1.38	38	
				S	1.42	7.7	1.36	20	
				E	1.31	15.0	1.08	8	

(continued)

APPENDIX 1   Properties of European Scale inks (continued)

Code	Colour	Viscosity (Poise)	Yield Value (dynes per cm^2)	Paper	Optimum ink film thickness (g/m^2)	ΔE CEI	Density	Gloss (%)	Transparency (%)
C	Yellow	224	9,600	N	0.70	3.0	0.79	70	86
				T	0.70	4.6	0.84	70	
				U	0.88	5.2	0.88	43	
				S	1.31	7.8	0.83	19	
				E	1.31	9.8	0.77	12	
	Magenta	248	11,300	N	1.19	5.8	1.40	69	88
				T	1.08	4.4	1.36	78	
				U	1.27	4.8	1.38	54	
				S	1.50	21.0	0.98	18	
				E	1.38	28.0	0.90	11	
	Cyan	168	5,000	N	1.27	3.8	1.56	56	88
				T	1.20	4.4	1.58	65	
				U	1.23	5.4	1.44	37	
				S	1.46	10.5	1.30	15	
				E	1.38	21.0	0.94	7	
D	Yellow	346	9,100	N	0.83	1.6	0.92	65	84
				T	0.65	4.4	0.88	77	
				U	0.88	4.6	0.85	43	
				S	1.42	3.4	1.05	24	
				E	1.42	8.7	0.87	11	
	Magenta	331	7,300	N	1.04	4.9	1.30	65	90
				T	1.08	6.0	1.33	70	
				U	1.19	5.0	1.36	46	
				S	1.54	1.4	1.12	18	
				E	1.54	20.0	1.08	13	
	Cyan	510	1,200	N	1.31	2.8	1.55	54	91
				T	1.31	3.6	1.64	59	
				U	1.31	4.8	1.58	42	
				S	1.50	8.5	1.24	16	
				E	1.70	15.0	1.13	8	

APPENDIX 2A Ink film thickness (g/m^2), ΔE_{min}, and density at various ink film thicknesses
Paper: APCO II/II
Inks: 4 commercial ink sets (A - D)

Ink	Yellow (g/m^2)	Yellow ΔE_{min}	Yellow Density	Magenta (g/m^2)	Magenta ΔE_{min}	Magenta Density	Cyan (g/m^2)	Cyan ΔE_{min}	Cyan Density
A	0.72	2.3	0.81	0.66	22.2	0.95	0.74	16.7	0.93
	0.85	2.8	0.88	0.80	18.9	1.02	0.85	12.7	1.05
	0.94	3.0	0.91	0.94	7.0	1.20	0.92	10.1	1.19
	1.06	7.3	0.97	1.02	3.2	1.26	1.02	8.6	1.27
	1.17	8.3	1.00	1.17	3.4	1.35	1.11	8.0	1.34
	1.31	7.4	1.05	1.26	5.2	1.39	1.26	6.6	1.43
	1.43	8.0	1.11	1.42	8.9	1.45	1.38	6.0	1.48
B	0.69	5.2	0.74	0.72	14.9	1.00	0.68	6.8	1.24
	0.80	2.9	0.80	0.86	9.3	1.12	0.82	3.7	1.46
	0.89	3.1	0.84	1.02	6.3	1.24	0.91	1.9	1.68
	1.00	5.1	0.94	1.11	6.6	1.34	1.03	2.4	1.72
	1.17	6.4	0.99	1.34	10.6	1.42	1.12	4.8	1.79
	1.32	7.4	1.03	1.42	10.9	1.42	1.29	5.9	1.86
	1.42	8.4	1.06				1.38	6.7	1.89
C	0.66	3.0	0.79	0.77	17.5	1.03	0.62	12.8	1.00
	0.82	3.5	0.86	0.86	11.9	1.15	0.75	9.8	1.10
	0.92	4.4	0.90	0.98	9.1	1.18	0.91	6.9	1.26
	1.02	5.7	0.96	1.14	5.8	1.36	1.05	5.1	1.35
	1.11	6.7	1.00	1.25	7.0	1.43	1.14	3.8	1.46
	1.17	7.7	1.05	1.37	8.5	1.48	1.23	3.6	1.55
	1.26	8.7	1.08				1.34	3.9	1.57
D	0.66	5.1	0.81	0.82	15.3	1.06	0.75	9.8	1.07
	0.75	2.8	0.88	0.86	12.0	1.08	0.78	8.7	1.13
	0.86	1.8	0.94	0.94	8.1	1.20	0.86	6.8	1.26
	0.98	3.4	1.02	1.05	5.0	1.30	1.00	5.1	1.36
	1.11	4.9	1.09	1.14	6.0	1.34	1.09	3.8	1.47
	1.22	5.7	1.11	1.28	7.9	1.40	1.23	3.2	1.53
	1.32	7.1	1.09	1.38	9.0	1.42	1.31	3.1	1.55

APPENDIX 2B Ink film thickness (g/m^2), ΔE_{min}, and density at various ink film thicknesses
Paper: Double blade-coated (Uno-cote)
Inks: 4 commercial ink sets (A - D)

Ink	Yellow (Blue filter)			Magenta (Green filter)			Cyan (Red filter)		
	(g/m^2)	ΔE_{min}	Density	(g/m^2)	ΔE_{min}	Density	(g/m^2)	ΔE_{min}	Density
A	0.74	5.0	0.78	0.72	27.7	0.88	0.71	19.2	0.85
	0.85	3.0	0.87	0.82	19.1	1.01	0.82	14.5	1.02
	0.91	3.1	0.93	0.95	16.4	1.06	0.94	13.2	1.05
	1.02	3.9	0.90	1.05	10.0	1.16	1.05	10.2	1.23
	1.12	4.2	0.92	1.14	2.4	1.36	1.12	9.8	1.26
	1.29	6.6	1.05	1.25	3.9	1.42	1.28	8.6	1.40
	1.40	7.1	1.11	1.37	6.5	1.48	1.38	8.0	1.49
B	0.71	9.6	0.67	0.69	26.1	0.94	0.72	11.6	0.97
	0.78	6.4	0.75	0.83	19.1	1.02	0.82	6.8	1.32
	0.91	4.3	0.82	0.98	12.3	1.12	0.95	6.1	1.34
	1.02	4.7	0.93	1.08	9.3	1.19	1.03	5.7	1.41
	1.14	6.1	0.96	1.18	9.2	1.24	1.17	5.3	1.50
	1.28	6.7	1.02	1.35	6.2	1.32	1.31	7.6	1.91
	1.43	6.4	1.00				1.38	7.8	1.84
C	0.69	6.0	0.75	0.65	25.2	0.98	0.74	13.6	1.00
	0.85	5.1	0.86	0.83	18.0	1.05	0.80	11.2	1.07
	0.98	5.8	0.94	0.97	9.1	1.20	0.92	8.4	1.23
	1.06	5.7	0.99	1.06	8.0	1.28	1.00	7.1	1.29
	1.17	7.0	1.00	1.14	5.9	1.34	1.12	5.8	1.40
	1.28	7.1	1.05	1.26	5.0	1.38	1.26	5.6	1.45
D	0.43	9.8	0.65	0.66	32.5	0.78	0.66	15.1	0.91
	0.52	9.6	0.68	0.89	18.0	1.03	0.80	12.0	0.98
	0.74	5.3	0.80	0.96	10.5	1.18	0.92	9.7	1.06
	0.83	4.5	0.82	1.02	7.2	1.23	1.02	6.3	1.31
	0.88	4.9	0.85	1.18	4.9	1.36	1.09	5.0	1.36
	1.02	5.1	0.98	1.31	7.1	1.44	1.29	5.0	1.47
	1.03	5.8	1.06						

APPENDIX 3 European Scale Inks, commercial printing

Colour	Ink code	Paper	Print density	Dot area F_T(80%)	Dot area F_M(50%)	Density full tone	ΔE
Yellow	B	APCO	0.80	84%	38%	0.75	7.9
			0.85	81%	40%	0.84	4.8
			0.90	86%	31%	0.90	5.7
		Art	0.80	82%	38%	0.75	
			0.85	83%	28%	0.82	
			0.90	86%	30%	0.91	
		Natur	0.80	82%	45%	0.64	
			0.85	85%	43%	0.70	
			0.90	88%	48%	0.71	
Yellow	D	APCO	0.80	82%	36%	0.80	5.2
			0.85	81%	38%	0.81	3.7
			0.90	82%	40%	0.86	4.3
		Art	0.80	82%	36%	0.82	
			0.85	82%	38%	0.82	
			0.90	82%	39%	0.86	
		Natur	0.80	84%	40%	0.64	
			0.85	85%	45%	0.66	
			0.90	85%	45%	0.67	
Magenta	B	APCO	1.60	94%	53%	1.25	9.3
			1.70	96%	58%	1.34	7.0
			1.80	97%	62%	1.34	6.5
		Art	1.60	95%	53%	1.22	
			1.70	98%	57%	1.24	
			1.80	97%	62%	1.21	
		Natur	1.60	93%	57%	0.93	
			1.70	94%	58%	0.92	
			1.80	97%	62%	0.97	

(continued)

APPENDIX 3 European Scale Inks, commercial printing (continued)

Colour	Ink code	Paper	Print density	Dot area $F_T(80\%)$	Dot area $F_M(50\%)$	full tone	ΔE
Magenta	A	APCO	1.60	90%	47%	1.30	5.8
			1.70	92%	47%	1.38	5.6
			1.80	93%	49%	1.55	8.5
		Art	1.60	91%	47%	1.32	
			1.70	93%	45%	1.42	
			1.80	93%	50%	1.49	
		Natur	1.60	92%	50%	0.93	
			1.70	93%	53%	0.99	
			1.80	93%	51%	1.02	
Cyan	B	APCO	1.70	90%	50%	1.59	2.7
			1.80	92%	52%	1.69	1.6
			1.90	94%	52%	1.69	1.6
		Art	1.70	89%	48%	1.58	
			1.80	92%	51%	1.67	
			1.90	92%	52%	1.73	
		Natur	1.70	92%	52%	1.00	
			1.80	92%	55%	1.03	
			1.90	93%	53%	1.07	
Cyan	C	APCO	1.70	94%	56%	1.54	5.8
			1.80	96%	59%	1.61	4.5
			1.90	100%	66%	1.66	7.1
		Art	1.70	93%	55%	1.50	
			1.80	98%	57%	1.59	
			1.90	100%	67%	1.66	
		Natur	1.70	95%	59%	0.99	
			1.80	97%	62%	1.02	
			1.90	99%	65%	1.04	

APPENDIX 4: Working procedure for DETERMINATION OF COLOUR STRENGTH OF LETTERPRESS AND OFFSET INKS based on measurement of reflection values for proof prints

1. INTRODUCTION

This method does not lead to absolute values for colour strength, but only gives a value for relative colour strength of printing inks with approximately the same colour hue.

2. EQUIPMENT AND MATERIALS

2.1 The printing inks to be tested.

2.2 Standard paper (APCO II/II, see CEI 14-74) or samples of the paper, which will be used in the commercial printing.

2.3 Proof print press, on which prints with known ink film thickness can be produced.

2.4 Reflection photometer (e.g. Elrepho Remissionphotometer).

2.5 Millimeterpaper.

3. PROCEDURE

3.1 Preparation of proof prints

Using the proof print press (2.3), at least 5-6 proof prints with known ink film thicknesses equally distributed in a relevant area of ink film thicknesses are produced, using the inks to be investigated (2.1) and the paper in question (2.2).
The proof prints are dried for at least 24 hours at ca. 23°C. The drying of the prints shall take place in a relatively dark room, since printing inks with low light fastness can change their colour hue (fade or darken) on exposure to light.

Note 1 When the colour strength of letterpress and offset inks is checked in routine control, the preparation of proof prints should be made on standard paper. If the colour strength determination is carried out in connection with a special delivery of inks (or current delivery of inks for a special purpose), the colour strength determination (and the preparation of proof prints) should be carried out using the same type of paper to be used for the commercial printing.

Note 2 A relevant area of ink film thickness can normally be chosen as between 70% and 200% of the ink film thickness used in the commercial printing.

Note 3 The method described for determination of the colour

strength of printing inks is, in principle, applicable for all printing methods, depending on the possibility of preparing proof prints with known ink film thicknesses. For all printing methods (e.g. heatset, news print in letterpress and offset, gravure, etc) areas of ink film thicknesses are selected according to note 2.

3.2 Measurement of reflection values for proof prints

The reflection values of the proof prints are measured using a reflection photometer (tricolorimeter, see 2.4) and the red, green and blue filters (R, G, B), i.e. the filters normally used for determination of colour coordinates in C illumination (for Elrepho the filters FMX/C, FMY/C and FMZ/C are used). The reflection values measured can be used directly for the calculations described in the following, and calculation of X, Y, Z-values is not necessary. The reflection values measured are entered in the table indicated in this appendix.

Note 4 For multiple determination of colour strength compared with a standard printing ink, the originally produced set of proof prints of the standard ink can be used as long as there exists a certainty that no change of the standard prints has taken place and no changes concerning the adjustment and calibration of the reflection photometer have occurred.

Note 5 It is not possible in the determination of colour strength to use densitometers, since these instruments only measure the contrast ratio of the prints, and this contrast ratio is to a very high degree dependent on other print properties than the colour of the print (e.g. the content of black in the print).

3.3 Calculation of relative colour strength

In a diagram (graphpaper) the reflection values measured are plotted as a function of the ink film thicknesses of the prints. This is done individually for each of the three filters used for the measurements (see 3.2). Each sheet of graphpaper shall include reflection/ink film thickness-measurements using one and the same filter for the measurements. A smooth curve is drawn through the depicted points, disregarding if a single point is outside the curve (this is normally caused by an error in determination of the ink film thickness of the print). Similar curves are made for all the printing inks to be compared on the same sheet of millimeter paper.

Vertical lines at 75%, 100% and 150%, respectively, of the standard ink film thickness selected (for letterpress inks, e.g. at 0.75 g/m^2, 1.0 g/m^2 and 1.5 g/m^2 when selecting an area of ink film thicknesses from 0.7 - 2.0 g/m^2) are drawn from each individual curve.

A horizontal line is drawn through the point of intersection between the vertical line at 75% of the standard ink film thickness and the reflection curve for the standard printing ink. The other reflection curves then indicate the ink film thickness which is required for the print in order to obtain the same reflection value as for the corresponding standard print, the reflection values for the standard print at 100% and 150% standard ink film thickness, and the corresponding ink film thickness values of the other prints are read, and these values are entered in the table as illustrated in this appendix.

For the ink film thicknesses determined (3 sets of ink film thicknesses (75%, 100% and 150%), each including 3 values measured with R, G and B filters, respectively) one calculates the relationship between the ink film thickness of the standard ink and the corresponding ink film thickness for the ink(s), whose relative colour strength shall be determined.

Example The standard ink film thickness for the standard print (1.00 g/m^2) measured with filter R over the ink film thickness of ink No. 2 necessary to obtain the same reflection value measured with filter R (e.g. 0.90 g/m^2). The colour strength of ink No. 2 relative to the standard ink is for the R-filter:

$$F = \frac{1.00}{0.90} = 1.11$$

The calculated ratios between the ink film thicknesses necessary to obtain the same reflection value are entered in the table (see appendix).

4. EVALUATION OF RELATIVE COLOUR STRENGTH

For each of the printing inks investigated, the measurements and calculations carried out lead to 9 values for relative colour strength (3 ink film thicknesses, at which the colour strength has been determined for 3 filters (R, G and B)).

If the values for colour strength for each of the 3 filters indicate a gradual increase with increasing ink film thickness, the relative colour strength at each of the 3 ink film thicknesses can be indicated as the mean value of the colour strength measured for R-, G- and B-filter. For some inks (e.g. yellow inks) results based on measurements for only 2 filters should be used (for yellow inks: filter G and B), since reflection values measured with the third filter will be almost constant in the total area of ink film thicknesses.

Generally, the colour strength ratio for different colours can be indicated on the basis of the following mean values, referring to the colour strength ratio determined with different filter (R, G, B):

Yellow:	Filters G + B
Orange:	Filters R + G + B
Red:	Filters R + G + B
Magenta:	Filters R + G + B
Violet:	Filters R + G + B
Blue:	Filters R + G + B
Cyan:	Filters G + B
Green:	Filters G + B

Note 6 It generally occurs that the colour strength ratio for two printing inks strongly depends on the ink film thickness at which the relative colour strength is determined. This appears with the reflection curves (reflection values as a function of ink film thickness) crossing each other. Relative colour strength shall, therefore, always be indicated in connection with an indication of the ink film thickness at which the colour strength has been determined. Preferably, the relative colour strength shall be indicated at at least 2, or better, at 3 ink film thicknesses in the neighbourhood of the standard ink film thickness.

5 REPORT

Referring to this working procedure, the following shall be indicated:

Data for the standard printing ink.

Data for the printing inks compared with standard.

Paper used for the printing.

Proof print press used.

Reflectometer used.

Standard ink film thicknesses selected.

Ink film thicknesses of the prints (see table).

Reflection values for the proof prints (see table).

Calculated ink film thicknesses of prints to obtain same reflection values as for the standard print at standard ink film thicknesses (see table).

Calculated colour strength ratios at standard ink film thicknesses based on measurements with different filters (see table).

Colour strength ratios at standard ink film thicknesses determined as mean values of colour strength ratios determined from measurements with 2 filters $(\frac{G + B}{2})$ or 3 filters $(\frac{R + G + B}{3})$ (see table).

DETERMINATION OF RELATIVE COLOUR STRENGTH BY REFLECTION MEASUREMENTS

TABLE 1A Ink film thicknesses and reflection values

Ink	No. 1 (standard)				No. 2				No. 3			
Print No:	g/m^2	Reflection R	G	B	g/m^2	Reflection R	G	B	g/m^2	Reflection R	G	B
1												
2												
3												
4												
5												
6												
7												

TABLE 2A Film thickness of prints to obtain same reflection values

Thickness of standard print	Reflection values of standard print at standard ink film thickness			Film thickness of proof prints to obtain same reflection values as for standard print					
Ink No. 1	Ink No. 1			Ink No. 2			Ink No. 3		
	R	G	B	R	G	B	R	G	B
75% = g/m^2									
100% = g/m^2									
150% = g/m^2									

TABLE 3A Relative colour strength

Thickness of standard print	Relative colour strength (thickness of standard print: thickness of print compared at which same reflection value is obtained)					
Ink No. 1	Ink No. 2			Ink No. 3		
	R	G	B	R	G	B
75% = g/m^2						
100% = g/m^2						
150% = g/m^2						
Mean value						
75% = g/m^2						
100% = g/m^2						
150% = g/m^2						

DISCUSSION

W. D. Schaeffer (1) What laboratory printing device did you use?

(2) Would you agree that a letterpress printing instrument produces prints with different 'ink lays' than does an offset litho press and that these differences can produce colour differences?

(3) There is considerable reluctance among suppliers in the U.S.A. to produce 'standard materials', i.e. paper and ink, because of the cost and concern for Government attitudes. Is this the situation in Europe?

W. Hansen The laboratory prints were made using an IGT tester. I agree that the ink layers produced differ from those obtained when printing in offset, which explains some of the differences between laboratory prints and commercial printing of the inks tested.

The attitude among European printing ink manufacturers, and I believe also among paper manufacturers, is very much against 'standard materials', being inks, paper or other products, since 'standard materials' always mean a loss of some degree of freedom, especially concerning the development of new and better materials. The European Standard for letterpress and offset, therefore, only 'standardize' the colour of the print obtained on a specified type of paper, including reasonable tolerances.

M. H. Bruno What densitometer did you use for your measurements of density? The cyans seem very high and yellows low compared with U.S. inks.

W. Hansen The densitometer used was an ordinary MacBeth reflection densitometer. The density values measured are at the same level as generally found for European inks. Differences from U.S. inks could be caused by deviations in pigment types generally used for four-colour printing inks in Europe and U.S.A., and the colour hues of the prints.

J. Fetsko My question has to do with experimental procedure:
(1) In Table 1, what shear rate was used to compute viscosity? (In respect of shear force, the same shear rate should be used for comparing inks).

(2) Since ink sequence is usually controlled by roller-type tack readings, did you consider such measurements?

W. Hansen The viscosity measurements were not considered important in this investigation, and were only made to establish eventual large differences between the inks. The measurements were carried out in 'the old fashioned way' on a Laray viscometer using intervals of shear forces between 1200 N/m^2 and 11000 N/m^2. Shear rates varied from ca. 20 sec^{-1} to ca. 400 sec^{-1}. According to investigations carried out at an earlier state, the tack measurements are, in our opinion, very dependent on the instrument used, and, therefore, tack measurements were not carried out, since the ranking of tack of the inks will depend on the tackmeter that happens to be used.

R. Ginman In Table 13 you report some gloss values. Your results show that the gloss usually decreases with increasing ink film thickness. Our experiences in test printing coated paper and board is the opposite. Was your paper or ink very special?

W. Hansen For high quality inks, you would not expect greater differences in the gloss of the printed film, when changing the ink film thickness from 1 g/m^2 to 1.5 g/m^2. In this case, the inks examined were all based on experimental binders, which are not commercially used and, therefore, abnormal behaviour of gloss values can be expected. Gloss deviations of 1 - 2% are considered as experimental errors. In the cases where a lower gloss is found when the ink film thickness is increased, the lower gloss is probably caused by the poor wetting of the pigment and a high content of mineral oil in the ink, leading to a kind of 'powder drying' of the ink film at increasing ink film thickness. For the pigment in question, the gloss values measured are generally very low, indicating poor wetting of the pigment.

INFLUENCE OF CHEMICAL COMPOSITION AND SURFACE STRUCTURE ON THE QUALITY OF COMMERCIAL OFFSET PLATES

W. K. Sobotka, Federal Training and Research Institute of Graphic Arts, Vienna, Austria.

ABSTRACT

The suitability of different metals for offset is discussed. Oleophilic and hydrophilic behaviour of different alloys were investigated. The pre-eminence of aluminium for offset plates is pointed out and different ways of manufacturing special surfaces of aluminium plates for offset printing are described. Surfaces of commercial offset printing plates were investigated by means of a scanning electron microscope and with a surface profile measuring instrument. Quality criteria for printability and runability in relation to surface structure are discussed.

INTRODUCTION

Offset plates are produced worldwide in a great variety, differing in composition of alloys, plate surface, method of use and price. A survey of the suitability of these different plates may be interesting information for printers, plate manufacturers and scientists. For this purpose the main physical and chemical parameters to investigate are oleophilic and hydrophilic behaviour, surface structure, durability and corrosion phenomena.

These characteristics have to be ideally balanced to develop plates specially designed for all kinds of offset printing, from duplicating to web offset plates.

The description of an ideal model for the surface structure as well as information on the best suited material and surface treatment for offset plates provides a useful survey for the plate manufacturer. Finally, the commercial aspect is also an important consideration in finding a plate for a special purpose.

2. METALS AND METAL SURFACES FOR OFFSET PRINTING

Many metals and alloys are technologically well known, but only few of them fulfil the quality parameters for an offset plate. The following 6 points are very important:

(1) Ductility of the metal or alloy
(2) Good durability and hardness
(3) Fine grain structure
(4) Thermal stability
(5) Ability for etching
(6) Low costs

These factors reduce the number of materials to zinc, copper, aluminium, magnesium, and their alloys. Brass, steel or iron, chrome and other special alloys are often used for bi- and trimetal plates.

2.1. Metal surface structure and behaviour

The metal behaviour and surface structure of an offset plate are crucial for a good working product. The following properties of the plate surface are very important:

(1) Good adhesion of the photopolymer coating and ink film to the image area.

(2) Strong bonding between plate surface and lacquer during plate development and subsequent fixing.

(3) Good wettability of the whole non-image area by water.

The criteria for mechanical and chemical behaviour of metals and their alloys depend on their crystallography. The crystallographic system for aluminium, copper and chrome is the cubic one, for zinc hexagonal. Furthermore, the behaviour is determined by the chemical composition and grade of deformation caused by manufacture.

Figure 1 illustrates the orientation of the crystals after the rolling process. Different solubility, corrosion resistance, hardness and elasticity of a crystal in different directions can be measured. Smaller and less orientated crystals are less anisotropic. It follows that the crystals for offset plates should be very small and deorientated. The method of manufacturing the plates is one reason why the ideal structure cannot be achieved. Information about mechanical properties of the different metals is also very important for the printer, because these properties determine the runability of the plate.

Table 1 shows how the physical properties differ with the composition of the raw material. The rolling procedure is also very important. Stretching of the metal is a method of increasing the elasticity. A part of the work done during the stretching procedure is stored as energy in the metal. The values are: Aluminium 10%, iron 15%, brass 16%, zinc 20%. The higher energy level increases elasticity and hardness.

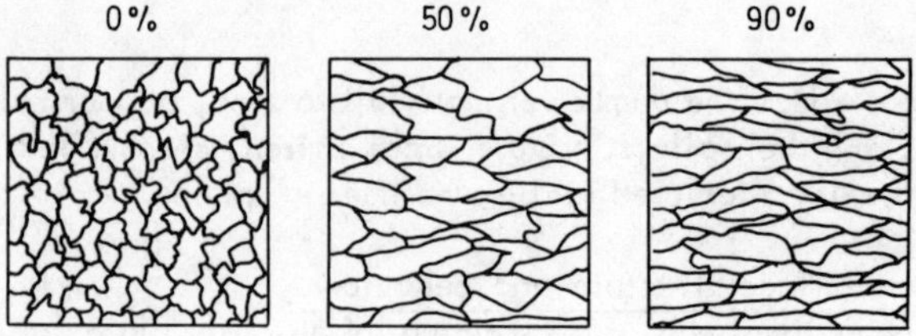

Fig. 1. Determination of crystal structure.

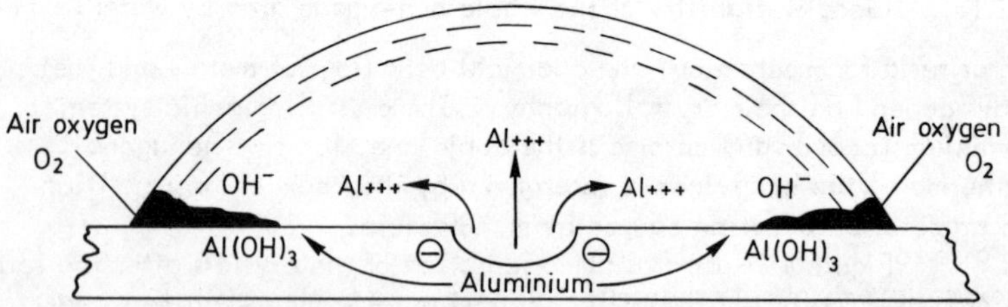

Fig. 2. Corrosion under a water drop.

Sensitivity to corrosion can also be raised by stretching. This effect, however, can be prevented by temperature treatment of the metal.

TABLE 1

Metal	Tensile strength σB(kg/mm^2)	Brinell hardness HB(kg/mm^2)
Zinc, pure, rolled	10-20	32-34
Aluminium, purest	3-6	12-15
Aluminium pure Al 99.5 (Alloy F 13) rolled and annealed	7-9	18-23
Al 99.5 (Alloy F 13) hard rolled and stretched	13-18	35-40
Aluminium alloys (1% Mn)	9-16	20-40
Steels	35	-
Zinc-alloys ZnCu 1	18	50
ZnCu 4 Pb 1	27	70
ZnAl 4 Cu 1	35	80

2.2. Corrosion

Aluminium plates are covered with a dense alumina (Al_2O_3) 1 micron thick layer, due to atmospheric oxidation. The thickness of this layer can be increased anodically. Although the alumina layer is a dense one, it contains a lot of fine pores that are responsible for good water balance and also for fixing the photopolymer image areas. Zinc plates also carry an oxide layer. Its mechanical strength is, however, much lower and it is very sensitive to chemical attack.

Corrosion is defined as the deterioration of a substance - metal or alloy - through reaction with its environment. Oxidation is the most common reaction leading to deterioration of metals and the bulk of corrosion theory is concerned with this aspect. For discussion purposes corrosion of metals may be classified according to type of reaction:

(1) By electrolyte solutions, usually aqueous, where the corrosion reactions involve electrochemical oxidation of the metal and a separate electrochemical reduction of some oxidizing species.

(2) By gaseous environments, usually at elevated temperatures.

(3) By solvents such as liquid metals or fused salts.

In the following section some forms of corrosion of importance for offset plates are described.

Electrochemical corrosion in electrolytic solutions   Corrosion of metals in aqueous solution involves the electrochemical oxidation of the metal and a separate electrochemical reduction of an oxidizing agent. Phenomenologically, corrosion can take many forms and may range from a single pit or perforation in a large container to the complete uniform dissolution of entire pieces of metal. The pH of the solution, its conductivity, the amount of dissolved oxygen, the presence of trace amounts of impurities, temperature, the degree of agitation, composition of the metal, the mode of fabrication, welding, riveting, or cold working, are all factors which will influence the rate, the extent and type of corrosion. These many diverse factors, however, can all be interrelated by viewing the corrosion process as a simple galvanic cell according to the scheme illustrated in Figure 2.

Stress corrosion   Stress corrosion is the accelerated corrosion of a metal under sustained stress. Its effects are greater than those of the corrosive environment or the stress alone. Methods of avoiding stress corrosion include design, choice of alloy, heat treatment and use of inhibitors, coatings or cathodic protection.

High purity - high temperature water corrosion   This form of corrosion has many applications. The corrosion resistance of metals in water is quite different. A number of factors affect the corrosiveness of the system, including temperature, pressure, water purity, dissolved solids and gases. Corrosion rates in high-purity water generally increase with temperature but the rate of increase is quite small up to a critical temperature (Table 2).

TABLE 2

	Corrosion resistance			
Material	94°C	148°C	204°C	260°C
Aluminium	A	A	B	C
Chromium	A	A	A	B
Copper	A	A	C	C
Magnesium	B	C	C	C

A = highest resistance B = intermediate resistance C = low resistance

2.3. Oleophilic and hydrophilic behaviour

Offset plates should have special qualities in their relative receptivity to water and ink beside general qualities like strength, fine grain structure, resistance to temperature and sensitivity to etching. Normally, offset plates are coated with a special lacquer for ink receptivity. The main property is the hydrophilic behaviour of the plate combined with the rejection of ink. The wettability of metals for the offset process can be investigated by measuring the contact angle measured in the liquid placed on it.

Measurements For the measurements a Lorentzen & Wettres contact angle measuring device was used. Drops of size 0.004 ml ± 10% of either castor oil or distilled water were applied from a pipette to the polished and degreased surface of the plate. Measurements were made after 10 seconds. The averages of 10 measurements are given in Table 3.

TABLE 3

Metal	Contact angle in degrees Distilled water	Castor oil
Zinc	68.6	35.1
Magnesium	54.3	39.3
Copper	71.9	38.7
Chromium	33.1	45.9
Aluminium	33.2	41.5
Iron/steel	27.3	29.7
Molybdenum	48.4	33.2
Tungsten	42.7	31.3
Zirconium	41.2	36.9
Hafnium	46.7	38.4
Beryllium	45.84	34.1
Titanium	49.7	26.9
Niobium	38.3	26.0
Tantalum	37.4	33.2

Discussion Copper is the most hydrophobic metal, followed by zinc and magnesium. These metals are also oleophilic and suitable as ink receptive metals. Chromium, aluminium and iron or steel are metals having smaller contact angles against water than against oil. These metals are hydrophilic. Chromium and aluminium also show good hydrophilic properties and therefore they are water receptive metals. Only molybdenum and titanium show oleophilic and hydrophobic properties. Commercial aspects apart, they could be used as ink receptive metals. The other metals are not suitable for offset plates.

2.4. Ink transfer

Ink transfer provides the main information for offset printing. In the offset process, ink is not directly transferred to the paper. Ink rollers bring the ink to the plate, the plate transfers it to the blanket and from there to the paper. The ink transfer is characterized by the ink transfer coefficient. In this investigation the influence of the different metal surfaces on ink transfer was measured. Accordingly, measurement was done with ink as a constant factor. The amount of ink after inking and the ink transfer-coefficient to the blanket were determined as material parameters for different metal surfaces.

Procedure With an ink-pipette a known amount of ink was put onto an inking roller. The ink was then distributed very carefully and the metal surface inked for a constant time. After inking, the ink was transferred to a blanket roller, with atmospheric conditions constant in all measurements.

Measurement 1: The metal surface was polished and degreased.

Measurement 2: The cleaned metal surfaces were used for the ink transfer test seven days later without protecting the metal surface.

Measurement 3: The metal surface was re-polished and degreased. This second attempt also showed an order for the durability of the metals as measured by the weight lost by polishing.

Measurement 4: Immediately after the third measurement the metal metal surface was cleaned and protected with a thin film of gum arabic to stop the oxidation. After seven days the gum arabic was removed and the last measurement made.

Between the different measurements the metal plates were stored in a desiccator. The results are shown in Table 4.

TABLE 4

Metal	Amount of ink (g/m^2)	Ink transferred to blanket roller (g/m^2) x	Ink retained by metal surface (g/m^2) y	Ink transfer coefficient $\frac{x}{y}$
Measurement 1:				
Copper	4.68	2.43	2.25	1.07
Aluminium	4.55	2.75	2.88	0.99
Chromium	4.13	2.13	2.00	1.06
Zinc	3.93	1.88	2.05	0.91
Steel	3.80	1.75	2.05	0.85
Measurement 2:				
Copper	3.65	1.73	1.93	0.90
Aluminium	3.88	1.90	1.98	0.96
Chromium	3.88	1.85	2.03	0.91
Zinc	3.83	1.80	2.03	0.89
Steel	3.93	1.95	1.98	0.99

TABLE 4 (continued)

Metal	Amount of ink (g/m^2)	Ink transferred to blanket roller (g/m^2) x	Ink retained by metal surface (g/m^2) y	Ink transfer coefficient $\frac{x}{y}$
Measurement 3:				
Copper	4.34	2.19	2.15	1.01
Aluminium	4.30	2.15	2.16	1.00
Chromium	4.28	2.01	2.26	0.89
Zinc	4.13	2.03	2.10	0.96
Steel	4.38	2.16	2.22	0.92
Measurement 4:				
Copper	4.35	2.19	2.15	1.02
Aluminium	4.41	2.16	2.25	0.96
Chromium	4.39	2.11	2.28	0.92
Zinc	4.26	2.08	2.18	0.95
Steel	4.15	1.99	2.16	0.97

Discussion The results of these ink transfer investigations are in agreement with the results described in Section 2.3. Only aluminium, as a hydrophilic metal, also shows a good ink transfer coefficient.

It has to be pointed out that of the ink transfer measurements, three were executed from high polished, non-oxidized metal surfaces. Only measurement 2 was made on an oxidized metal surface. This result shows quite different ink transfer coefficients. The hydrophilic and oleophilic behaviour changes rapidly with oxidation of the surface.

ALUMINIUM

All the investigations in Section 2 show that aluminium is the metal best suited for a single metal offset plate. Today, aluminium plates are used for offset printing as pure aluminium or as an alloy with manganese. The thickness of the plates varies from 0.1 - 0.5 mm. Therefore aluminium is the key material for offset plate manufacturing. Beside the properties pointed out in Section 2, aluminium has a low specific gravity, a bright colour (control of the copy on the plate) and a passive surface.

Pure rolled aluminium is, however, not suitable for offset printing because the surface is too smooth and even wetting of a smooth surface is impossible. It is therefore necessary to give the metal surface a special structure which is called 'grain'. Graining gives the plate better adhesion to surface materials. The fact that a rough surface has better adhesion than a smooth surface is well known. Lacquer and photopolymer coatings also have to have a good bond with the plate surface. Furthermore, the resolving power and the sharp edges of a print are very important. For the latter aspect the surface should be very smooth. To produce a good offset plate in regard to the surface treatment, wettability, adhesion, resolving power and image-sharpness have to be considered. The statement that the surface treatment of an offset plate made of aluminium is the most important step in producing a good working plate must be emphasised.

3.1. Methods of manufacturing aluminium offset plate surfaces

To make offset printing possible it is necessary to roughen or 'grain' the surface either mechanically, chemically or electrolytically.

Ball Graining is done in a machine which consists of a tub with a rotary motion. Small steel marbles rotate over the surface of the plates. Water is added and then abrasive material. Problems with this type of graining are inconsistency from plate to plate, scratchiness, dirt and imbedded abrasives.

Sandblasting The plates are mounted on a rotary drum and a dry abrasive impinges on the surface and normal to the direction of rotation of the plates. Imbedded abrasives can also be a problem.

Dry Brush Graining can be done with brass or steel wire brushes. The advantage is that dry brush graining can be done in line with the treatment and coating of presensitized plates. It has the disadvantage that the grain is oriented in one direction. The use of these plates must be such that brush direction is parallel to the cylinder axis of the printing machine. Otherwise it is impossible to get good moisture control.

Rotating and Oscillating Plastic Brush Graining Advantage of this surface is the absence of a grain-direction.

Ball Brush-Graining - a combination of rotary tub and wet brush graining. These plates have the texture for good quality printing and the depth for good moisture control, especially for large presses.

Chemical and Electrochemical Graining is used primarily for treating plates prior to coating in the manufacture of presensitized plates. The most widely used method is the electrochemical treatment of aluminium in a solution of hydrofluoric acid. This produces a fine grain, which is used as a base for wipe-on and presensitized plates. It is also used as a preliminary treatment for anodizing.

Anodized Aluminium Anodizing is a process by which a uniformly controlled thickness of oxide is produced electrolytically. The grain surface is very fine and better than all mechanically treated plates. The process of anodic oxidation is by growing the alumina on the electrochemical roughened base and therefore it is very strongly bound to the metal. The alumina consists of lots of small cells which make the surface denser and also acts as a protective coating against the normal atmospheric oxidation.

Furthermore it is hydrophilic, which is very important for offset printing. The electrolytical roughening enlarges the surface area and improves water receptivity. This effect is also enhanced by the pores produced during the anodic oxidation. The disadvantage of these fine pores lies in the fact that particles of emulsified ink vehicles or fillers from the paper can be caught in them, so reducing the water adhesion to a level when scumming can occur. Gum arabic can insulate against this effect as well as optimizing the dampening of the plate surface.

3.2. Influence of corrosion on hydrophilic behaviour

As described in Section 2.2, corrosion is an important factor for an aluminium plate surface. Using the contact angle method, a pure aluminium plate with a smooth surface was studied after various treatments.

The treatment and resulting contact angles are summarised in Table 5.

TABLE 5

Aluminium plate		Contact angle in degrees, with water
Untreated		87.0
75 min in distilled water at 20°C		84.9
75 min in boiling distilled water		15.9
Drying cupboard at 200°C		
	after 24 hr	60.4
	after 48 hr	62.8
	after 72 hr	49.8
	after 120 hr	49.1
	after 144 hr	44.2
	after 168 hr	41.9

Discussion The aluminium treated in the drying dupboard became more hydrophilic with longer heating time. The highest hydrophilic properties were obtained after 75 min boiling in distilled water. The cause of this high figure can be explained by the increased number of oxidation pores resulting from the high temperature water corrosion described in Section 2.2.

4. INVESTIGATION OF COMMERCIAL OFFSET PLATE SURFACES

The offset plates listed below were investigated and are described in Sections 4.1. - 4.5.

KODAK Polymatic	LNL	Anodized
	LNM	Anodized
	LP	Anodized

KODAK Polymatic	M	Anodized
	S	Anodized
OZASOL	P7	Anodized
	N4	Mechanically grained
KRAUSE	Maximal	Mechanically grained
	VSP 7	Mechanically grained
	VSP 15	Chrome surface
	VSP 31	Chrome surface
	VSP 41	Chrome surface
	Bio	Chrome surface
	M-plate	Chrome surface
	Neumetall	Chrome surface
ALUCOPRINT PA-S		Mechanically grained
Lithal		Anodized
Polychrome		Mechanically grained
POLYCHROME PAP		Mechanically grained
FUJI FPM		Mechanically grained

4.1. Hydrophilic behaviour

Discussion These investigations showed that the chrome surfaces of bi- or trimetal plates are the most hydrophilic. The mechanically grained plate surfaces are very different in their hydrophilic character. The anodized plates are not more hydrophilic than the others. The reason may be that the surface tension of water prevents the water penetrating immediately into the fine pores. Therefore dampening water additives are used to wet the plate surface more easily and more quickly. The contact angle measurements are shown in Table 6.

TABLE 6

Offset plate		Contact angle in degrees, with water
Kodak	LNL	64.7
	LNM	60.7
	LP	68.0
	S	69.4
Ozasol	P7	64.0
	N4	56.0
Krause	Max.	56.1
	VSP7	69.4
	VSP15	69.9
	VSP21	50.8
	VSP31	60.1
	VSP41	53.5
	Bio	54.4
	M	43.8
	Neumetall	57.3

4.2 Water evaporation

This test was carried out to find a correlation between the kind of surface graining and the speed of water evaporation on the plate surface after dampening.

Test Procedure Conditions 21 °C, 55% relative humidity.
Plate samples 10 x 13 cm with coating removed and degreased. The plates were marked 10 cm from the edge, completely immersed in water, taken out and left tilted in the air for evaporation. The time taken for the water line to recede to a marked position on the plate was measured. Table 7 gives the average of three timings.

TABLE 7

Offset plate		Time (minutes, seconds)
Kodak	LNL	11.22
	LNM	11.39
	LP	11.46
	M	11.46
	S	11.48
Ozasol	P7	15.32
	N4	11.33
Krause	Max.	13.10
	VSP7	12.55
	VSP15	13.19
	VSP21	10.39
	VSP31	10.39
	VSP41	11.10
	Bio	10.47
	M	11.03
	Neumetall	10.34

Discussion The chrome surfaces showed the shortest water evaporation time. This value confirms that bi- and trimetal plates are only able to hold a small amount of water during printing. The anodized plates showed at least similar values. The mechanical grained plates showed the longest evaporation time. These plates are not as fine grained and therefore considerably more dampening water is necessary to prevent scumming during the printing process. It was very interesting to note that the evaporation time of the OZASOL P7 plate, which is an anodized plate, is the longest. This phenomenon might be explained by the fact that this plate has a much rougher surface compared with the anodized KODAK plates.

4.3 Water receptivity

This test was carried out to investigate the ability of an offset plate to absorb water. A water soluble blue writing ink was used instead of water.

Test Procedure The ink was put onto the surface of the plate with a pipette. A constant sized drop was used. The drop was then distributed with a squeegee under constant pressure in one pass. The amount of ink absorbed by the plate was determined. A subjective scale from 1 - 5 depending on the density and length of the ink stripe was used as a measure of water receptivity, where 1 stands for poor ink receptivity, i.e. no water absorbed, and 5 stands for good water receptivity.

The dried samples were investigated with a microscope to observe the amount of inking on the grain surface of the plates (Table 8).

TABLE 8

Offset plate		Absorbency	Plate surface	After cleaning
Kodak	LNL	4-5	Anodized, uniform dark blue surface	Invisible
	LNM	4-5	as LNL	Invisible
	LP	4-5	as LNL	Less visible
	M	4-5	as LNL	Less visible
	S	4-5	as LNL	Less visible
Ozasol	P7	4-5	Anodized, little defects, dark blue surface	Very visible
	N4	2	Mechanically grained, less ink on the surface	Less visible
Krause	Max.	2	Mechanically grained, less ink on the surface	Less visible
	VSP7	2	as Krause Max.	
	VSP15	3	Mechanically grained, ink very visible	Invisible
	VSP21	1-2	Chrome, uneven ink of the surface, less ink	Invisible
	VSP31	1	Even inking	Less visible
	VSP41	1-2	as VSP31	
	Bio		as VSP31	
	M		Chrome, uneven	Invisible
	Neumetall		Uneven, less ink	Less visible

Discussion The comparison of water receptivity and water evaporation on the plate surfaces showed that the rough mechanically treated plate types can carry more water than the smooth anodized plates. This fact may be because more water is necessary to fill the holes of the plate surface from the bottom to the top of the grain. The anodized plates are much smoother, therefore they carry less water than do the mechanically grained ones. On the other hand, they are much more absorbent because of their porous structure.

The OZASOL P7 plate combines both properties.

4.4. Scanning electronmicroscopical investigations

The plate surfaces listed in Section 4 were investigated on a JEOL JSM 35 C Scanning Electron microscope. The ultrasonically cleaned plate surfaces were then sputtered with gold.

In the Appendix Figures 1 - 30 show the results of these investigations. Figures 1 - 26 show unused commercial offset plate surfaces, while Figures 27 - 30 show plate surfaces of used, printed plates.

The plate runs vary from 80,000 - 200,000 prints and the damage to the grain structure is very clearly demonstrated. Especially does Figure 29 show the typical compressed tops of the grain structure.

The remaining figures afford a good explanation for all the surface structure problems discussed in Section 4.

ACKNOWLEDGMENT

My sincere thanks to Dipl.-Ing. Dr. Schreiner of the Inorganic Institute of Technology of the Technical University of Vienna for carrying out the electron microscope investigations.

APPENDIX

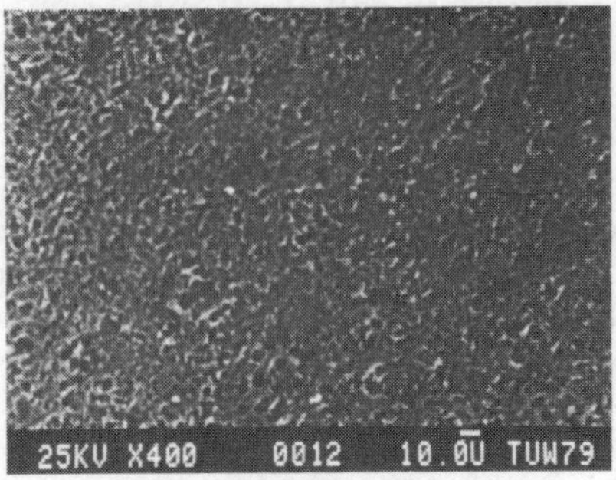

1 KODAK POLYAMTIC LN-L,
Magnification 400 ×

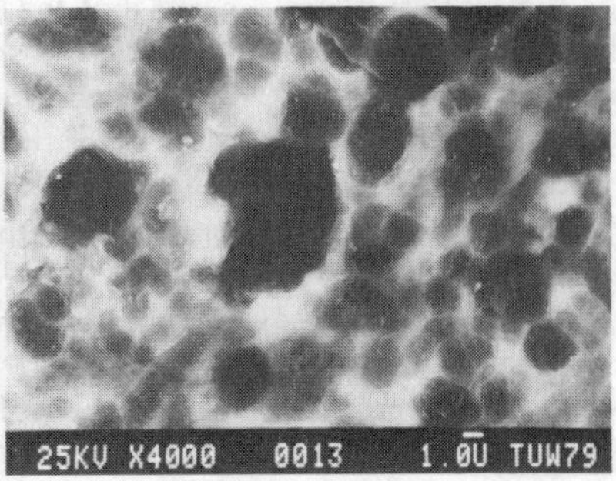

2 KODAK POLYMATIC LNL,
Magnification 4000 ×

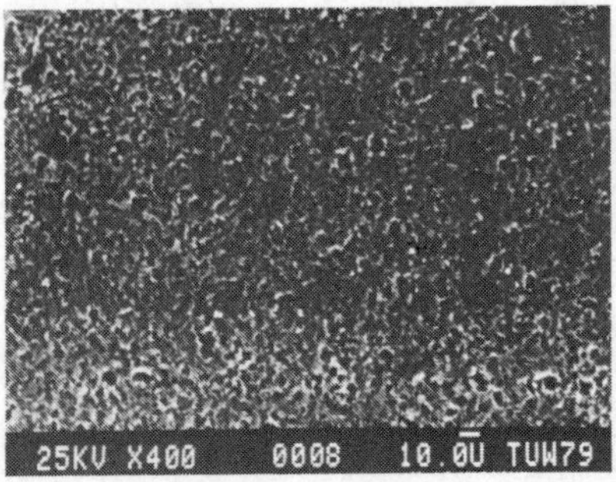

3 KODAK POLYMATIC LP,
Magnification 400 ×

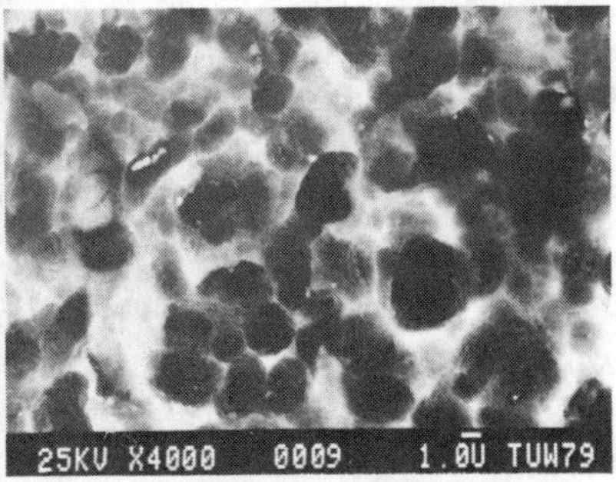

4 KODAK POLYMATIC LP,
Magnification 4000 ×

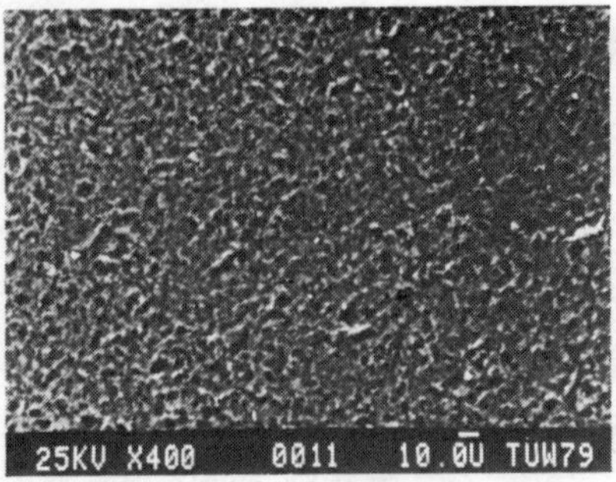

5 KODAK POLYMATIC M,
Magnification 400 ×

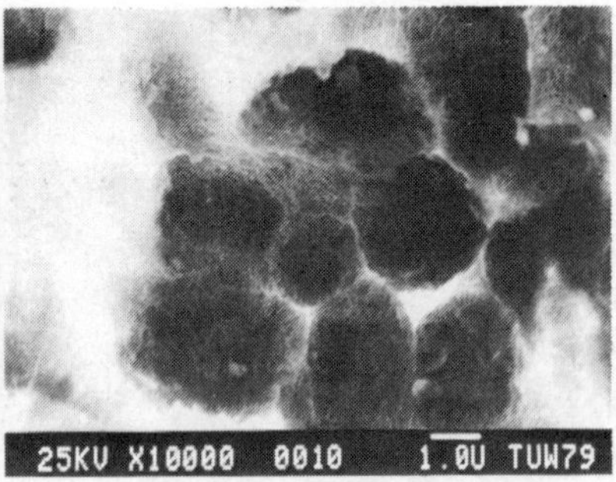

6 KODAK POLYMATIC LP,
Magnification 10 000 ×

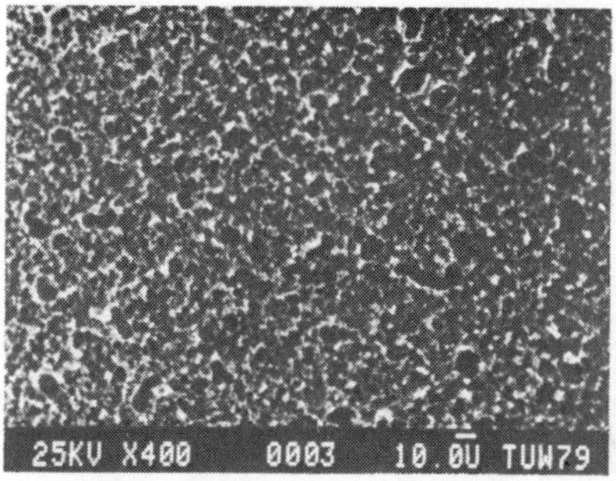

7 *OZASOL P7, Magnification 400×*

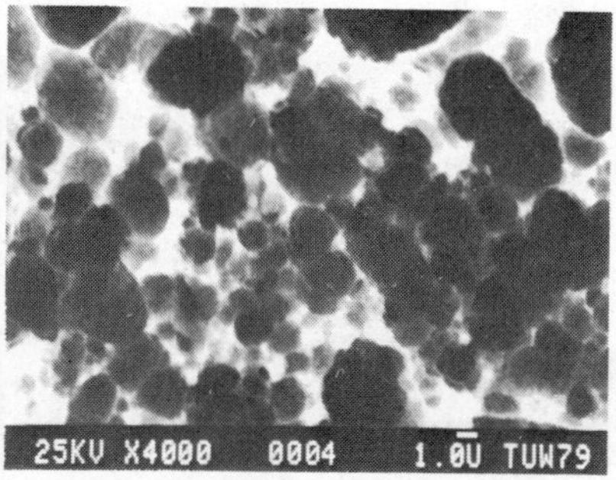

8 *OZASOL P7, Magnification 4000×*

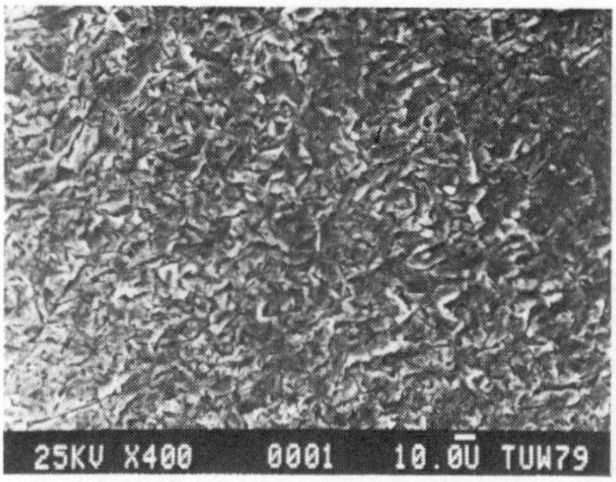

9 *OZASOL N4, Magnification 400×*

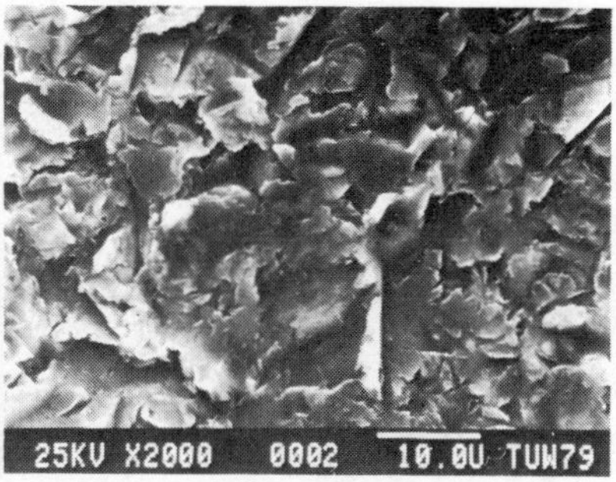

10 *OZASOL N4, Magnification 2000×*

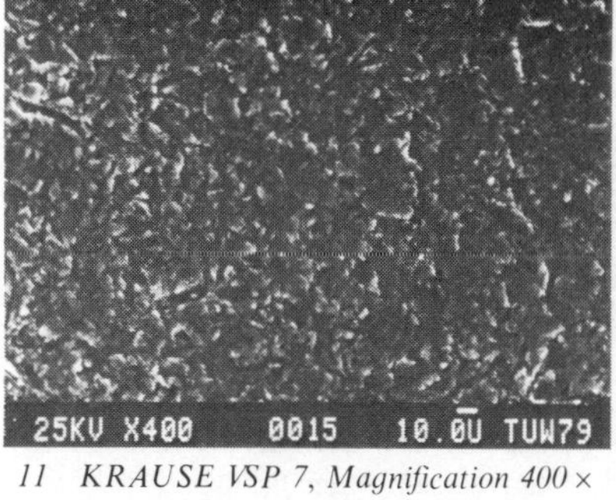

11 *KRAUSE VSP 7, Magnification 400×*

12 *KRAUSE VSP7, Magnification 2000×*

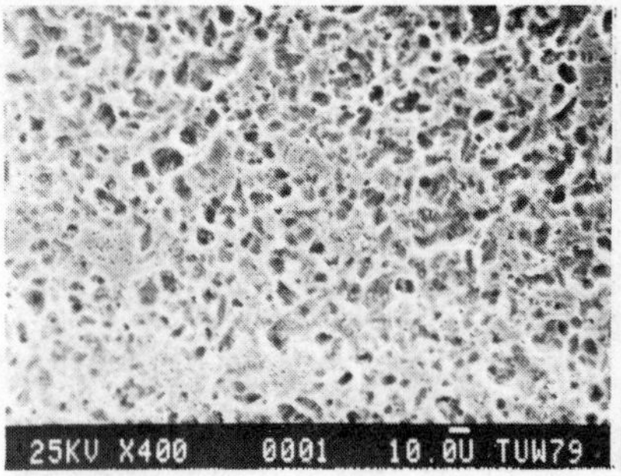

13 *ROTOPLATE, Magnification 400 ×*

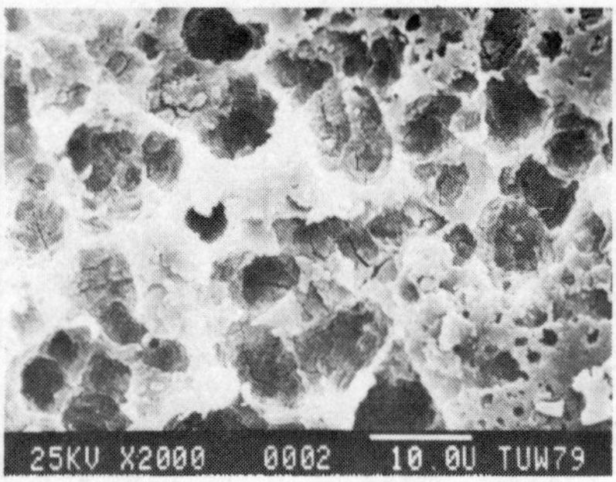

14 *ROTOPLATE, Magnification 2000 ×*

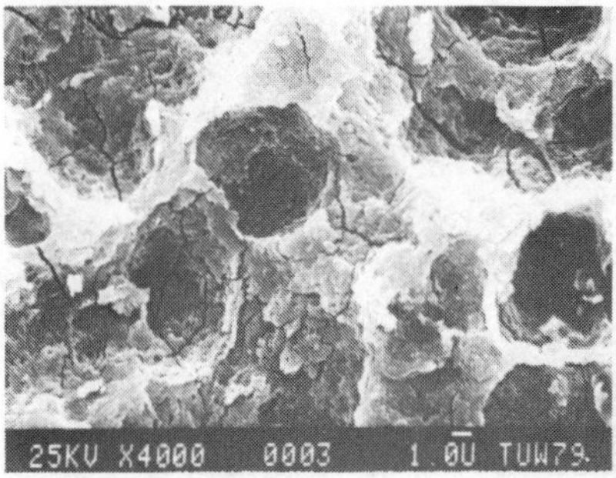

15 *ROTOPLATE, Magnification 4000 ×*

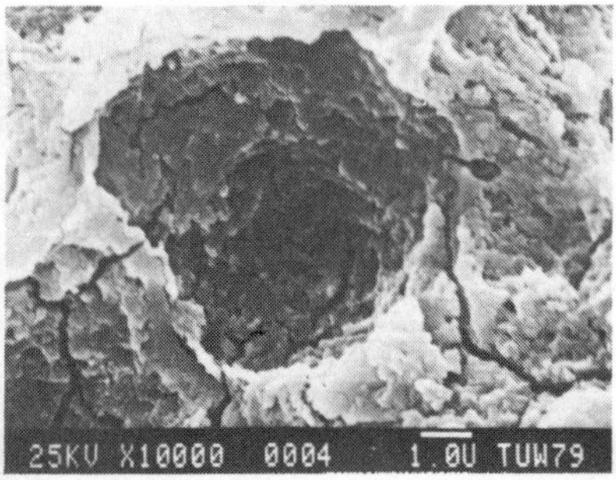

16 *ROTOPLATE, Magnification 10 000 ×*

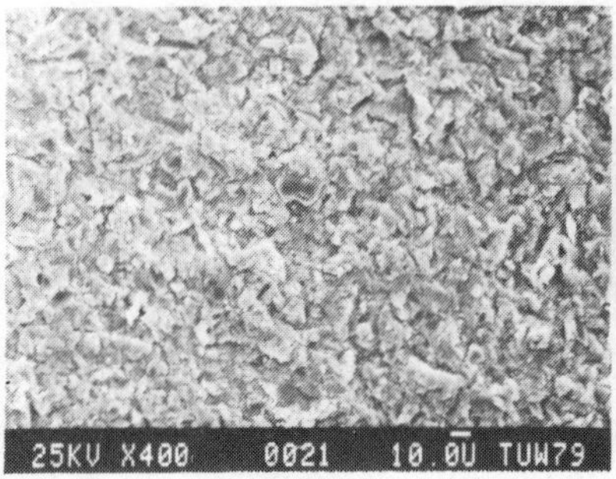

17 *MIKRAL, Magnification 400 ×*

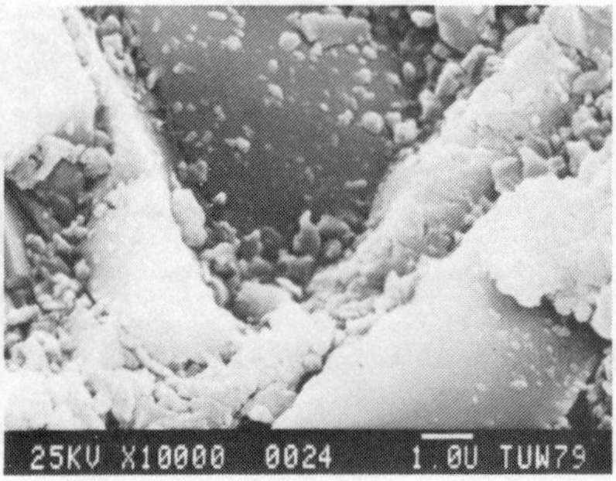

18 *MIKRAL, Magnification 10 000 ×*

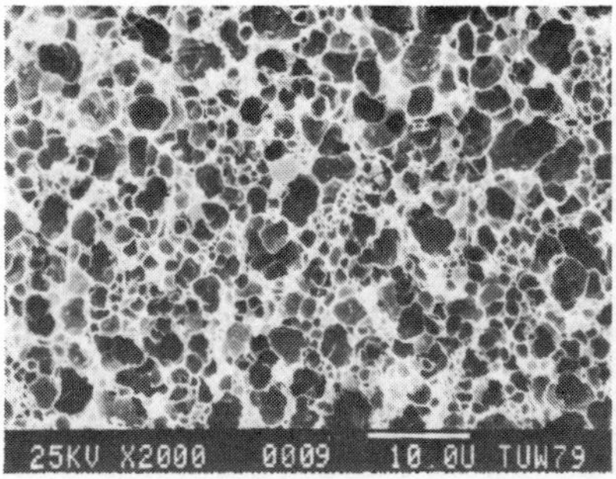

19 OZASOLP3, Magnification 2000 ×

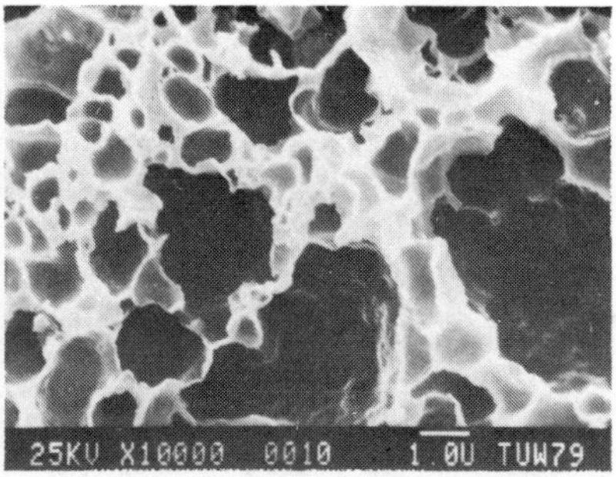

20 OAZSOLP3, Magnification 10 000 ×

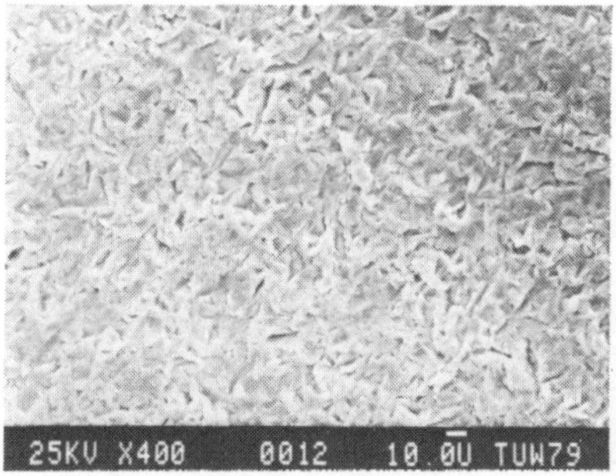

21 ALUCOPRINT PA-S, Magnification 400 ×

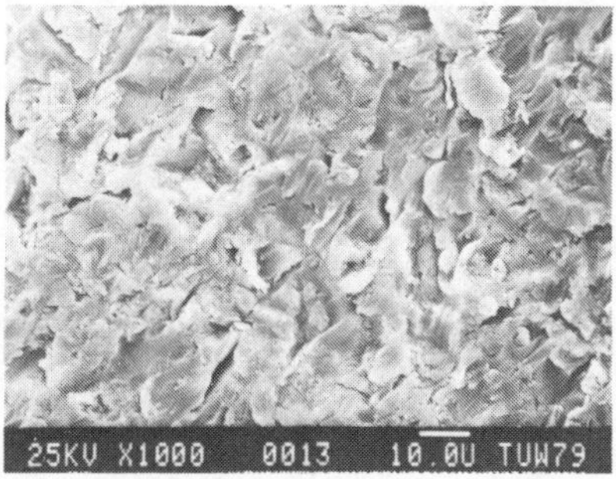

22 ALUCOPRINT, Magnification 1000 ×

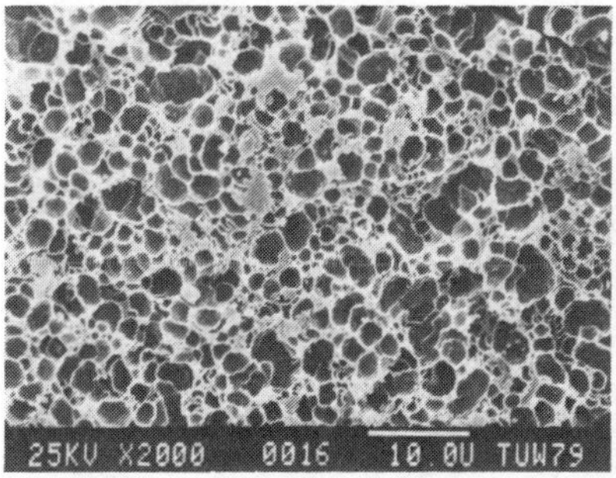

23 LITHAL, Magnification 2000 ×

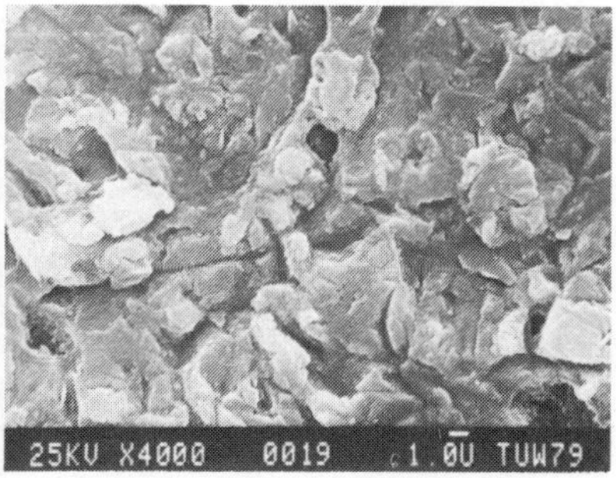

24 POLYCHROME, Magnification 4000 ×

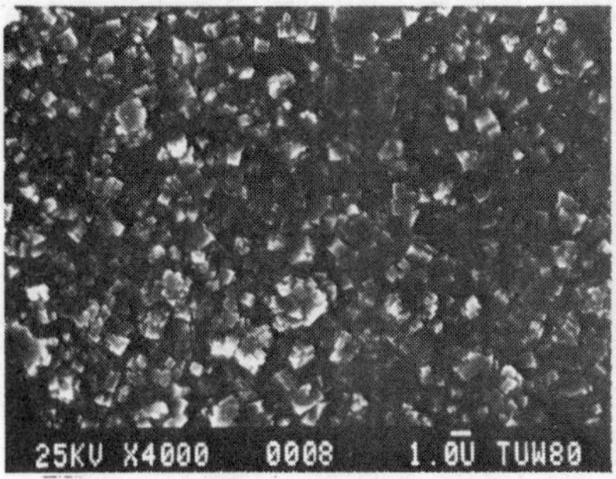

25 *KRAUSE BIO-PLATTE, Magnification 4000 ×*

26 *KRAUSE BIO-PLATTE, Magnification 10 000 ×*

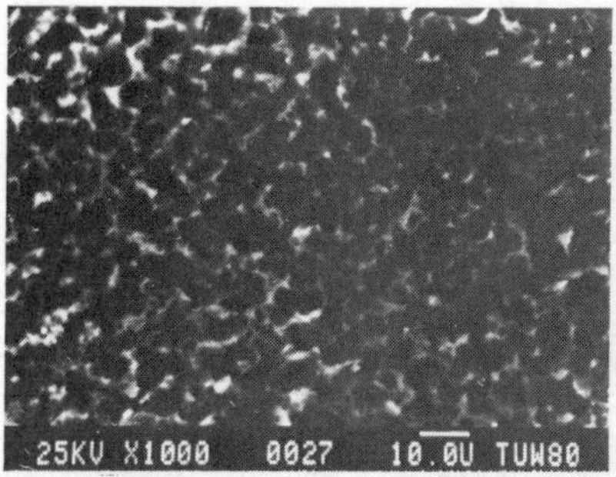

27 *KODAK-N PLATTE, Magnification 1000 ×, after 80 000 prints*

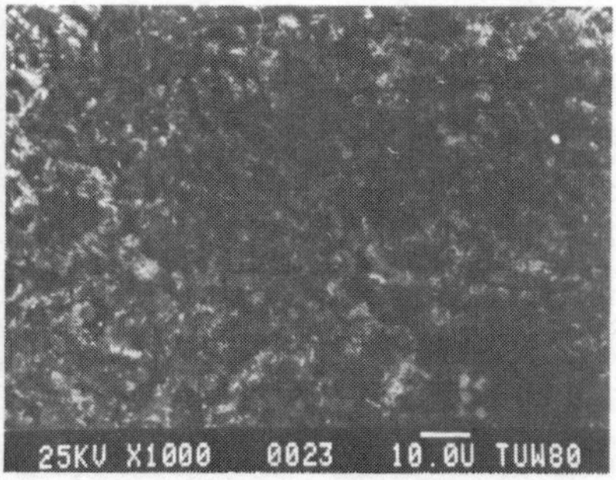

28 *POLYCHROME PAP, Magnification 1000 ×, after 200 000 prints*

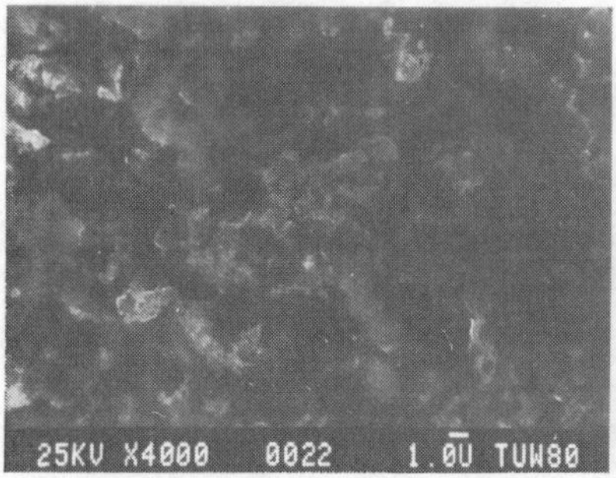

29 *PAP POLYCHROME, Magnification 4000 ×, after 200 000 prints*

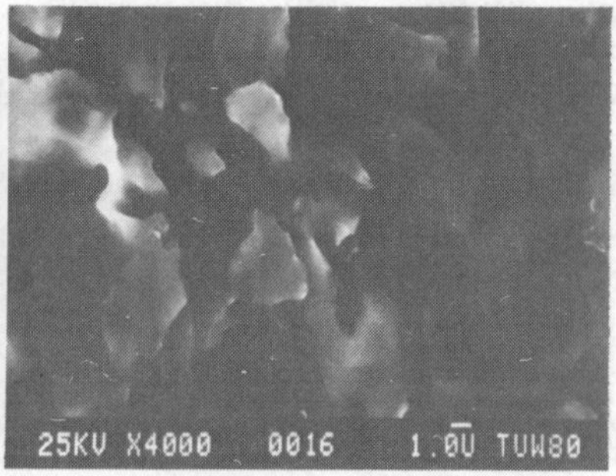

30 *FUJI FPM, Magnification 4000 ×, after 170 000 prints*

DISCUSSION

J. Fetsko Your finding that corrosion had no influence on ink transfer is not surprising. Our work showed that different types of metal plates did not affect transfer, but there were dramatic trends in transfer, solid optical density, and dot enlargement as a function of rubber plate hardness. At some time in the future, a study of blanket hardness and surface structure might be fruitful.

J. Aspler (1) Would you expect any significant difference in the wetting of an aluminium surface if you substituted an acidic or an alkaline fountain solution for distilled water?

(2) What was the surface tension of the blue ink used in the water receptivity measurements?

W. Sobotka (1) We did not investigate different pH fountain solutions. But I think in terms of longer range wetting of grained aluminium surfaces there will be some difference, especially the corrosion phenomenon is more affected than with distilled water.

(2) We did not measure the surface tension of the blue ink but we will carry out measurements soon.

J. L. Poujade You have studied the effect of corrosion on the non-printing areas of the plates after runs of various lengths. Have you also investigated the same effect on printing areas and in various conditions, more especially fountain solutions of different pH?

W. Sobotka We did not investigate the influence of different pH fountain solutions in relation to the grain structure. We are starting now to investigate the image areas of offset plates. At least the pH of the fountain solution is a very important factor concerning corrosion on aluminium plate surfaces and it will be investigated as an on-going project.

F. Andresen The manufacturers have strengthened their research departments for some years. Do the results of your studies correspond with the findings of the manufacturers?

W. Sobotka We have used all literature available on plate manufacturing from the manufacturers. Especially you can find many useful reports on plate manufacturing at the last TAGA meeting. These reports were used as a source of information.

P. J. Mangin Would you say more about the definition of your ink transfer parameter. It seems to me that you can only define an ink transfer coefficient with the same amount of ink on your plate. It would also be worth defining two transfer coefficients in percentage, i.e. from the ink roller to the plate, then from the plate to the blanket.

W. Sobotka The different amounts of ink on the different metal surfaces used is a good material coefficient for this type of metal. But I agree that for ink transfer investigations the transfer procedure has to be done with the same amount of ink on the plate surface.

INTERPRETATIONS OF INK-STAIN TESTS ON COATED PAPERS

J. Anthony Bristow and Hans Bergenblad, Swedish Forest Products Research Laboratory, Stockholm, Sweden.

ABSTRACT

A number of ink-stain tests are used for quality control purposes in the production of coated papers. The best known of these is probably the K & N Testing ink, but other inks such as the Lorilleux Porometrique ink and the Croda Red ink are also widely used.

The modes of behaviour of these inks and their interactions with coated papers are critically analysed and their merits as measures of paper absorptivity are discussed. The influence of surface roughness is demonstrated and the use of Lorilleux Microcontour ink for the assessment of roughness is described. The use of a special ink designed to permit the simultaneous measurement of both absorptivity and roughness is also referred to.

The value of the information derived from an ink-stain test in predicting the nature of the interaction between an ink and the paper in a printing operation is discussed. It is emphasised that tests should be carried out with different contact times to facilitate a better interpretation of the tests. In addition, the use of an ink-stain test to predict the tendency of a paper to give a mottled printing result is illustrated. Problems connected with the calibration of different batches of ink are discussed.

INTRODUCTION

The use of ink stain tests to study printability properties is not new. K & N ink which is probably the best known of the inks used for this type of test was put onto the market in the 1930s. Other test inks have since made an appearance. Some of these have been developed for special purposes, but often the aim seems to have been to emulate the K & N ink rather than to introduce an ink with radically different properties.

These test inks are widely used for quality control purposes and in monitoring technical developments in the paper and board industries, but the interpretation of the result of an ink-stain test is not a trivial matter. This paper describes in general terms a number of aspects related to the use of these inks and indicates various factors which should be borne in mind if the results of an ink-stain test are to be correctly interpreted. The detailed results presented relate more specifically to the K & N ink, but reference is also made to Lorilleux Porometrique ink, Croda red draw-down ink and Lorilleux Microcontour ink and briefly to Ault & Wiborg's MCA ink.

This paper deals with a series of general topics of importance in this connection. It does not provide the final answers - indeed it cannot pretend to do so, since the interaction between ink and paper is very dependent on the particular properties of the paper being tested - but it seeks to emphasise and illustrate the manner in which different factors may affect the results of an ink-stain test and to give guidance as to how the results should be interpreted.

It is assumed that these test inks will be used primarily for the evaluation of coated papers. There is in general no hindrance to their use on uncoated papers except that the method becomes insensitive when the intensity of the stain is high.

GENERAL PROCEDURE

The general routine for an ink-stain test on a paper involves the following stages:

(1) an excess of ink is applied to the surface of the paper

(2) the ink is allowed to remain on the surface of the paper for a given contact time

(3) the excess of ink is removed from the surface of the paper by wiping

(4) the intensity and appearance of the resulting stain are assessed.

The procedure for the K & N test is described in TAPPI RC-19 and is standardised in a British Standard[1] which specifies a contact time of 2 min. This time is usually also taken as standard for the other inks.

One point which we wish to emphasise is the value and importance of carrying out tests at different contact times in order to ensure that a low absorptive capacity is not mistaken for a low absorption rate or vice-versa. This report shows data for contact times up to an hour or more. In general, measurements in the range $\frac{1}{2}$ - 8 min are usually sufficient. The results are presented on square-root-time scales, as is customary in the study of absorption processes which may be expected to take place under the action of capillary forces.

Some comment deserves to be made about the manner of removal of the excess ink. In many cases it may be sufficient to remove the excess by hand using, e.g. cellulose wadding or a suitable tissue, but such a routine is clearly liable to be influenced by individual wiping characteristics. Where greater reproducibility is required, a mechanised wiping is clearly desirable. The British Standard includes the supplementary information that the error can be reduced if an automatic instrument such as that produced by GFL in Sweden[2] is used and this instrument is in fact widely used in Scandinavia.

QUALITATIVE OR QUANTITATIVE?

It is a historical fact that K & N ink was introduced solely for qualitative comparative tests and it is worthy of note that this laboratory continues to market the ink with such comparative tests in mind.

It has, however, become increasingly common for the ink to be used as a quantitative test on a single paper material, the result of the test being characterised by the K & N-value for the paper. Unfortunately there is some lack of agreement as to how this value should be calculated. In the USA a K & N-value is sometimes calculated according to:

$$K \& N = 100\, R_F/R_\infty \qquad (1)$$

where R_F is the reflectance factor of the stain and R_∞ the reflectivity of the paper.

According to the British Standard, however, the K & N-value is computed as follows:

$$K \& N = 100(R_\infty - R_F)/R_\infty \qquad (2)$$

This equation is accepted in Europe and is to be preferred since it gives a value which increases with increasing stain intensity. The measurements are carried out with a FMY/C filter and with an opaque pad of the paper as background.

An analogous ink-stain value may be calculated for any ink Q:

$$Q = 100(R_\infty - R_Q)/R_\infty \qquad (3)$$

where R_Q is the reflectance factor for the stain of ink Q. The ink-stain value is hereinafter referred to as the Q value.

The interpretation of this Q-value is not a simple matter. Many opinions have been expressed as to how such a value should be interpreted and as to whether or not it deserves to be called an absorption value[3, 4, 5, 6, 7]. In the simplest and most unequivocal terms, it shows the percentage decrease in reflectivity of the surface resulting from the retention of ink by the paper. This retention can depend on different factors, and the Q-value should certainly be interpreted in different ways in different circumstances. The British Standard designation of the K & N-method as an 'ink absorption test' is something of an oversimplification.

INK COMPOSITION

Clearly no ink-stain test can be properly interpreted unless the composition of the ink and its mode of interaction with paper is correctly understood. When the merits of a new ink are being assessed it is worthwhile to carry out a few simple tests to investigate the types of interaction occurring. In the first place, it is for instance essential to know whether or not the intensity of the stain is dependent on the time during which the ink is in contact with the paper before the excess is removed.

Figure 1 shows the Q-value as a function of contact time for four different test inks on four different grades of coated paper. A cursory glance at the figure is sufficient to show that these four inks behave in different ways. To describe in detail exactly how the respective inks interact with the paper is a considerably more difficult task. One or two features are, however, apparent and these deserve comment.

Microcontour ink

This ink gives a value which is virtually independent of time and the curves for the different papers are virtually horizontal and parallel. The Microcontour value is evidently primarily a surface roughness parameter. A visual inspection confirms that the blue colour of the stain is associated with a solid-phase pigment which is retained in the surface cavities while the colourless oil component is absorbed. This oil does gradually lead to a slight darkening of the stain so that a Microcontour assessment should preferably be made after a short contact time to eliminate the risk that the Microcontour stain value will include a contribution from this oil. The behaviour of Microcontour ink is an indication that it is to be expected that the result of any ink-stain test will be dependent on the surface roughness.

K & N ink

This ink is basically a non-drying ink consisting of a coloured oil with a white pigment extender. The value increases with increasing contact time and thus clearly gives a measure of the absorptivity, although it is to some extent affected by the surface roughness. The fact that the curves intersect shows how important it is to carry out measurements at different times in order to obtain a clear picture of the pattern of behaviour. An additional aspect is apparent: after an initial rapid absorption the curves reach a plateau when the coating is saturated and thereafter there is only a slow absorption as the ink penetrates into the base stock.

Figure 2 shows one important result of a study of differences between different batches of ink. A correlation was found between the K & N-value on a given sample of paper and the viscosity of the mineral oil phase which had been separated by ultracontrifugation. This is an important factor to be borne in mind in the formulation of inks for this type of test and in the interpretation of the results. It is the viscosity of the liquid phase entering the paper which must be kept constant and not necessarily the viscosity of the ink as a whole. See also Table 1. The composition of the K & N ink is such that, if tests are carried out over a long period (ca 24 h), the mineral oil (a) produces a transparent dark stain

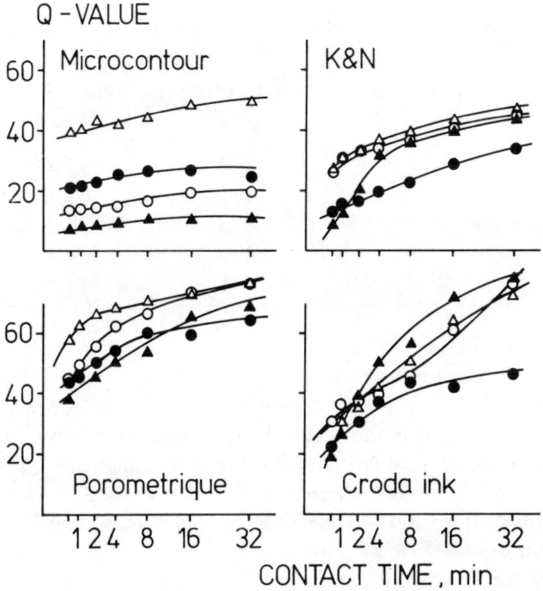

Fig. 1. *The ink-stain value as a function of time (square root scale) for four different test inks on four grades of coated paper.*

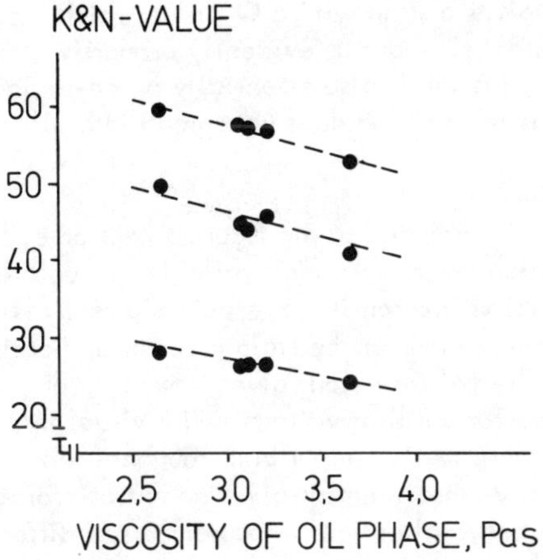

Fig. 2. *The K and N value (2 min) is dependent on the viscosity of the oil phase. Data for three different papers.*

and (b) begins to show lateral spreading away from the stain.

TABLE 1 The absorption depends on the oil phase viscosity

Ink sample	K & N-value	Absorption g/m^2	Total viscosity Pas	Oil phase viscosity Pas
K	52.6	4.0	55.4	3.64
L	57.0	4.1	63.8	3.19
M	56.9	4.3	57.8	3.10
N	57.2	4.3	62.0	3.06
O	59.4	4.8	48.9	2.63

Porometrique ink

This ink is also stated to consist basically of an extender in a coloured non-drying oil. The ink also gives a value which increases with increasing contact time and is thus essentially a measure of absorptivity. The stains are however very dark, and the curves appear to intersect the ordinate axis at a high level which suggests that surface roughness probably has a greater effect than in the case of K & N-ink. It is also noteworthy that curves 1 and 2 intersect at a much longer contact time than in the case of K & N ink.

The ink differs markedly from K & N ink however in its behaviour over very long contact times. The Porometrique ink gives a dark opaque stain with no tendency to exhibit lateral spreading.

Croda red draw-down ink

Croda red ink gives a stain with a Q-value which increases steeply with increasing contact time and is evidently primarily a measure of the paper absorptivity. This ink is also essentially a non-drying ink. It also gives a stain which is opaque and does not spread laterally.

A & W MCA ink

Although it is not included in the figure, mention can here be made of another ink which has been made available for such tests. This Micro-contour-Absorption ink consists of a blue pigment in a yellow oil and in this case the blue colour of the stain gives an indication of the surface roughness and the yellow colour gives a measure of the absorption. This ink is attractive for qualitative tests and a visual inspection of the stain gives valuable information about both roughness and absorption, but a single quantitative measurement gives no extra information. It is necessary to carry out several optical measurements at different wavelengths in order to separate the two components and to extract the information which the stain offers.

General comments

It is clear that the Q-value for any given ink will depend on both the

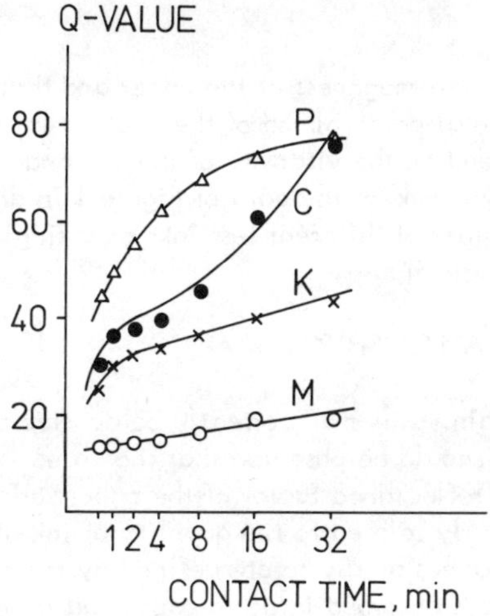

Fig. 3. *The ink-stain value as a function of time for four different inks on a single coated paper. P = Porometrique, C = Croda, K = K and N, M = Microcontour.*

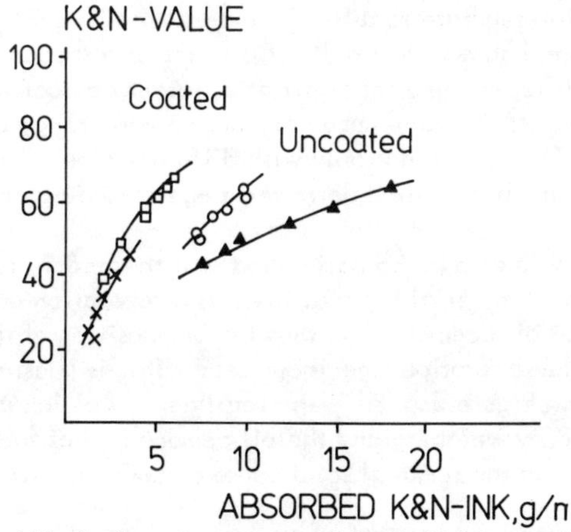

Fig. 4. *The K and N value for a given quantity of absorbed ink is strongly dependent on the type of paper tested.*

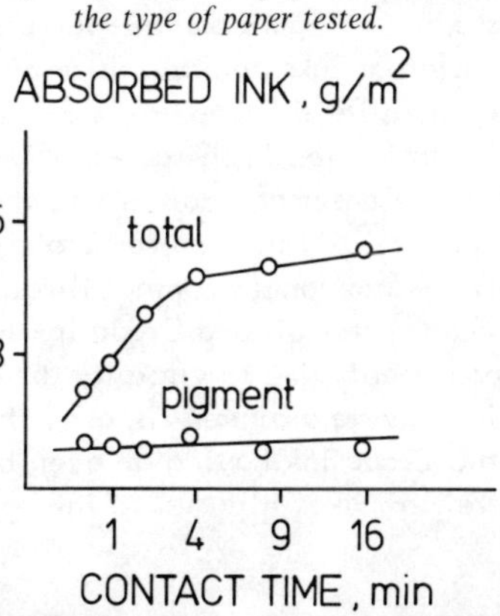

Fig. 5. *The quantity of K and N ink absorbed by a coated paper as a function of time, before and after extraction of the oil phase.*

absorptivity and surface roughness of the paper and that it may in some cases indicate the total pore volume of the coating layer. Clearly the value will also depend on the viscosity of the ink and on its optical intensity. Figure 3 shows some of the data of Figure 1 in another way, the curves for the Q-values of different test inks on a single grade of paper being shown in a single figure.

GRAVIMETRIC RELATIONSHIPS

Although the ink stain value may correctly be considered to be a measure of ink absorption it should be observed that the value is an optical measure of a change in reflectance factor of the paper surface and that the Q-value is not linearly related to the quantity of ink absorbed. The optical value approaches an asymptote defined by the colour of the ink. In addition, the ink is absorbed into the paper and interacts with the fibre network or coating pigment and the relationship between Q-value and ink quantity is not independent of the paper. Furthermore, the various components in the ink separate to different extents in different papers.

Figure 4 shows the results of a series of tests carried out with K & N ink on different types of paper at different contact times. It is evident that the K & N-value may vary over a wide range at any given ink quantity. This is true not only with different types of paper as shown in the figure, but also within a narrower range including only different qualities of coated paper.

It should also be borne in mind that the K & N ink is initially a multi-component material but that there is a separation and preferential absorption of the oil component so that the composition of the ink absorbed changes as the absorption time increases. This is illustrated for one grade of paper in Figure 5. Here the total amount of ink absorbed has been determined by weighing and the oil component has subsequently been extracted and the residual solid phase component determined by reweighing.

The differences between the different test inks can also be clarified by gravimetric studies. Figure 6 shows the uptake of ink as a function of time for three different inks and for castor oil, on a single grade of coated paper. The curve for this vegetable oil is typical of the absorption of oil in paper, the inflection in the curve being related to the saturation of the coating and the onset of sorption into the base paper. It is evident that the curves for K & N ink and for Croda red ink are similar but that there is a shift towards longer contact times, i.e. K & N ink is abosrbed more slowly than castor oil and Croda ink more slowly than K & N ink. This is presumably due to viscosity differences, the oil component of the K & N ink having a higher viscosity than the castor oil and the oil component of the Croda ink having an even higher viscosity. An awareness of this difference makes it easier to interpret the variation in behaviour in Figure 1.

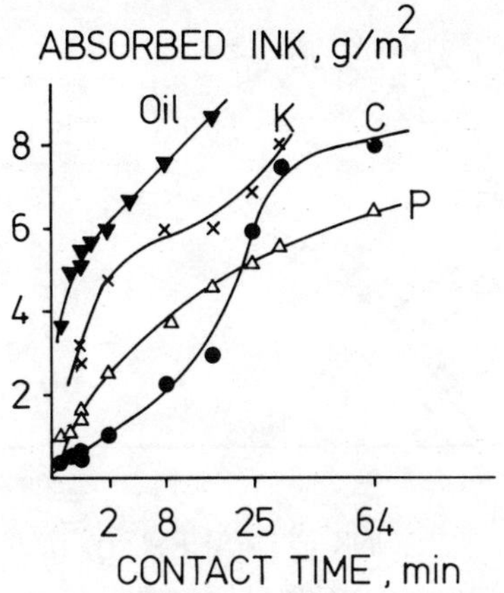

Fig. 6. The quantities of different test inks absorbed as a function of time.

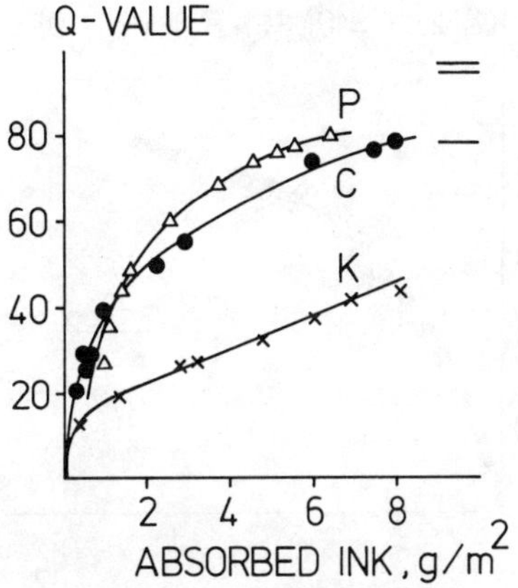

Fig. 7. The ink-stain value for three test inks as a function of the quantity absorbed.

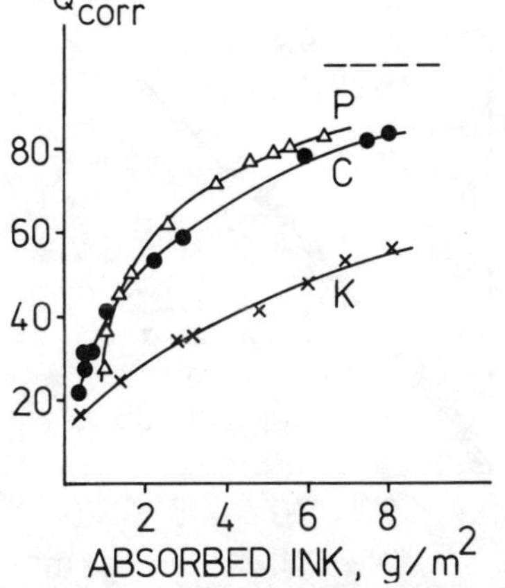

Fig. 8. The data for Fig. 7 corrected according to Equation 4.

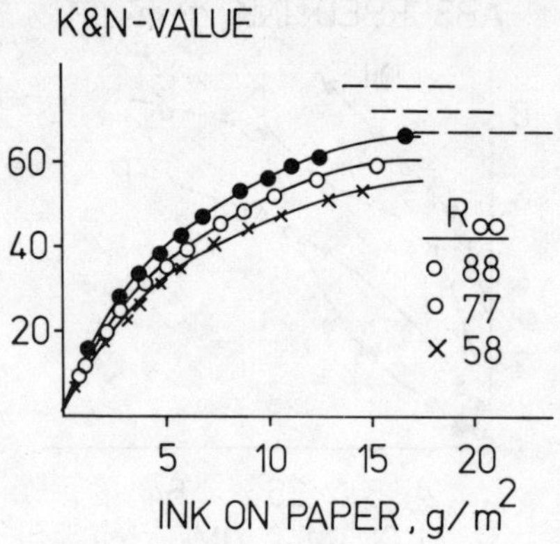

Fig. 9. The K and N value as a function of the quantity of ink transferred to a single grade of paper coloured to different reflectivity levels.

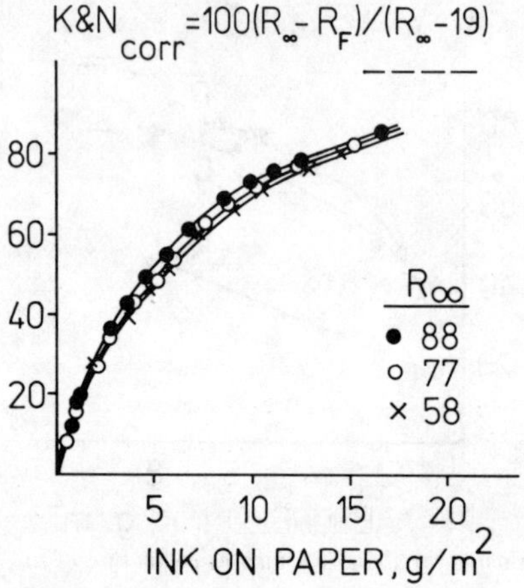

Fig. 10. The data of Fig. 9 corrected according to Equation 4.

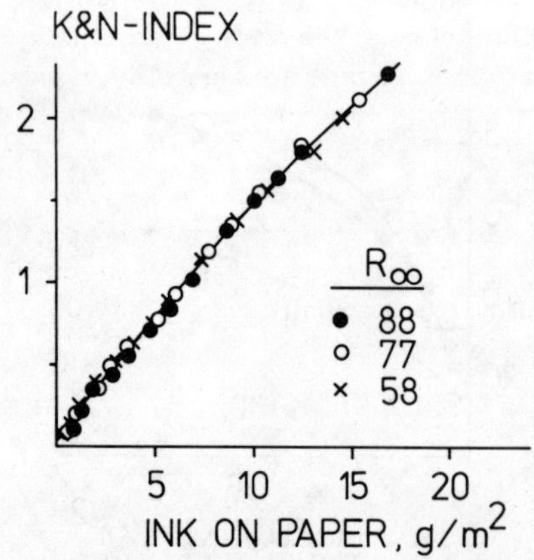

Fig. 11. The data of Fig. 9 calculated to give the K and N Index defined by Equation 5.

The curve for the Porometrique ink in Figure 6 intersects the curve for Croda ink and has in general a different form from that of the other inks so that it is not possible at the present time to give a clear explanation of how this ink behaves.

This gravimetric information provides a partial explanation of the differences in Figure 3.

It is evident in Figure 7, where the Q-value is plotted against the quantity of ink absorbed in g/m^2, that the inks have different intensities and in particular that K & N ink is much lighter in shade than the other inks. The curves tend towards different asymptotes depending on the saturation intensity of the ink, but it is not possible to define this asymptote unequivocally since the saturation intensity is paper dependent.

Figure 8 shows an attempt at normalisation based simply on values for the saturation intensity obtained from thick layers of the ink. The corrected ink-stain value, Q_{corr}, has then been calculated according to:

$$Q_{corr} = 100(R_\infty - R_Q)/(R_\infty - R_{Q\infty}) \qquad (4)$$

where $R_{Q\infty}$ is the reflectivity of the ink itself. The K & N ink is clearly lighter than the other two inks even when these corrected values are compared.

INFLUENCE OF PAPER REFLECTIVITY

If the Q-value is to be taken as a measure of the ink absorptivity of the paper it may be important when comparing papers to take into consideration the reflectivity (Y-value) of the paper itself.

Figure 9 shows how the K & N-value for a given quantity of ink on the paper is influenced by the Y-value of the paper. In this study[8] the K & N ink was transferred to the paper by a printing procedure on an IGT AC2 laboratory press. Samples of paper were stained by dipping them in solutions of Sudan brown in a solvent which was inert towards the paper structure and did not alter the ink transfer characteristics.

The primary reason for this reflectivity dependence lies in the fact that the ink is not black and that the curves approach maximum values which are distinctly different from 100. The easiest way to eliminate the dependence is to calculate a corrected value as in Equation (4). The Q-value is then the percentage change in reflectance factor within the limits represented by the reflectivities of the paper and of the ink and the absorption curve approaches the limiting value of 100 independent of the reflectivity of the paper.

Figure 10 shows the data of the previous figure corrected in this way using a value $R_{F\infty} = 19$ for the K & N ink. In practice even this value of $R_{F\infty}$ may be paper dependent since the ink absorbed varies in composition according to the roughness and absorptivity of the paper as previously indicated.

It has also been shown that a K & N-index X which is independent of the reflectivity of the paper may be calculated according to:

$$X = 100(R_\infty - R_F)/R_\infty R_F \qquad (5)$$

$$= 100(1/R_F - 1/R_\infty)$$

This index has the advantage of being an almost linear function of the quantity of ink transferred to the paper. A further advantage is that it is not necessary for $R_{F\infty}$ to be known.

Figure 11 shows this K & N index corresponding to the date of Figure 9.

INFLUENCE OF SURFACE ROUGHNESS

One of the basic problems of all absorption evaluation is that the surface roughness almost always affects the test results. This is true of ink-stain tests in general. The wiping action leaves ink in the surface craters and a rougher surface thus retains more ink than a smoother surface.

This is essentially the mode of operation of the Microcontour test in which it is the retention of the coloured solid phase which creates a stain. It also affects, for example, the K & N test because the white extender pigment is retained in the surface. Although this pigment masks the ink - as can readily be seen if the test is carried out over a period sufficient to allow penetration of the ink to the reverse side of the paper where a darker stain is observed - the pigment does not lie on the paper as a white layer but is itself stained and gives a positive contribution to the K & N-value.

The fact that the K & N-value is influenced by the surface roughness was recognised by Chatwin[9] who proposed that the K & N-value should be corrected for surface roughness by subtracting the value obtained by extrapolation of the K & N curve to zero time. Although a reasonable suggestion for obtaining a roughness value, the subtraction is mathematically unsatisfactory since the absorption and surface roughness components cannot be expected to be additive.

In general terms, the roughness component Q_r is related to the reflectance factor at zero time R_{Qo} by the relationship:

$$Q_r = 100(R_\infty - R_{Qo})/R_\infty \qquad (6)$$

Instead of correcting for surface roughness by subtracting Q_r from Q, it is better to determine R_{Qo} by extrapolation to zero time and to consider this value the starting point for the absorption process. The following expression should then be used to compute an absorption component Q_a:

$$Q_a = 100(R_{Qo} - R_Q)/R_{Qo} \qquad (7)$$

This expression appears to be mathematically more defensible than Chatwin's proposal but the assumption of additivity in reflectance values is still of doubtful merit. It can only be approximately valid for very low roughness and absorption values where the reflectance factor is a linear function of the ink quantity on the paper. The merits of the Q-index are here worth noting.

In the first place, the Q-index is an almost linear function of the transferred ink quality, and, in the second place, the index is additive.

$$X = X_a + X_r \quad (8)$$

where the roughness and absorption values are calculated as follows:

$$X = 100(R_\infty - R_F)/R_F R_\infty$$

$$X_r = 100(R_\infty - R_{Fo})/R_{Fo} R_\infty$$

$$X_a = 100(R_{Fo} - R_F)/R_F R_{Fo}$$

Figure 12 shows for a number of different grades of paper that there is a reasonable correlation between the X_r-value and the Parker-Print-Surf roughness value.

COATING SATURATION

Reference has already been made to the fact that one important consideration when using an ink-stain test to characterise a coated paper is that the absorption curve passes through a plateau when the coating is saturated with the ink. This may occur within the 2-min contact time normally adopted and erroneous conclusions may be drawn from the results of a test unless this possibility is taken into account.

This has been illustrated[10] for samples of coated paper with different coating weights and dried under different conditions. In the case of the samples dried under mild conditions, the curve rapidly levels out at a saturation level and the 2-min value is thus solely a measure of the available pore volume in the coating. In the case of the material dried under extreme conditions, the K & N-value increases steadily and the 2-minute value is an indication of the absorption rate.

VISUAL ASSESSMENT OF DIFFERENCES - SUBJECTIVE THRESHOLDS

These inks are not always used in circumstances where a quantitative Q value is readily obtainable. Some people rely solely on a qualitative assessment and for this reason a simple test has been carried out to provide some indication of the extent to which an observer can visually discern a difference between the intensities of two stains.

By varying the contact time by small intervals on a single grade

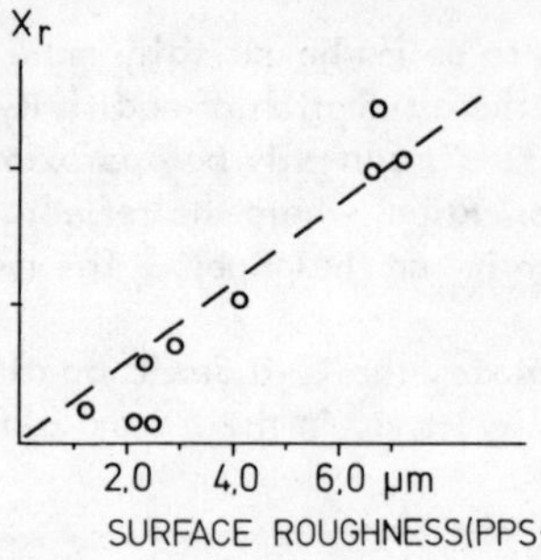

Fig. 12. The surface roughness component of the K and N Index plotted against the Parker-Print-Surf roughness value.

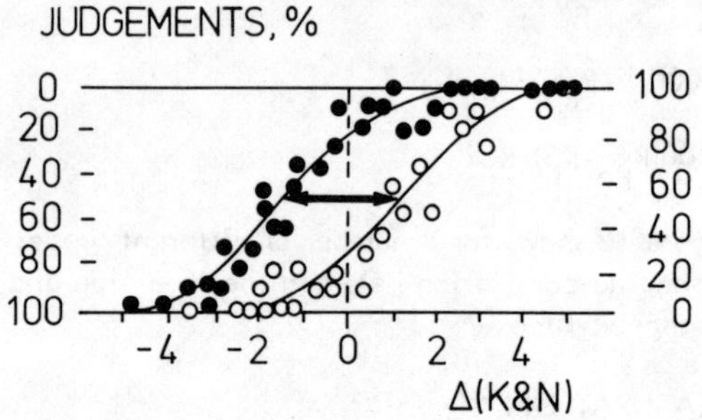

Fig. 13. The proportion of judges who consider one K and N stain to be lighter than (left-hand curve) or darker than (right-hand curve) a reference stain plotted against the measured difference in K and N value.

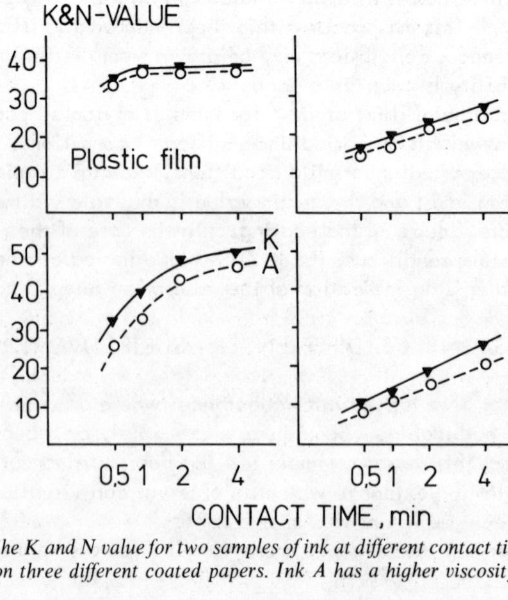

Fig. 14. The K and N value for two samples of ink at different contact times on a plastic film and on three different coated papers. Ink A has a higher viscosity than ink K.

of coated board, a series of stains was prepared with only very small differences in intensity between adjacent members in the series.

A panel of observers was presented with sub-series of these stains in random order and was asked to compare the stains in pairs and to judge whether or not there was any difference in intensity between the stains. The K & N-values of the stains were measured and the data were analysed as indicated in Figure 13, the fractions of the panel correctly discerning, failing to discern and wrongly discerning a difference being plotted against the known difference. This permits the evaluation of any desired threshold L_X at which X % of the persons on the panel correctly discern the known difference.

In the range of 20-40 K & N units, the L_{50} threshold was 1 unit and the L_{90} threshold about 3 units. A detailed analysis of the data showed, however, that the responses of the judges were not random but clearly showed different thresholds for different persons.

This test was carried out under uniform conditions of illumination. The nature of the illumination may be expected to influence the result and the perception threshold may be different in other mill or laboratory locations.

STAIN NON-UNIFORMITY - MOTTLING, SPECKLE

Although it can be an advantage to measure the reflectance factor of an ink stain and to compute a quantitative ink stain value, it should be emphasised that there is information to be gained from a subjective investigation and that it is unwise to neglect this aspect.

A single optical measurement provides only a mean value for the area of the test whereas the stain may show deviations from uniformity which are important warnings of non-uniformity in printability.

The Microcontour test, for instance, may yield a non-uniform stain which is an indication of a roughness variation to which other roughness gauges are insensitive. Such a non-uniform roughness may be critical for speckle in gravure printing and a subjective assessment of a Microcontour-ink test may provide important information in this respect.

Similarly, an absorption test such as the K & N test may give a stain of variable intensity which can be a valuable indication of a non-uniformity which can give rise to serious mottling problems in an offset process. It is not possible to state with certainty that any of the test inks is better than another in this respect. In some series of tests, the K & N ink has apparently shown a better ability to detect a tendency to that type of mottling, whereas in other tests the Porometrique or Croda red inks seem to have a greater potential. Much depends on the surface roughness level of the samples and on the level of absorptivity since at a given contact time these different inks display different regions of the absorption curve. It can be said in favour of Croda red ink that it is darker than K & N ink and shows good contrast and is less affected by surface roughness than Porometrique ink. The fact that it is absorbed more slowly and

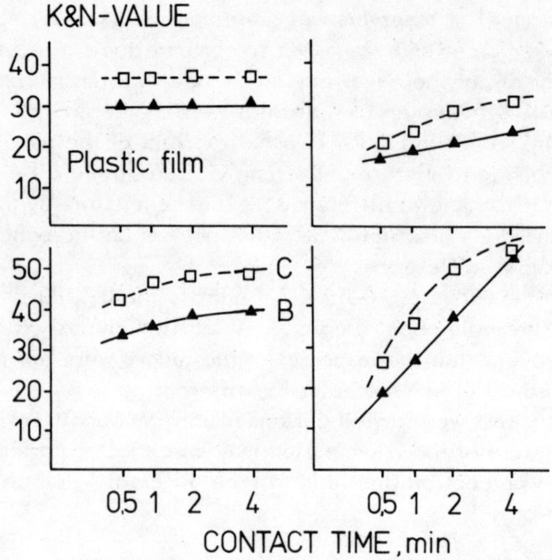

Fig. 15. The K and N value for two samples of ink at different contact times on a plastic film and on three different coated papers. Ink C is darker than ink B.

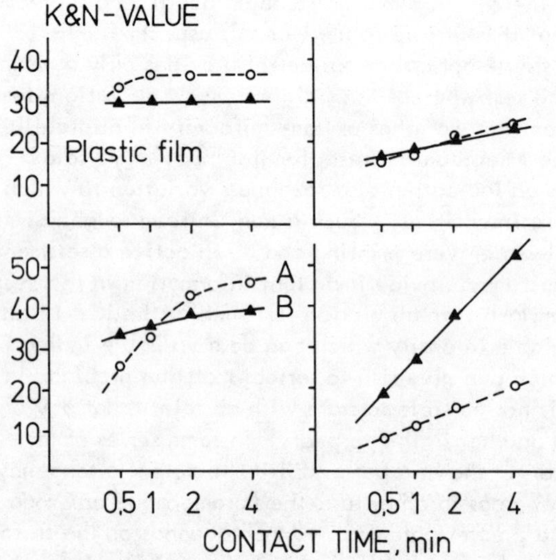

Fig. 16. The K and N value for two samples of ink at different contact times on a plastic film and on three different coated papers. Ink A is darker and has a higher viscosity than ink B.

does not spread laterally also means that it is potentially more sensitive to local variations.

CALIBRATION OF INKS

Anyone using these test inks for any considerable time knows that there are differences between batches. The manufacturers are aware of this and cautiously refrain from claiming reproducibility. A statistical multivariant analysis involving different laboratories and different batches has shown significant differences between batches [11] and this difference can mathematically be reduced by the adoption of a calibration term between one batch and another, but further analysis shows that the interaction between papers and inks is so complicated that it is not possible to assign to a batch any simple generalised calibration term or calibration factor of universal applicability.

Differences between batches arise not only from differences in viscosity but also from differences in colour intensity and perhaps also from differences in the content or structure of extender.

The following illustrates an extreme case of the behaviour of different samples of K & N ink. There is good reason to believe that K & N Laboratories Inc. have now improved their calibration routines, so that these results are presented for their illustrative value and not as any comment on the quality of K & N ink. It should also be mentioned that these were residual samples from almost empty jars and not fresh samples. The results do however illustrate the manner in which two different factors give conflicting calibration situations. Each figure shows K & N-values for two different jars of ink as a function of contact time on four different materials, a coated plastic film (3M Scotch Print)[12] and three different coated papers. The plastic film is rapidly saturated and gives a comparison between the intensities of the ink. The three papers are listed in descending order of surface roughness.

Figure 14 shows two inks which evidently have the same intensity but differ in viscosity so that ink A is absorbed more slowly than ink K.

Figure 15 shows two inks which differ greatly in intensity so that it is not easy to determine whether or not there is also a viscosity difference.

Figure 16 shows a case where ink A is darker than ink B but also has a higher viscosity. The absorption curves therefore intersect and the results observed depend on where this intersection lies in relation to the contact times employed. This figure shows clearly that it is not possible to label the ink with any simple calibration term which enables the 2-minute values for one ink to be transformed to 2-minute values for the other ink.

Theoretical calculations indicate that it might be possible to introduce a calibration system based on the specification of a batch-dependent contact time in conjunction with an intensity term. This would, however, involve a departure from the standard 2-minute test and the use of different times for each batch and this is hardly feasible.

It thus appears desirable and correct that each company should seek to ensure continuity in its own measurements by calibrating each new batch of ink in relation to its own usage, although it is to be hoped and expected that an awareness of these problems will lead to improvements in the manufacturers ability to control their own products. Unfortunately the problem is not merely in the hands of the manufacturers themselves - they are in turn subject to variations in the raw materials supplied to them.

CONCLUSIONS

These ink-stain tests will continue to be an important part of the arsenal of quality control and product development in the coated paper and board field. Although a standardised ink contact time is essential if quantitative ink-stain values are to have any value, it is recommended that tests be carried out at different times to ensure that the pattern of behaviour is understood and to facilitate a correct interpretation of the reasons for variations which may be observed.

REFERENCES

1. British Standards Institution BS 4574:1970
2. Carlsson, G.E., GFL Medd., No 58 (1967)58
3. Karttunen, S., Paperi ja Puu, 53 (1971)617
4. Kelly, J.M., Russell, P.J., Graham, F.H., Paper, 1 (1972)407
5. Fetsko, J.M., Tappi, 42 (1959)110
6. Aschan, P.J., Tappi, 56 (1973)4:78
7. Austin, M., Proc. PIRA, IARIGAI Conf. Applied Litho Tech. 1970, PIRA (1971)
8. Bristow, J.A., Sv. Papperstidn., 78 (1975)249
9. Chatwin, C.E., Printing Technology, 16 (1972)2:45
10. Fineman, I., Pauler, N., Rylander-Hällen, B., STFI Report B512 (1978)
11. Bristow, J.A., Bergenblad, H., STFI Report B321 (1975)
12. Suggested by Hultgren, K., GFL

DISCUSSION

M. N. Bruno Have you noticed any effect of the pigment in coated papers on the readings? When Preucil was working on the Paper Surface Efficiency factor we ran into one lot of paper which gave very misleading results and discovered that the paper contained satin white in the coating which appeared to bleach the K & N stain.

J. A. Bristow I have no comment to add to this. There may be interactions. Incidentally, the manufacturers have indicated that if the K & N ink is used on uncoated papers, the colour may be affected by the pH of the paper.

R. Ginman At FPPRI we have noticed that for certain coating pigments the K & N test gives misleading results in comparison with the ink absorption which takes place in printing.

M. B. Lyne Are there differences in the chemistry of the ink vehicles and are there volatile components in any of the inks?

J. A. Bristow I have not enough information to be able to discuss the chemistry of the ink vehicles. I have never had cause to suspect that the results may be affected by the loss of volatile components and I do not believe these are present.

ELECTRONIC PRE-PRESS SYSTEMS SURVEY

Peter Daciuk and Harvey F. George, Gravure Research Institute, Port Washington, New York, U.S.A.

ABSTRACT

Gravure printing in North America is entering an era of computer-assisted methods for pre-press operations. There are a number of compelling reasons for adopting computer-assisted methods including lower operating costs, greater control of operations, increased processing speed, and pressure from competing printing processes. Several approaches are being taken by gravure printers including custom design, integration of existing equipment, and purchase of pre-arranged systems which have complete capabilities from copy input to cylinder engraving.

While many previously labor-intensive functions are handled by newer systems, programming capabilities and the complete documentation of operating systems and application programs are essential in order to retain full control over the computer-assisted methods used in preparatory operations. The acquisition of capabilities needed to change programs, work routines, and to incorporate new types of terminals or engraving devices, is likely to be time-consuming and expensive. However, these skills are necessary for the printer who intends to advance his system beyond developments offered by vendors.

Proper planning will require careful systems evaluation on a regular basis. Management will be faced with rapid development in various technologies which may impact on their particular operations.

This paper includes a survey of available systems and their operational characteristics. The development of systems approaches in North American gravure plants are described and the advantages and disadvantages of two systems' approaches are discussed. Challenges that management will face such as work flow control, data management, and the general integration of preparatory requirements and computer processing are examined. Future trends in computer and data communications technology, which may have a significant impact on gravure preparatory operations, are described.

INTRODUCTION

Rationale for Incorporation of Digital Image Process Systems into Publication Gravure Printing Operations

For the past several years Gravure Research Institute has been conducting techno-economic studies of North American gravure printing operations. These studies have centered around cost performance of existing technology and theoretical examinations of new or soon to be introduced digital process technologies. This work has been known to our members as Op Gravure 90.

Initial Op Gravure 90 work centered around finding present operating costs for gravure operations on an average basis in the publication field. Data was collected from nearly all companies operating in the United States and Canada with a standard questionnaire. Several GRI reports were written which documented the findings - included were average man-hour costs for the production of positives and gravure cylinders[1].

Comparisons were made between costs of producing various types of pages using scanners and photomechanical reproduction techniques. Similar comparisons on a man-hour per operation basis were made with cylinder making operations. Data was collected and comparisons were made of electronic engraving systems and chemical etching techniques.

During the course of this study much information on methods of operation and problem areas became known. For instance, color scanners are today quite popular for inplant reproduction of color copy. Typically 70 to 90% of all incoming copy is scanned. Retouching of positives is reduced by at least 30% when scanners are used. However, productivity on scanners is low compared to the potential capacity which this type of equipment should have. Even with low actual productivity, scanners are dramatically faster than older photo-mechanical methods. In order to increase productivity on scanners, job routing techniques which include copy and job analyzing functions, were recommended. In cylinder making operations, present techniques require large amounts of time to be spent on cylinder correction for color adjustment. Long lead times and higher capital investments are direct results of cylinder correction requirements. There is reason to believe that two recently introduced hard copy pre-proofing systems can help reduce cylinder lead times and the number of press proofing tests[2, 3].

Op Gravure 90 studies of gravure press operations indicate productivity which is quite impressive[4]. According to our figures, when scheduled press times were examined, publication gravure presses printing on various paper grades, produce acceptable copy close to 80% of the time they are scheduled to operate. New press speeds in gravure for high quality work are typically 20 to 25% faster than speeds of 1400 to 1600 ft/min which are currently being obtained in the fastest web offset presses. Low paper waste in gravure is also impressive when compared to other processes. As paper represents a growing part of total printing cost, low waste in gravure tends to favor the gravure process with each new paper price increase.

Since 1968, when the last GRI study of gravure operations was made, gravure printing plants producing publication products have made some dramatic cost improvements[5, 6]. Average press productivity, not including increased speed, has grown by 15%. In addition, positive making man-hours have been reduced by 40% on the average.

However, even with the large reduction in positive making costs, gravure preparatory operations are still considerably higher in cost when compared to lithographic preparatory operations. The cost difference between preparatory operations, which use halftone films and those which use continuous tone films as in gravure, is probably the single most influencing factor for the rapid introduction of digital image processing techniques to the gravure preparatory process.

The successful incorporation of digital image processing into gravure publication operations promises to maintain the growing market share of gravure printed publication products. This will primarily be possible by a resulting decrease in the breakeven point where gravure printing becomes economical, due to decreased fixed preparatory costs with digital image processing. The high speed and low waste of gravure press equipment, coupled with digital preparatory methods, which will probably be similar in operating cost for all printing processes, pose an exciting challenge to gravure printing enterprises.

However, the many difficulties imposed by this radical change in operating methods needs much further study. The final phase of Op Gravure 90 addresses itself to these difficulties.

Interview of Several Available Electronic Preparatory Systems for Gravure

What follows are short descriptions of the available graphic arts based image processing systems:

Sci-Tex Response 300 Sci-Tex Corporation was one of the first companies to offer full color page makeup and color correction systems for process color work in the graphic arts. Sci-Tex is a relatively new company founded in 1968 and based in Israel.

The Response 300 System is based on extensive software development undertaken through the 1970s. Before introduction of the Response 300 System, Sci-Tex was able to obtain graphic arts experience through their Response 200 family of systems which are line color, digital image processing systems, intended for the packaging and specialty printing fields.

Response 300 functions in the following manner. All copy is input by scanners. Sci-Tex does not manufacture this equipment. However, interfacing with any modern scanner is said to be possible. Scanners which scan all colors at the same time are preferred as scan time is shortened. At the present time type is input through scanners, and computer interfacing with typesetters, is said to be under development.

Input, image processing (page make-up/correction), and output are controlled by three separate Hewlett Packard computers.

Digital picture information controlled by a computer and obtained from the input scanner is stored on disc memory. If page make-up and correction are scheduled soon after scanning, disc information from the scanner is used. If page make-up and correction is scheduled for another time, the image information can be transferred to magnetic tape for long term storage. Using data compression techniques, a full color $8\frac{1}{2} \times 11$ in. page requires approximately 40 megabytes of storage area on magnetic tape.

The page make-up and correction subsystem has in addition to separate computer control, a digital planning table, color television tube, a computer terminal, and a function console. The computer terminal serves as a subsystem master control device. The digital planning table has an electronic stylus for graphic communication. The function console contains commonly used function commands. It also controls display of function lists called menus. The color television tube is a 19 in. diagonal with 256 vertical and 320 horizontal visible lines.

Since one station is used for page make-up and color correction, layout operations can be carried out in unity with retouching. Position accuracy is related to resolution requirements. Copy size can be changed. However, to maintain image quality, the range of size change depends on input resolution at the scanning station. Picture areas can be rotated. Special images and logos can be stored for repeat use. Image lapping is available with adjustable overlap. Through the function console, geometrical shapes and borders can be made automatically and only end coordinates need to be indicated. For non-geometrical shapes, several points can be indicated and the system will generate required shapes.

For color correction, ink and paper characteristics, which conform to plant requirements, can be stored and are used to produce the color television image. Any number of ink and paper sets can be stored. During correction two hundred and fifty six gradation levels can be addressed. Local correction is possible in that specific areas of an image can be corrected without affecting others. The planning table is used to locate areas for this activity. For correction on small areas, zoom is possible on the television screen. Density measurement can be made with display directly on the screen. If dramatic changes in color are desired, electronic airbrushing is used. Image areas can also be deleted and filled with adjacent color information by electronic airbrushing. Separate color records, as well as progressive colors, can be viewed. Black and white records can also be viewed. By using the function console, tints can be created. Additional stations can be added to the system for increased production capability.

For output from Sci-Tex Response 300, gravure users have two possibilities. Continuous tone film output is said to be possible through their laser exposure light source 60×40 in. film plotter. It was reported that this equipment is presently being installed.

Data in a composed and impositioned format can also be used to output through electronic engraving equipment. Sci-Tex supplies interface and control equipment for this sort of application. However, engraving equipment is not supplied by Sci-Tex.

Crosfield 570 System   The Crosfield 570 System is comprised of a Magnascan 550 digital scanner, DEC PDP11/04 image processing minicomputer, layout digital planning table with CRT monitor, and a disc drive for 300 megabyte discs.

To produce final 'stripped' films the 570 System operates in the following manner:

All copy including type matter is scanned to size. For each original one scaled 'plotting' film is made for subsequent positioning purposes. Digital color image information is forwarded to 300 megabyte disc storage. Once all copy for a pagination job has been scanned, the scanner section can be used in normal mode for conventional film output. A mechanical (layout) in the form of a positive film transparency or precision line drawing is needed to position plotting films. This is done at the digital planning table with an electrical sensing cursor. Plotting films are registered with the mechanical using register marks previously exposed by the scanner. Image placement positions are shown on the CRT. The CRT also confirms or rejects functions and placements.

For cutting irregular shaped images, a program is used to smooth cutting with the cursor. The digital planning table has a menu for frequently used functions needed for stripping and cutting of standard shapes such as squares and circles. The menu can also be used to create tint values. Once all positioning and cutting is done the minicomputer is used for digital image movement and manipulation. This enables output of each color page fully composed. Processing takes about 10 minutes between layout completion and start of scanner exposure. Crosfield has carefully considered operator interfacing.

This system has been reported to work well with former conventional layout personnel. In part this is due to the use of planning films made to final size. Many of the operations familiar to a layout person are still present.

As the 570 System is presently arranged, computer memory is only used for temporary storage. Work flows through the system on a page basis with output to film when a page is complete. For a typical page size, all four separations are output at the same time.

Plans for the 570 System include a high resolution color CRT and software for local color correction. This development was introduced at IPEX 80, Birmingham, England.

Hell Chromacom   The Hell Chromacom System was shown in Chicago at Print 80. The first system was installed at a lithographic printing plant in Germany[7]. Four additional systems have since been installed in Europe.

Input to Chromacom is from the Hell DC300, DC350 or CP340 color scanners with intermediate storage on disc memory. There is a computer with video terminal at the scanning station which is used to organize image data on the disc memory and code each picture element for identification purposes later. Size scaling is done during scanning. However, in addition, a range of size enlargement dependent on scanning input pitch, will be available later. Image size reduction is also available at a later time at the Combiskop station. With the present system type is also input by scanner. However, plans are underway for digitized font storage in the future.

The Combiskop section of Chromacom is used for page make-up and color correction. This station is an offline operation. The Combiskop station includes a color CRT, digital planning table with electronic cursor and light pen, function console, and a video data terminal with operating keyboard. In addition, there is a digital computer for image processing and two disc drives for program and image storage by means of 300 megabyte disc packs. One disc drive is used for input and the other for output of finished pages.

Combiskop operates in the following manner: A page layout, which is a line drawing showing position for all image elements, is superimposed on the digital planning table. Identification codes are used to address images for placement. As the operator traces layout lines, coordinate data is input to the computer for image placement. Features available for page make-up include positioning, cropping, geometric and irregular shape generation, placement of one image into another image, and recurring logo or special image storage. Color tints can be created at the planning table with color densities input through the computer terminal.

The Combiskop station also is used for local and overall color correction. Airbrushing is possible by using the tracer pen. Image contrast can be altered for printing conditions. All color correction operations are shown on the color CRT screen.

Output with the Chromacom system can be to film, or directly to gravure cylinder, using Hell K201, 202 electronic engraving machines. Before output to large format recorders, CP340 or electronic engravers, imposition operations are performed digitally. Hell markets an offline sorting station for imposition of pages. The sorting station consists of two disc drives, a computer, and video screen with terminal. Floppy discs are used to imposition pages according to preset imposition requirements. Variable data such as page allocation are input by the operator. Pages are impositioned according to printing color on to disc packs.

Optronics Pagitron System Presently, Optronics has a black and white page composition system installed at one American gravure printer. The Pagitron System has tone correction available in that tonal curve shape can be specified. Calibration for highlight, shadow, and range is automatic for each image individually.

As recently delivered, Pagitron differs from the other available systems in one major area. Layout operations are specified before input scanning. A digital planning table is used for layout. Up to 0.005 in accuracy is possible. To compose the page, each piece is mounted on a sheet of prepunched mylar which fits over pins on the digital planning table and the scanner. The prepunched mylar sheet is placed over a layout mechanical on the planning table. The operator indicates, by means of a cursor, the spacial relationship of his input copy to its position on the final page. Enlargements can be made. However, it is extremely rare that enlargements are needed for black and white work. To identify each copy piece, identification codes are input. Each piece, type or continuous tone for a full page, is treated in this manner at the planning table.

After layout is done for the whole page, each piece is scanned into the system by means of an Optronics Scanner. When all pieces are input the system arranges all data for composed page output. At this point the operator has interaction in that he can check and adjust layout at an editing terminal which has a CRT screen. If changes are needed he may make them at this point.

Output is done with an Optronics recording device. Film output can either be continuous tone, halftone, or variable density halftones. Scanning and plotting resolutions for the system are 625.23 lines/in. for type and line work and 156.31 lines/in for continuous tone work.

Comtal 3M Twin Image Display System Recently Comtal was acquired by 3M Corporation. Comtal 3M is involved in the development of color correction and layout systems for graphic arts applications. The company is also regarded as a leader in the field of remote sensing.

Comtal 3M's Twin Image Display System is basically a standalone system which has color correction and image placement capabilities. A primary and a reference display are used so image processing operations can be somewhat increased in speed. The primary display holds the picture for which processing is required. The reference display is used to work on specific areas from the primary display which need manipulation. Enlarged portions of the primary display are shown on the reference display. The enlarged portions shown on the reference display are equal in spacial resolution to the printed page. By using two displays the time needed to go back to the original image is eliminated since the original is always on display. The reference display is viewed for color correction, airbrushing, and image masking. Image masking can be used to do local correction. Layout operations can also be done. Specific areas from different originals can be used to compose a page. For operator communication a digital planning table with electronic cursor and a function console are used.

Comtal 3M is working closely with some of the large printers integrating Twin Image Display capabilities with completely digital copy to cylinder processing systems. They are also working with other equipment vendors on integrated systems for offset and gravure preparatory operations.

Lasergravure System 700 The Crosfield Lasergravure System 700 is intended to be used with input from an imposition unit which arranges page information from the 570 system. The imposition unit arranges pages for each color on a separate disc, via terminal keyboard for operator instructions.

For cylinder manufacture, the undersized cylinder base is electrostatically sprayed with plastic which is later temperature cured and hardened. The finished cylinder is turned to obtain a smooth surface of correct size. After engraving the cylinder is coated with electroless nickel.

For cylinder imaging Crosfield is using a 200 watt CO_2 laser which is modulated to create helical grooves with variable depths on the plastic cylinder surface. The laser light source engraves at a speed of 4 in/min, and the cylinder turns at 1000 RPM. Cylinders with a face width of approximately 60 and 100 in have been successfully engraved. Under test conditions one million impressions have been run without detectable cylinder wear. Field trials under production conditions are scheduled for early 1981.

GRI Model Printing Plants for Evaluation of New Electronic Preparatory Systems

From cost information collected by GRI, two hypothetical gravure printing plants were developed[8]. One plant prints magazine type products while the other prints newspaper supplements on uncoated paper grades. Primarily these plants were developed to study the effects of new technology on costs and operating methods in gravure. Working with the equipment vendors and manufacturers, new systems are being tested by GRI using the models.

One study was recently completed by Crosfield Electronics who were interested in the cost effectiveness of their Lasergravure system. Using the GRI Magazine Plant Model, the study concluded that a lasergravure system could save up to 35% of the reproduction and cylinder making costs over a film based system in the first year of operation[9].

The model plants provide insight into memory requirements for daily operation of a gravure publication plant. The magazine model plant processes 909 color pages and 219 monotone pages each month. Of this total 455 color pages are processed in-house, 300 supplied from service houses, and 154 are repeat color pages.

For digital storage and processing, the image is divided into small finite areas called pixels. The tone value of each pixel is divided into 256 density steps (to avoid false contrast) which can be represented by an 8 bit binary number called a byte. Each color (yellow, magenta, cyan and black) needs one byte reserved in storage, or 4 bytes per pixel on a color page. At 300 pixels per linear inch, referred to output, 90,000 pixels per square inch, an $8\frac{1}{2} \times 11$ in. color page requires over 33 megabytes of storage capacity if data compression is not used.

Image information usually contains much redundancy, i.e., data compression through encoding should be effective in reducing memory requirements at the cost of some programming complexity. The fact that engraving screens are usually known in advance, will allow for additional reductions in memory requirements for electronic engraving. At the front end, storage requirements are increased by provisions for radical post-scanning enlargement and unanticipated cropping. The storage requirements for one week's work of the model plant has been estimated to be in the 16 billion byte range without data compression. At present, magnetic tape provides the most economical means to store masses of data of this order of magnitude. Because of the slow access on magnetic tape, magnetic disc drives will be required at most stations where the image data are entered, manipulated or used.

Direct cylinder engraving from digital storage may require retention of the digital data for several weeks to complete all the preparatory operations for a full 48-page signature. Also, it may be wise to retain digital data until the press run is completed in case of damage to a finished cylinder. This will greatly increase the total storage requirements. Careful consideration should be given to alternatives for temporary storage of fully composed pages. Initial cost, physical storage and movement in and out of disc drives represent a cost burden. Storing images on film for later re-scan may be more cost effective and should be examined.

The movement of data within and between the electronic equipments for each preparatory work station requires the handling and storage of vast amounts of data. The scope of this data problem can be gauged by reference to the Op Gravure 90 magazine model plant. We need to manage and control over 60 billion data bytes for our model plant when we include enlargement of the film transparencies supplied each month. Improved methods to do this will utilize new technologies that offer more powerful computers, better data communications using fiber optics, a broad mix of memory types for archival or short term storage and better methods of computer input/output.

TABLE 1 Reproduction Area Peak Month Operating Costs in Magazine Production Plant Model

	Photographic Operations $	Cylinder Making Operations $
Labor	209,181	76,889
Equipment	45,623	63,438
Supplies and Materials	19,751	10,125
Light and power	350	700
Rent and Heat	4,500	5,100
	$279,405	$156,252

Table 1 shows operating costs for reproduction areas, photographic and cylinder making operations. These costs are for one month's peak operation based on 1979 equipment and labor costs. For photographic operations, labor cost equals close to 75% of total department costs. This further illustrates the importance of new technology introduction into gravure preparatory operations. The decline in available skilled craftsmen, as well as the rapid rising costs of maintaining such a labor pool, are substantial concerns in the United States. New technology promises an easing of these problems while shifting part of this cost to the equipment area.

Planning Considerations for Pre-Press Computer Systems

Although details of the internal operation of specific computer based image processing systems are presently not generally available, these systems are subject to the general capabilities and limitations of computer systems.

It is realistic to expect that newly introduced image processing systems will require changes to adapt to unforeseen requirements and novel technology. Many of these changes will be made by the vendors as experience is gained in actual day-to-day preparatory operations. It is also likely that the printer will have to initiate changes to expand his capabilities, to re-configure hardware for back-up purposes in the event of failures near deadlines and - to have control over his own preparatory operations in the absence of tangible film intermediates.

The printer who desires the same degree of control over his digital preparatory system that he now has with tangible film intermediates, will need to have some insight into the computer and software capabilities which he will have to manage. These objectives can be attained if selected members of reproduction management have a basic understanding of computer hardware and software. This will allow them to more fully work with vendors and computer professionals as they initially assess computer based preparatory systems available and adapt them to particular plants and future needs.

Another important requirement is systems documentation. The hardware of standard computers is extensively documented by the manufacturers. Industry specific equipment such as scanners and engravers are usually documented to levels needed for maintenance and repair. The major difficulty is in the interface area, especially when custom made interface modules are used. Besides such basics as description of operating principles and circuit diagrams, listings of data and control register codes, interrupt (vector) addresses and interrupt priorities should be obtained. In the past a skilled engineer or technician could get much of this information from tracing signal paths on the physical module. With the increasing use of microprogrammed devices, this approach will not be possible any more.

A work flow based on film intermediates will closely resemble traditional preparatory operations with distinct color separation, retouching, page make up and engraving departments. The level of documentation that is required has to satisfy equipment adjustment, maintenance and repair criteria. This type of specific function equipment is likely to need only documentation that satisfies adjustment, maintenance and repair criteria. The potential for later changeover to a less film dependent digital preparatory system that allows data to flow between specific function equipment, should be anticipated by review with the vendor of possible equipment input/output modification. Standard computer communication protocols would be needed to avoid design and construction of customized interfaces for this purpose.

A preparatory system that is largely without film intermediates requires thorough documentation in the programming area. In many dedicated systems, programming or software represents the largest cost item. Programs in general fall into three categories, namely, operating system, service programs and application programs. The operating system, part of which is always in working memory, acts as a general housekeeper and performs subsidiary functions (pre-programmed requests) when ordered to do so by other programs. A variety of standard operating systems are usually available from the computer manufacturers and these systems are extensively documented. These operating systems are usually very complex. Provisions are included to run programs written for earlier versions of the systems and to allow for a wide range of applications. In the interest of simplicity and speed, computer applications groups often write their own customized operating systems for dedicated computer applications. Thorough documentation of the operating system is needed to assure that program changes and new programs are supported by the system.

The same considerations apply to application programs and the way image information is encoded. Service programs such as compilers and linkers can be run off line with standard operating systems. Care must be exercised that the modified or novel application program section contains no requests that are not supported by customized operating systems.

Future Trends Which May Influence Pre-Press Technology

The trend has been established for a total electronic preparatory system that will greatly reduce today's labor intensive film based preparatory systems to one that is capital intensive and largely filmless.

The system will move pages and sections of pages to and from page composition. Typesetting capability, including font selection, will be added in the future for a complete electronic preparatory system. Faster scanning operations will overcome present data rate limitations. Color correction and color manipulation of selected sections of a photograph, together with video color displays, will gain greater acceptance. Direct cylinder engraving from digital inputs, without intermediates, is becoming practical and cost effective. Electronic transmission to the printing plant of complete pages, sections of pages, color input copy, etc. from major metropolitan areas, where advertising agencies and printing buyers are located, is likely to commence as broad channel communications and other technological innovations become more prevalent at all levels.

The computer industry has been offering more and more powerful computers, together with substantial reductions in size and cost to performance ratios. This trend continues and by the end of the decade computers and their application will have greatly altered industrial processes such as printing.

Paced by rapid and continuing changes in semiconductor technology, these improvements include the entire computer spectrum from small microprocessor based microcomputers, to the minicomputer to the large main frame computer. These changes are starting to blend the traditional lines between computer types.

Most computers today execute only a single set of instructions that can act simultaneously on multiple streams of data. Recently 'supercomputers' capable of processing multiple streams of instructions on multiple data streams, were introduced. These techniques are likely to be used in the future in lower end systems to increase computer power.

Thus, it can be seen, that computers are growing more powerful, smaller, and less costly. This trend continues and, based upon IBM experience over the last 25 years, will further improve computer cost/performance at a 25% compounded annual rate.

Data communication bottlenecks involving data rates, bandwidths, data errors due to noise and cost of data transmission, will be overcome by the rapidly developing fiber optics technology which is the fastest growing technology in the communications field. New fibers, new low loss splicing methods, new diode lasers with matched spectral emissions and higher bandwidth fibers were announced recently. In fact, optical components have now improved to the point that General Telephone Company of Indiana and the Bell System are proceeding to install fiber optic systems for real rather than experimental usage. This trend is illustrated by the decision of AT&T to develop a 611 mile system to link Boston, New York, Philadelphia and Washington by a laser powered lightwave system. This $79 million system is expected to be extended to Chicago, Atlanta and Miami later in the decade with Boston, New York and Washington completed by 1983. This all digital system will carry voice, video and data operating at 44.7 megabits per second. This allows 80,000 calls per second in a $\frac{1}{2}$ inch optical cable, using amplifiers (due to the long distance) at appropriate spacings.

Integrated semiconductor devices that perform both optical and electronic functions are receiving significant attention. Recent research advances are bringing us closer to the goal of commercially available, monolithic, integrated optoelectronic circuits. Using these circuits could bring a significant simplification in equipment for communications between computers and other high speed electronic devices.

The prospect of optoelectronic integrated circuit chip to chip communications over optical fibers at data rates in the gigahertz range represent very significant future improvements over present information flow rates available today. Realistically, one can conceive of future optoelectronic integrated devices such as optical transmitters, receivers, multiplexers, demultiplexers and, with a good imagination, logic circuits and signal processors.

Future impact of fiber optics on telecommunications is likely to correspond to the impact of semiconductors on vacuum tubes. As electromagnetic spectrum congestion increases in the 1980s, limiting expansion of satellite communications and terrestrial microwave systems, fiber optic links will begin to replace them, becoming a major factor by the 1990s. Lower costs for broadband transmission, due to fiberoptics, should result in expanded private facsimile networks, electronic mail, etc.

Two Distinct Systems Approaches for the Gravure Printer

There are essentially two approaches to computerized preparatory operations. The equipment, used for page makeup or color correction, can be treated in an 'off-line' fashion, Figure 1, performing specific functions and - in most cases - producing a tangible film intermediate, which acts as input to the next operation. The second approach involves completely computer integrated image manipulation with the use of film reserved for image input and most likely for high quality pre-proofing purposes, Figures 2 and 3. The choice of one or other alternative, off-line use of equipment or fully computerized integration will require different degrees of computer skills by preparatory management. For off-line operations, with film intermediates, equipment and program documentation is required to the point needed for checks for proper operation, adjustments, maintenance and repair. If magnetic disc packs or tapes are used as intermediates, storage formats, interface specifications and input and output protocols should be documented to facilitate the use of magnetic media prepared by equipment from other vendors. With fully integrated systems, it is, in our opinion, imperative that the printer acquire the capabilities to understand and possibly alter and adjust programs. The pieces of film that are manipulated and transferred in traditional preparatory operations are tangible objects that can be located, inspected and measured. Image data on disc are not tangible objects in the same sense. They can usually be read even when there is a partial system's failure, but the reading will be in terms of numbers from which an image has somehow to be synthesized. More important, when a scanner or processor behaves erratically, the offending equipment is readily identified and can be bypassed until it is fixed. Given enough equipment redundancy, the same applies to computerized equipment where faults are located through the use of diagnostic programs. The user should be aware of the mode in which the diagnostic programs operate.

The acquisition of capabilities needed to change programs, work routines, and to incorporate new types of terminals or engraving devices,

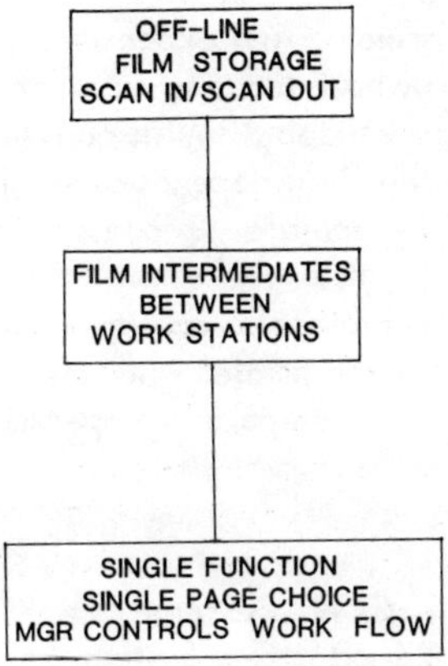

Fig. 1. Film based digital image processing system.

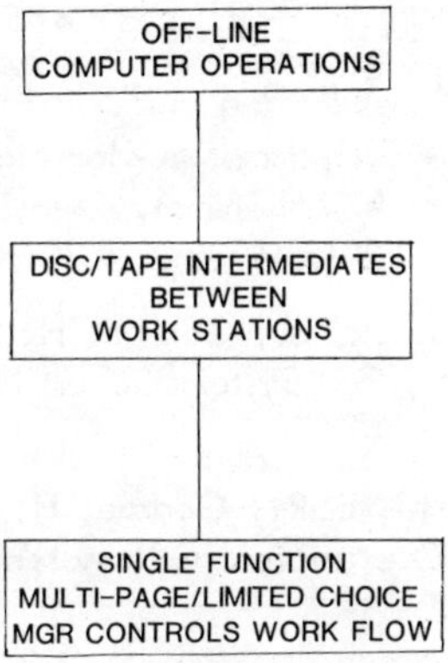

Fig. 2. Fully computer integrated digital image processing system.

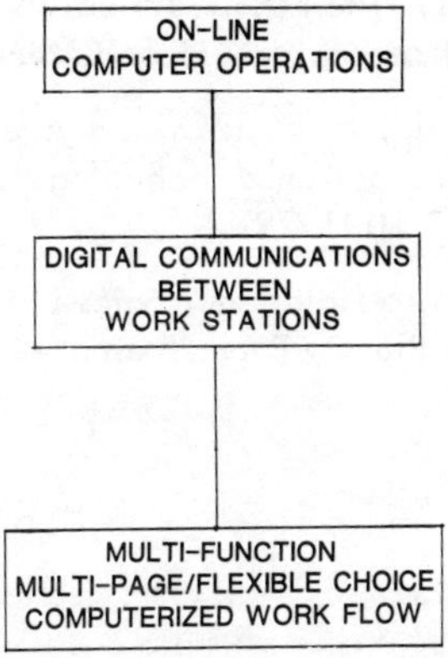

Fig. 3. Futuristic fully computer integrated digital image processing system.

is likely to be time-consuming and expensive. These capabilities are essential for the printer who intends to advance his system further than the developments offered by vendors. With conventional systems, the efficiency of a scanner in the field depends as much on the integration of the scanner into the overall preparatory process as on some peripheral capabilities a particular scanner may or may not have. Programming capabilities and the complete documentation of operating systems and applications programs are essential for the printer who intends to retain full control over the methods used in his preparatory operations.

REFERENCES

1. Daciuk, P., Iselin, C.W., George, H. F., 'Op Gravure 90 Preparatory Operations Report', GRI Systems Report No. S.22, November 8-11 (1978).

2. Warter, L. C., E. I. duPont de Nemours & Co. Inc., 'Cromalin Proofing Application for Gravure', GRI Report No. M.133, November 7-10 (1979).

3. Underthun, G., A/S Hjemmet, Oslo, Norway, 'Pre-proofing with Ektachrome R.19 at A/S Hjemmet, Oslo, Norway', GRI Report No. M.156, November 12-15 (1980).

4. Iselin, C.W., Daciuk, P., George, H. F., 'Op Gravure 90 Press Operations Report', GRI Systems Report No. S.23, November 8-11, (1978).

5. Bier, T. M., Camhi, E.R., George, H.F., 'Systems Study of Gravure Preparatory Operations', GRI Systems Report No. S.1, March 17 (1967).

6. Camhi, E. R., George, H.F., 'Systems Study of Rotogravure Press Operations', GRI Systems Report No. S.7, November 22 (1968).

7. Jordan, Dr. Klaus, 'The First Practical Experience with the Chromacom System', GRI Report No. M.168, November 12-15 (1980).

8. Daciuk, P., George, H.F., 'Op Gravure 90 Model Plants for Evaluation of the New Electronic Technology', GRI Systems Report No. S.27, November 7-10 (1979).

9. Purdy, D.R.A., Crosfield Electronics Ltd, 'Laser Engraving - New Developments and Plans', Paper Presented at the GTA Convention, April 6-9 (1981).

DISCUSSION

K. Schläpfer How do you judge the reliability of a CRT image compared with a hard copy?

P. Daciuk It is quite obvious that the CRT image looks different from a hard copy proof or a press sheet. Each vendor does request press samples for test targets for set up of the CRT. The actual techniques used by the vendor depend on the system and involve either hardwired circuits or software programming. It has been reported that tubes fatigue; however, recalibration is possible with most systems either on a daily basis or whenever visual correspondence with press sheets is in question. At this time most systems are based on densitometric values for operator interaction.

B. Corsham It may be that much of the technology you described is outside the conventional study areas for IARIGAI member institutes. However, these electronic filmless page make up and colour modification systems rely on efficient means of entering image data by conventional separation techniques which must at present be entirely by scanner. The use of expensive colour separation scanners for the input of all text and black and white originals is not efficient. There is a requirement for a high speed digitizing device for mono and line images. Even the efficiency of scanners for colour separation must be said to be far from efficient when only about one quarter of the production time is spent in scanning and most of the remainder in setting up for end points and also for adjusting controls to allow for the characteristics of different film emulsions which might appear similar to the human eye but different to the scanner. Any basic work which leads to more efficient setting of scanners will be most welcome.

PAPER PROPERTIES REQUIRED IN PRINTING WITH WATER-BASED GRAVURE INKS

Rolf Ginman, The Finnish Pulp and Paper Research Institute, Helsinki, Finland

ABSTRACT

In the first part of the study, a large number of commercial papers were printed on the Winstone laboratory press with water-based gravure inks from different ink manufacturers; in addition, a comparison was made with a toluene ink. It became evident from the results that major differences in print quality existed between the various qualities of water-based inks, especially with regard to mottle.

In the second part, the influence of coating raw materials on the water absorbency of the coated paper was studied. The water absorbency can be lowered by sizing the base paper, and by increasing the amount of starch in the binder.

In the final part, several commercial papers were printed on a four-colour production gravure press with one set of water-based inks. In all the papers, wrinkles appeared and mis-register occurred, although considerable differences between the papers were apparent in this respect.

INTRODUCTION

In recent years, increased interest has been shown in water-based gravure ink for publication printing. The advantages to be gained from the use of water-based inks include:

(1) less atmospheric pollution

(2) lower concentration of solvents, possibly harmful to health, in the print shop atmosphere

(3) less risk of fires, with consequent lowering of insurance costs

(4) less odour in the print

(5) an abundant supply of 'solvent' during times of raw material shortages (as occurred in 1974)

(6) no need for a recovery plant

(7) easy wash-up of liquid ink with water (depending upon the type of binder used).

The disadvantages include:

(1) risk of paper distortion

(2) as a rule, lower print gloss (possibly this does not apply to the latest types of ink

(3) higher requirements of drying energy

(4) risk of corrosion of the printing press

(5) higher ink costs (at least until now)

(6) foaming with certain types of pumps.

The binders used in toluene-based gravure inks are unsuitable for water-based inks, as they are insoluble in water. Special water-soluble binders are needed. Other requirements are that the gloss of the print should closely approach that produced with toluene-based inks, and that the binders should possess good water-release properties, to ensure that the drying is not impaired. One of the most important groups of binders used today in water-based gravure inks is the alkali-soluble acrylic copolymer. Even if an ink is called water-based, it may contain considerable amounts of organic solvents, usually alcohols, such as propanol, which improve the wetting of pigment and paper, and accelerate drying. The expensive binders required in water-based inks mean that today the price of these inks is higher than that of conventional gravure inks.

One important difference between water and toluene is the speed of drying. At room temperature, the solvent evaporation rate for water is 1102 s, and that for toluene 180 s. In general, water-based inks will require more efficient hot-air ovens, with higher temperature, and larger volumes of air. The drying speed is also influenced by the content of alcohol in the ink, and by the absorbency of the paper. If the absorbency is high, less drying energy will be required, but other difficulties may arise, such as increased distortion of the paper web.

Apart from water, there is also another alternative to toluene as solvent, that is naphta-type solvent. In the U.S.A. it is quite common to use a gravure ink which, in the press-ready state, contains 20% of toluene and 80% of naphta; it has been stated that the naphta content can be increased, perhaps to as high as 100%.

Amongst others, Sauntson[5] and Burke[2] have studied the different types of binders which are usable in water-based gravure inks. Hutchinson[3] has recently reviewed the progress made in water-based gravure inks. Agebro and co-workers[1], in GFL, carried out test printing of water-based inks, and a toluene-based ink as comparison, on a laboratory scale. They found that the main difference between the two types of ink lies in the drying properties. Johansson and Konradsson[4] test-printed different papers with water-based inks from different suppliers, and with a conventional ink on a Rotova Press. They noted that some of the water-based inks induced paper deformation, and that as a rule the water-based inks gave higher print-through.

In The Finnish Pulp and Paper Research Institute, a study has been made of the requirements imposed on the paper in printing with water-based inks; some of the results obtained are presented below.

LABORATORY PRINTING TRIALS WITH DIFFERENT PAPERS

In the first part of this study, 23 samples of commercial papers were collected; these included uncoated and coated papers, and gravure and offset papers. These samples were test-printed in our Winstone laboratory gravure press, with water-based inks from four different ink manufacturers, and for comparative purposes with a toluene-based ink. The results indicated a very great difference in quality between the different water-based inks, especially with regard to mottle in the dark tones. It was found that the inks giving a high degree of mottle could be improved by the use of more alcohol when diluting the ink than the amount recommended by the ink maker.

The optimum ink viscosity of the water-based inks (23 mPas) differed from that of the toluene ink (19 mPas). In selecting the optimum ink viscosity, consideration was given to missing dots in the light grey tones, and to mottle in the solids.

In print gloss, a clear difference was evident between water-based inks and the toluene ink. Figure 1 illustrates the diminution in gloss of the paper samples tested, caused by printing with the toluene ink, and with black and red water-based ink acquired from the same manufacturer which did not give mottle. For the toluene ink, this diminution in gloss is negative and this accordingly means an increase in gloss. This varies between 1 and 18 units. For water-based inks, the print gloss is always lower than the gloss of unprinted paper. As regards the black ink, the diminution varies between 1 and 24 units, and for the red ink between 5 and 30 units.

Figures 2 and 3 illustrate how print gloss depends on the original gloss of the paper. For ink 1, the toluene ink, the correlation between print gloss and paper gloss is good; for water-based inks 2a and 2b, the tendency is similar, but more scatter is apparent.

For the study of the influence of the amount of water upon the diminution in gloss, some of the paper samples were printed in the Winstone press with water instead of ink. It is observable from Figure 4 that even the smallest amount of water, which corresponds to the lowest engraving depth, affects the gloss; when the amount of water is increased, this brings about only a relatively small additional drop in gloss.

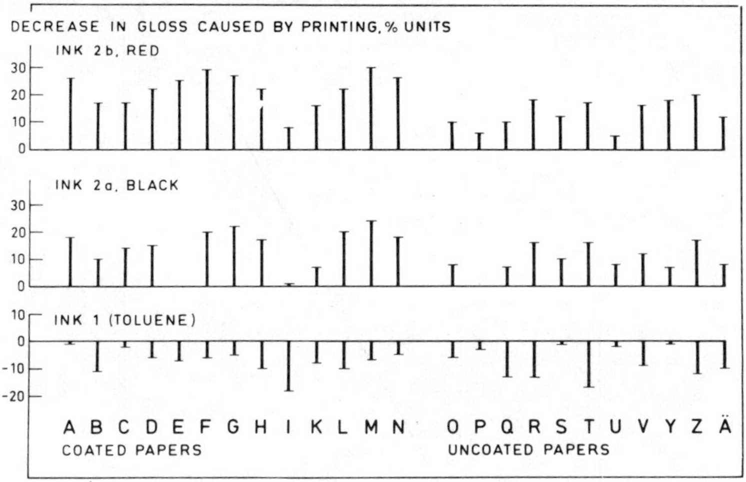

Fig. 1. *Decrease in gloss of different paper samples when printing with a black toluene-based ink and with two water-based inks, black and red.*

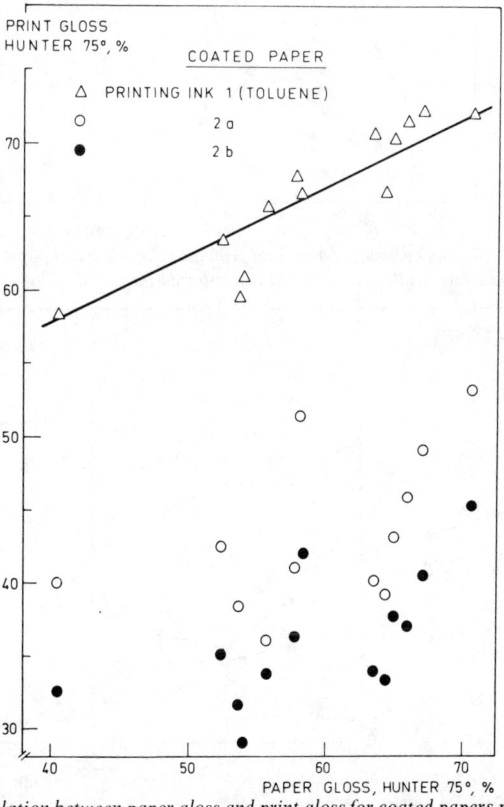

Fig. 2. *Correlation between paper gloss and print gloss for coated papers printed with a black toluene-based ink and with two water-based inks, black (2a) and red (2b).*

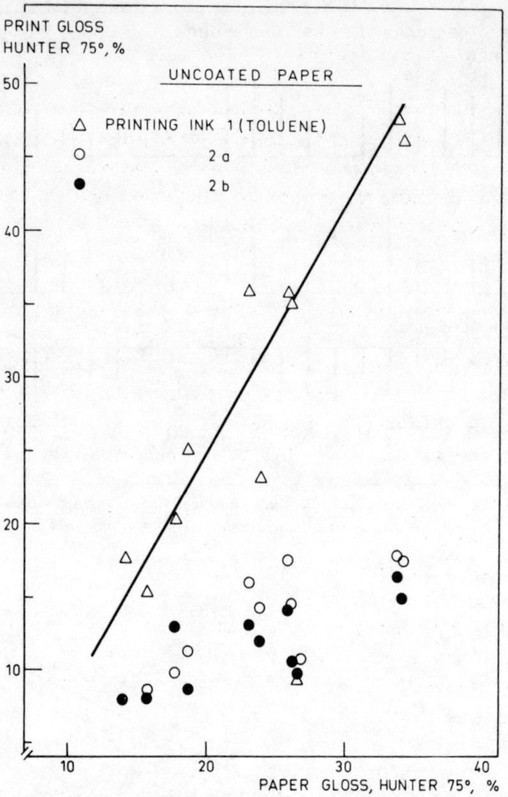

Fig. 3. *Correlation between paper gloss and print gloss for uncoated papers printed with a black toluene-based ink and with two water-based inks, black (2a) and red (2b).*

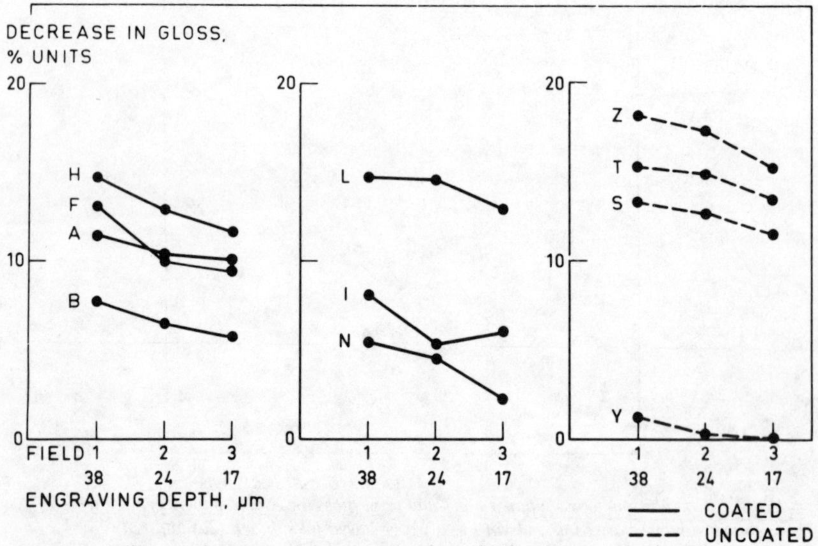

Fig. 4. *Influence of water application on the gloss of different paper samples. The water was applied by printing in the Winstone gravure press.*

COATING TRIALS

At this stage of the investigation, the results obtained from production trials with water-based inks showed that the main difficulty was paper distortion. It was decided that special attention would be given to this problem. Coating trials were made to study the possibilities of making a coated paper with properties such that the water would not penetrate to the base paper.

Influence of coating components on the contact angle

The influence of different binders, pigments and additives on water absorptivity and contact angle was studied by means of draw-downs made by hand on plastic, and measurement of the contact angle with water. In the first series, the influence exercised by the binder upon the contact angle was determined on films with binder containing no pigment, and no additives. The following types of binder were included: CMC, polyvinyl alcohol, different starch qualities, alginate, casein, Δ-protein, styrene-butadiene latex, and melamine formaldehyde. The contact angle (measured in the liquid) varied between 10° for a S/B-latex quality, and 80° for a starch quality and a PVA quality. The differences between different qualities of the same type of binder were quite large; for S/B-latex the lowest value was 10° (quality A) and the highest value 71° (quality B). It is well known that the binder manufacturer puts different additives into the binder; most likely, it is to a large extent these additives which determine the contact angle.

In the next series, a study was made of the influence of coating additives, 'Polysalz', NaOH, and Calgon, upon the contact angle, by means of draw-downs of coating paste. The contact angle varied only between 15° and 20°.

In the last series of draw-downs, the type of pigment was varied. The pigments studied were clay, talc, and calcium carbonate; of these, calcium carbonate had the smallest contact angle, 13°, and talc the largest, 24°.

To confirm the results obtained in the draw-down tests, trials were also made on a pilot coater. On the basis of the draw-down trials, four different coating paste formulations were used, representing combinations of binders and pigments that had extreme contact angles (Table 1).

TABLE 1

Coating paste	1	2	3	4
Talc	100	100	-	-
Calcium carbonate	-	-	100	100
CMC	2	2	2	2
S/B latex A	5	-	5	-
S/B latex B	-	5	-	5

The base paper was wood-containing, with a grammage of 47.5 g/m^2. Trailing blade coating was employed. The samples selected for testing had a coating grammage close to 10 g/m^2. The test results are listed in Table 2. The smallest contact angle, 58°, was obtained with coating paste 4. The contact angle of coating paste 3 was 68°, whereas the value was higher for coating pastes 1 and 2, 73° and 76°. On the basis of the draw-down results, one would have expected the following order: coating paste 3 smallest angle, coating paste 1, coating paste 4, coating paste 2 largest angle. The results of the pilot coater trials indicated that the pigment exercises a greater influence upon the contact angle than does the binder; this seems logical as there is more pigment in the coating than binder.

TABLE 2 Properties of paper samples coated on the pilot coater with different coating pastes

		Coating paste 1	Coating paste 2	Coating paste 3	Coating paste 4
		Talc CMC S/B latex A	Talc CMC S/B latex B	$CaCO_3$ CMC S/B latex A	$CaCO_3$ CMC S/B latex B
Grammage	g/m^2	55.9	56.7	56.9	55.8
Density	kg/m^3	934	915	948	962
Roughness, Printsurf S-20	µm	< 0.90	0.95	1.25	1.35
Ink absorptivity, K & N		29.6	30.7	46.5	45.8
Water absorptivity, STFI rapid moistener, 0.2s	g/m^2	4.8	4.7	6.3	6.0
Y-value, SCAN-P8:75	%	75.6	76.3	78.5	77.9
Opacity, SCAN-P8:75	%	93.5	94.2	93.6	93.2
Gloss, Hunter 75°	%	52.6	52.7	25.6	20.8
Wet rub, Taber		1.3	0.8	4.0	12.0
Contact angle (water)	deg.	73	76	68	58
Required etching depth, (D = 1.00)	µm	21.2	22.2	21.0	21.4
Print-through (D = 1.00)		0.045	0.038	0.051	0.050
Unevenness in light tones, U_{light}		0.077	0.088	0.110	0.113
Print gloss, Hunter 75°	%	63.1	64.5	42.5	36.6

In the draw-down investigation, it was assumed that the contact angle would describe how water would be absorbed if there were a sheet of paper under the coating layer. In general, this seems to be true; the less the contact angle, the larger is the water absorptivity determined by means of the STFI rapid moistener.

The print-quality values illustrate the well-known fact that calcium carbonate lowers the printing smoothness (the unevenness value U_{light}, which is a measure of missing dots, increases) and print gloss.

Influence of starch content in the coating on water absorptivity

It is necessary for coated offset papers to possess a certain degree of water resistance, which is obtained by the use of a higher content of binder in the coating than that of gravure paper. It was decided that a study would be made of the effect of the amount of starch upon the water absorptivity of the coated paper. The coating trials were made on the pilot coater with wood-containing base paper, and a grammage of 40 g/m^2. The pigment in the coating paste was clay, and the binder content varied between 4 parts starch and 4 parts S/B latex to 14 parts starch and 4 parts S/B latex to 100 parts of pigment. Thus the binder content was not constant, but increased in relation to the increase in starch content.

The results of the coating trial have been listed in Table 3. The water absorptivity (measured with STFI rapid moistener 0.2 s) diminished from 6.0 g/m^2 for sample A1, with a starch:S/B latex ratio of 4:4, to 3.4 g/m^2 for sample A6, starch:S/B latex ratio = 14:4, representing a reduction of almost 50%. Furthermore, the contact angle increased, although the change was much smaller, from 62° to 68°. With increasing content of starch the change in some other important paper properties was:

oil absorptivity	decreased considerably
ink absorptivity K & N	decreased
opacity	decreased
gloss	decreased
wet rub resistance	decreased considerably
roughness	increased.

The low wet rub resistance of coating layers containing large amounts of starch could introduce problems, but chemicals such as glyoxal are available that will increase the wet rub properties of starch.

Test printing in the Winstone press with a water-based ink demonstrated that the required etching depth and the print gloss increased with rising content of starch. The increase in print gloss was rather small (from 23.6 to 26.0%), but it is interesting to note that this occurred simultaneously with diminution in the gloss of the unprinted paper from 68.8 to 44.6%. As a rule, paper gloss and print gloss change in the same way, but in this case less ink binder penetrated into the paper, with increasing content of starch in the coating layer.

TABLE 3 Influence of binder content in the coating paste and of base paper sizing on paper properties

Sample description	Amount of coating (g/m^2)	Ink absorptivity K & N	Water absorptivity STFI rapid moistener 0.2s (g/m^2)	Contact angle (water) (degrees)	Y-value SCAN-P8:75 (%)	Opacity SCAN-P8:75 (%)	Gloss at 10 g/m^2 coating Hunter 75° (%)	Wet-rub Taber
Influence of binder content								
A 1 Starch/latex 4/4	11.6	26.2	6.0	62	73.6	91.8	62	12.0
A 2 6/4	10.7	25.1	5.0	64	73.8	91.5	59	27.5
A 3 8/4	10.5	21.6	5.0	63	73.4	90.5	57	37.8
A 4 10/4	12.0	18.6	4.1	66	73.6	91.0	49	55.5
A 5 12/4	11.5	8.7	3.2	67	71.7	89.1	50	60.3
A 6 14/4	9.9	7.4	3.4	68	69.5	86.9	45	75.8
Influence of base paper sizing								
B 1 Sized base paper	9.7	23.2	3.1	65	83.4	88.7	44.1	68.5
B 2 Unsized base paper	10.3	23.1	5.9	62	82.6	90.0	42.4	71.5

The influence of sizing of the base paper upon water absorptivity

Two base papers, with a grammage of 62 g/m^2, were produced on the pilot paper machine of The Finnish Pulp and Paper Research Institute. Both papers were typical base papers for lightweight coated paper, with a groundwood content of 50%. The only difference between the two papers was that one was unsized, and the other was beater-sized to a water absorbency of $Cobb_{60}$ = 20 g/m^2. Both papers were coated on the pilot trailing blade coater in the same way, and with the same coating paste, which contained clay with a binder of 8 parts starch and 4 parts S/B latex to 100 parts of pigment. The properties of these papers are indicated in Table 3. Sample B1, with the sized base paper, had a lower water absorbency, 3.1 g/m^2 (STFI rapid moistener 0.2 s) than sample B2 (unsized base paper), 5.9 g/m^2. Sizing the base paper thus lowered the water absorbency by almost 50%. The effect of sizing the base paper upon other paper properties was relatively small. This also applies to the effect exerted upon test printing results.

FULL-SCALE PRINTING TRIALS

In the final part of this investigation, ten qualities of commercial paper were printed on a four-colour production gravure press, with one quality of water-based ink (none of the qualities used in the laboratory trials). Five of the paper qualities were coated, and five uncoated. A general description of the paper qualities is given in Table 4.

TABLE 4

Sample		Grammage (g/m^2)	Ash content (%)
Coated samples			
1	Normal gravure paper	57	32.8
2	Normal gravure paper	60	34.7
3	Offset paper, starch in the coating	60	28.4
4	Offset paper, CMC in the coating	55	28.6
5	Normal gravure paper	72	34.6
Uncoated samples			
6	Normal gravure paper	61	24.2
7	Lightweight gravure paper	50	14.9
8	Coupon paper, beater-sized	70	11.1
9	SC News	56	13.2
10	Contains talc as filler and 10% TMP	56	28.6

The colours were printed in the following order: yellow, red, blue, black. The width of the paper web was 1300 mm. Some observations were made during the printing trial:

(1) The ink did not dry properly, but came off on the guide rollers on printing at the normal speed of this press, 15000 copies/h, and drying at an air temperature of about 58°C. After this initial test run, the air temperature was increased to about 63°C, and the speed was lowered to 8300 copies/h. All subsequent paper samples were printed under these conditions.

(2) The water-based inks foamed to a markedly greater extent than normal inks. At 8300 copies/h, this did not present any problem, but at 15000 copies/h the black ink foamed strongly. It proved impossible to run for more than a few minutes before the foam reached the edge of the ink duct. The foaming difficulties in the black unit arose because it was not possible to regulate the speed of the pump in this unit. The speed of the other pumps was adjustable.

(3) The inks tended to dry in the gravure cylinder cells when running at idleing speed during the changing of paper rolls. Once the ink had dried, it did not dissolve when printing at full speed, but had to be removed with toluene. The black ink introduced the main problem in this respect; perhaps the foaming exercised some influence. In the black unit, the ink also formed deposits on the underside of the doctor blade near the edges.

Figure 5 illustrates the layout of the print. The print consisted of two halftone pictures and a colour scale in the middle of the web; on both sides of the web there were larger fields with different combinations of the four colours and four engraving depths.

The two most important properties of the print measured were formation of wrinkles and mis-register. The value reported for the formation of wrinkles is a relative value, which depends upon the number and size of the wrinkles running over the pictures. Mis-register was measured from the register marks marked a1, a2, and a3. Under a microscope, the distances between the black register mark and each of the coloured register marks were measured in the cross direction of the web at points a1, a2, and a3. If the coloured register mark was on the right side of the black register mark, the mis-register figure was regarded as positive. It was then assumed that the web was moved sideways, so that the mis-register at point a2 was 0, and the mis-register figures at points a1 and a3 were corrected correspondingly. Following this, the difference between the mis-register at point a1 and point a2 was calculated for each of the colours yellow, red, and blue. The procedure for calculating the mis-register is indicated schematically in Figure 6. For each sample, three mis-register values were derived. From these the mean value was calculated and is the final measure of mis-register.

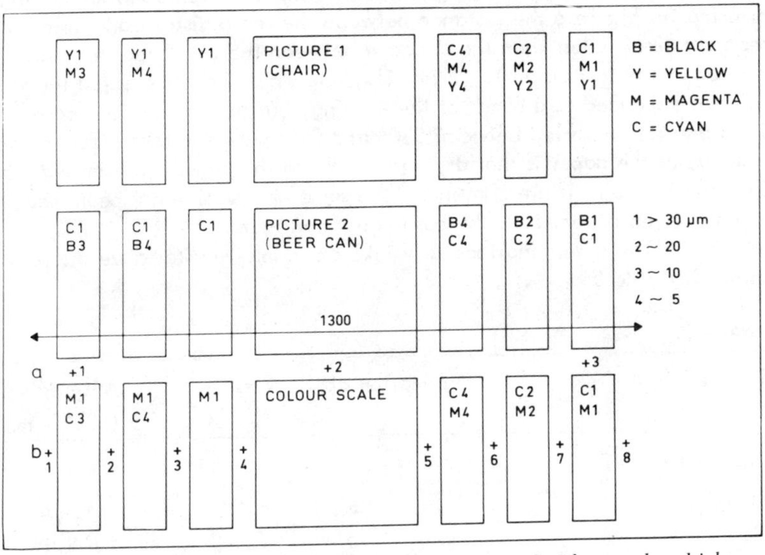

Fig. 5. Layout of print produced in full-scale printing trial with water-based inks.

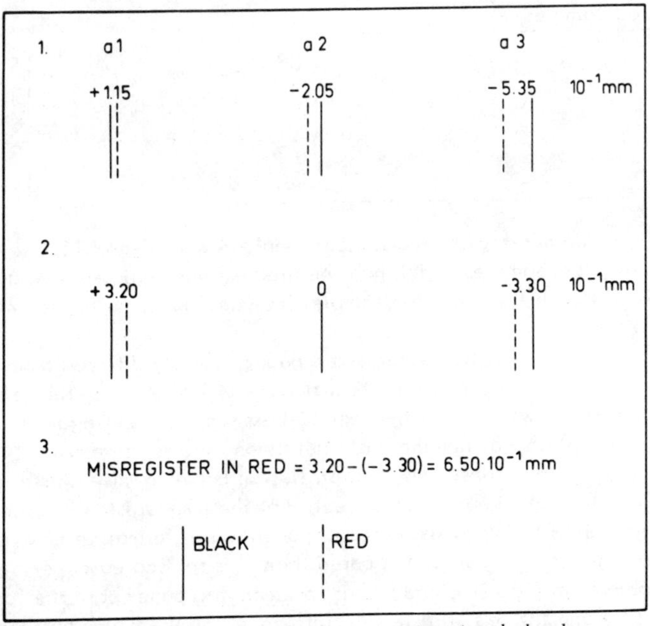

Fig. 6. Example showing how misregister is calculated.

The mis-register results indicate that the web contracts during printing (in Figure 6 the distance between the red register marks printed as second colour is less than the distance between the black register marks printed as the last colour). Primarily, one would have expected the web to have expanded as a result of the wetting, but this was not the case. The contraction varied between 0.06 and 0.17%. The reason for the contraction of the paper is that drying tensions in the paper are released with increase in the moisture content. The same observation has been made when coating paper, i.e. the base paper contracts.

The determinations of wrinkles and mis-register gave the results shown in Table 5.

TABLE 5

Sample	Formation of wrinkles	Mis-register ($10^{-1} \cdot$ mm)
Coated samples		
1	46	7.6
2	51	8.9
3	34	5.4
4	44	8.1
5	41	7.5
Uncoated samples		
6	38	15.5
7	49	13.9
8	8	6.5
9	25	9.0
10	30	14.2

The mis-register varied between 5.4×10^{-1} and 15.5×10^{-1} mm. The three paper samples which had the greatest mis-register were all uncoated, although the uncoated samples included papers with less mis-register.

Sample 8, the beater-sized paper, clearly differed from all the other samples with regard to the formation of wrinkles, 8 on the relative scale, compared with 25 for the second-best sample. In regard to mis-register, sample 8 was not the best, but among the top three.

Figure 7 illustrates the correlation between mis-register and the formation of wrinkles. It can be seen that the mis-register is larger when the formation of wrinkles increases. For the same formation of wrinkles, the mis-register is larger for uncoated than it is for coated paper. It is observable from Figures 8 and 9 that for uncoated paper both the formation of wrinkles and the mis-register are influenced by the water absorbency, measured with the rapid moistener. All of the printed papers had wrinkles

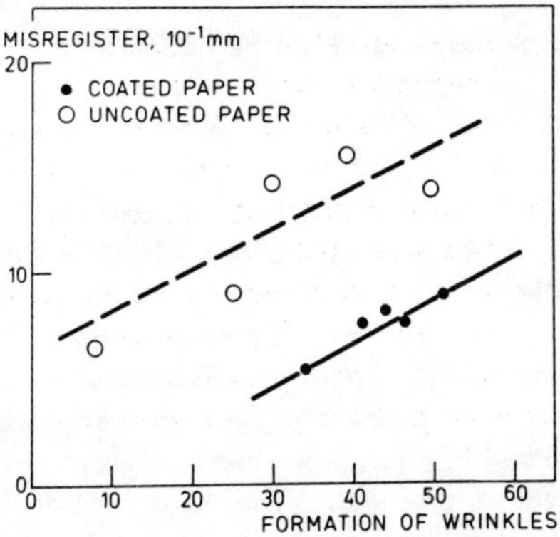

Fig. 7. Correlation between formation of wrinkles and misregister.

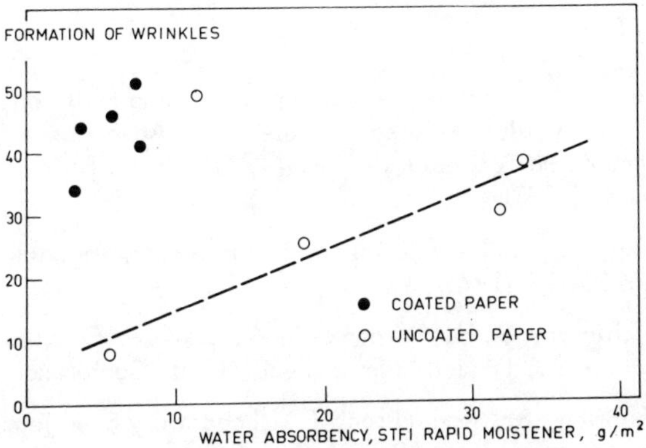

Fig. 8. Correlation between water absorbency and formation of wrinkles.

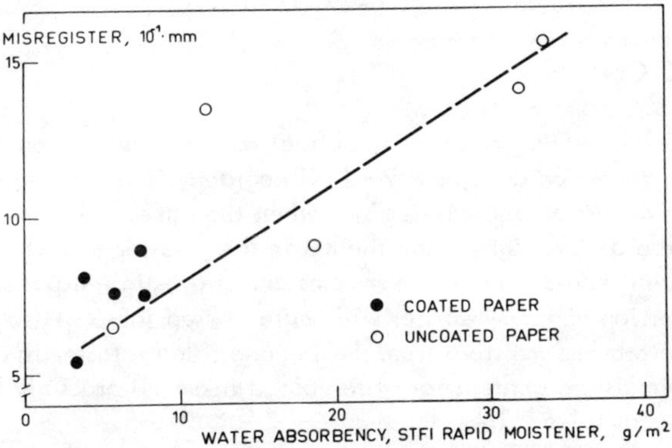

Fig. 9. Correlation between water absorbency and misregister.

to such a degree that they would not be acceptable.

CONCLUSIONS

If for some reason it becomes impossible to continue the use of toluene-based gravure inks, it seems very probable that the water-based inks will offer a concrete alternative in publication printing. A successful change-over to water-based inks requires development work, in close cooperation between printer, ink-maker, paper-maker, and press-manufacturer. The drying capacity of the press needs to be increased, and a quality of paper has to be developed which does not absorb water, as otherwise wrinkle formation will present a problem. The results of this investigation indicate two possibilities of increasing the water resistance of the paper, by beater sizing, and by increasing the binder content in the coating, which will produce a denser coating layer.

REFERENCES

1. Agebro, A., Arnamo, A., Enander, C. and Hultgren, K., 'Alternatives to Toluene-Based Gravure Inks', Advances in Printing Science and Technology, Vol. 14, Ed. W. H. Banks, Pentech Press, London (1979).
2. Burke, R. E., 'New Resins for Printing Inks', Am. Ink Maker, 54, 6:29-30, 32 (1976).
3. Hutchinson, G. H., 'Progress in Water-Based Gravure Inks', Pira Gravure 80, London, Nov. 27-28 (1980) Conference papers.
4. Johansson, B., Konradsson, T., 'Laboratory Experiences with Water-Based Gravure Inks', (in Swedish) Svensk Papperstidn. 78, 16:562-567 (1975).
5. Sauntson, B. J., 'Acrylic Copolymers for Water-Based Printing Inks', Am. Ink Maker, 18, 1:26-29 (1975).

DISCUSSION

M. H. Bruno Did you study the effect of the moisture content of the paper on distortion or register? In lithography it was found that there is minimum distortion and mis-register when the paper is in balance with an atmosphere 5 - 8% higher than the RH in the pressroom. This higher moisture content could relieve the stresses and strains in the paper and cause less distortion and mis-register with water-based inks. Also the paper would absorb less moisture from the ink and release faster in drying. The optimum moisture content for water-based inks will probably be different from that for solvent inks.

R. Ginman We did not measure the moisture content of the papers printed in the full-scale printing trials. As they were commercial qualities, I would think that the moisture content was around 6%. In further studies it is certainly worth while to look at the influence of moisture content on formation of wrinkles and on mis-register.

U. Lindqvist You mentioned the optimum viscosity level in the laboratory printing test to be 23 and 19 mPas for water and toluene based inks respectively. Does it hold true for all the different paper grades you tested? Also, what was the viscosity level in the full-scale tests; was it optimized and how was it controlled?

R. Ginman Our experience is that there is no significant difference between different types of paper with regard to optimum ink viscosity. The printer on the production press adjusted the ink viscosity to give optimum print quality and this was then kept constant during the whole printing run.

W. Hansen Replacing aromatics with aliphatics in solvent based gravure inks decreases the solubility of the polymers and leads to a decrease of print quality. In water-based inks, the wetting of the pigments, especially for black pigments, is not as good as for solvent-based inks, and this leads to a general reduction in print quality.

J. M. Fetsko Your results on the effects of water on paper gloss agree with our findings that high relative humidities (70-90% RH) during paper storage greatly reduced both paper and print gloss (with oil-based inks). However, gloss recovered with time after printing to the extent that after 30 days it surpassed gloss at 50% RH. These results suggest that control of time of measurements is required in experiments dealing with water/paper relationships.

R. Ginman We did not repeat the print gloss measurements at a later date, but it is an interesting observation that you made. We have made some tests on the effect of RH on paper properties and we found that if you condition uncoated paper containing groundwood at 65% RH or higher, you get an irreversible increase in roughness and decrease in gloss.

J. Aspler In Figures 8 and 9, what do you mean by the 'STFI rapid moistener'?

R. Ginman The weighed sample is placed on a rubber belt which transports it under a water shower which floods the sample with water, and then into a nip where a blotting paper is pressed against the sample and absorbs the free water from the surface of the sample. The sample is then weighed once more and water absorbency is reported as g/m^2. The time of contact between water and sample can be varied between 0.1 and 0.4 s by placing the shower closer to or further away from the nip.

P. J. Mangin In the figure showing the correlation between wrinkles and the water absorbed, are all the uncoated papers from the same type of furnish? If so, how do you explain the uncoated paper not included? What types of uncoated paper were used?

R. Ginman All the papers were commercial qualities manufactured by different paper mills. The coating composition may differ between the coated samples. The fibre composition is roughly the same in all the uncoated samples. More details are given in the section of my paper under the heading 'Full-scale printing trials'.

L. O. Larsson In your discussion of print quality, you are stressing 'mottle'. How did you assess mottle?

R. Ginman In this case it was done subjectively because the degree of mottle was very high. It would also be possible to measure mottle as the non-uniformity value U_{dark} using the scanning device built in our institute.

L. O. Larsson In Table 2 and in the text it is said that $CaCO_3$ in the coating increases the unevenness in light tones due to a increase in the number of missing dots. Do you have any comment on the occurrence of an increased print-through at the same time?

R. Ginman I cannot explain this. On the contrary, one would expect the paper containing $CaCO_3$ to have lower print-through.

L. O. Larsson Do you have any information on the price of the water-based inks as compared with the conventional ones, and what do you think the situation will be in the future?

R. Ginman It has recently been reported (Hutchinson[3]) that water-based inks are more expensive than conventional inks but that this cost difference is partly compensated for by the cheaper solvent used by the printer. Most likely, this situation will not change much in the future because special, more expensive binders are required in water-based inks.

L. O. Larsson Were there any particular reasons why some of the newest inks were not included in the study?

R. Ginman This investigation was done some time ago and since then some properties of the water-based inks have improved, at least gloss. With regard to the main problem studies in this investigation, namely the influence of water on the paper, I do not expect any difference between newer and older inks.

J. A. Bristow You indicate the advantages of reducing the water absorptivity and increasing the contact angle. If you extend this too far you may reach the area of water repellency where you will have ink-transfer problems. Did you notice such problems arising? Have you any idea where the critical (minimum) water absorptivity point is located?

R. Ginman In principle you can of course go too far in reducing the water absorbency of the paper. However, in our investigation we were very far from doing so.

W. Schaeffer Can you make a coating colour from a coated calcium carbonate? Coated carbonates are made from stearates or stearic acid and commonly are used as fillers for elastomers. The coating makes the carbonate hydrophobic to a degree comparable with that of a stearate multilayer.

R. Ginman It is an interesting idea, but I have no knowledge of this pigment.

GUIDELINES TO A CRITICAL CHOICE OF COPY PROCESSING SYSTEMS

Claes Enander and Jan Kullberg, The Graphic Arts Research Laboratory, Stockholm

ABSTRACT

This work is intended as a practical aid for decision-making before choosing a copy processing system. It applies to any integrated on-line type computerized composing system. It gives a purchaser an insight into all major selection criteria. Among the topics are scheduling of installation planning, system functions, hardware aspects, backup considerations, guarantee, education, documentation, service and maintenance requirements.

System functions, i.e. software properties, cover editing capabilities, file handling, hyphenation and justification, and various format systems, etc.

There is a thorough discussion on requirements to be met by all the usual hardware components of a copy processing system (except typesetters), i.e. mass memory, video display unit (or terminal) (VDT), printer, computer, etc.

How to evaluate a vendor's organisation is another vital part of the decision process which is also covered.

1. A GUIDE LINE FOR SYSTEM PLANNING

Undoubtedly a lot of work is put into choosing, buying and installing a text processing system. Since the work is often carried out by a project group, it is necessary to start the project early, at least one year before the projected installation time. This will enable the group to meet often enough to be able to understand and influence the installation and at the same time feel enthusiasm about it, which is very important. It also means having time to study the market before the decision, analyse the system development and last but not least the different vendors.

Typically, an outline for the project could be as follows:

(1) The project group. It is important for the project group that all

the different personnel groups whose work in some way will be affected by the system are represented in the group. This goes especially for those who will be working directly with the system (typographers, editors, advertising department staff).

When the different systems are to be studied in detail, visiting reference installations and attending demonstrations are practicable if done by small groups (3 - 4 persons) who can report to the project group. It is often difficult to try to gather a larger group for different study tours and the work will in this way suffer both in time and smoothness.

(2) Description of the present production situation. The different vendors have to know the production situation and development plans to be able to present a suitable system for a company. The following information is necessary for the vendor:

(i) copy flow (both editorial and advertisements)
(ii) the maximum work load
(iii) the number of work stations (terminals)
(iv) the required storage capacity.

(3) Planned future. It is important to look ahead - how will the production change within a five-year period? Can the system expand to meet a growing production which can be predicted, and not cost too much to adapt?

(4) System specification. The project group has an important task in defining what will be required of the new system and documenting it in a specification. This specification should be based on the actual need for the present production and also the various desires for a new system coming from different groups. In this phase, the wishes should not necessarily be limited to what is thought can be delivered today by the different vendors. It is, though, advisable to give different priorities to the desired requirement.

(5) Request for proposals. The system specification will be part of the request for proposals which will be sent to the vendors who could deliver a system that fulfils basic requirements and who is considered to have a sound organisation. A thorough system specification will help all serious vendors to make offers which will include adequate information on the issues that are thought important. The offers should include the points that are included in the system specification and in the same order - this will help in comparing the systems. In addition to comments regarding the system specification, the vendors should be asked to give a thorough documentation (in the appropriate language) of the total possibilities of the system.

(6) Preliminary selection. Based on high priority qualities of the system, in the first selection phase, the number of systems should be

reduced down to a few. A high quality system which is thought too expensive should be left among the few, since it is easier to negotiate prices rather than system qualities. Do not make the cost of the system a high priority matter in the first selection.

(7) Visiting reference installations. The few remaining systems must be thoroughly studied by attending demonstrations and visiting reference installations, preferably to companies of the same size as the project group's own company. Seeing the offered systems work in similar environments will guarantee that it can also be used for ones own production. One should be aware that it is a gamble to allow a vendor to install a pilot system on site; that is installing a new system which has not been tested previously in a production environment. It is a gamble especially when it comes to meeting the promised installment time and promised software functions.

(8) Final selection. The final selection will be done through negotiating prices and other terms with the remaining vendors. In the final offers, guaranteed terms, training, technical service, options, delivery terms, etc., should be well defined. These aspects must, along with price and technical specifications, be carefully considered when making a total judgment of the different systems proposed.

(9) Planning installation and training. When ordering a system, installation and training plans should be discussed with the vendor. The customer and the vendor should agree on a preliminary date for production start on the new system. The production start should be preceded by an acceptance test, which also should be agreed. It is also wise to document in the contract the responsibilities of the vendor, should installation be delayed.

In this phase the vendor should also define the environmental requirements of the system for the customer site planning.

(10) Signing the contract. Look for national standard contract terms for the purchasing of computer systems (hardware and software).

(11) Installing the system. The responsibilities of the project group should not end at the contract signing. The group (or a part of the group) could be very useful in supervising the installation, having taken part in the system project from the beginning. The original installation plans will almost always have to be changed and/or delayed. The group should in every new situation accept changes in work so leading to an efficient system that is operated by a well-trained staff.

2. GENERAL SYSTEM ASPECTS

2.1 Large or Small System

Generally speaking, there are (in Scandinavia) two main categories of newspaper systems - systems where computer and mass storage units are separate units (to which a number of terminals and peripherals are connected), and secondly, mini systems where computer and data storage are often mounted together in a small cabinet with a video display terminal (VDT), using a rather limited and standardized software. One must carefully balance the low price of the mini system with the limitations regarding maximum number of terminals, storage capacity, number of input/output units, overlapping activities (parallel typesetting and VDT work, etc.), available software, etc. If aiming at a larger system, it will often be more costly to start off with a mini system and gradually add on equipment than it would be to start off with the larger system in the first place.

2.2 Flexibility and Modularity

Systems that seem alike regarding price and capacity may be very differend when it comes to flexibility. What limitations are there regarding:

job length
number of jobs
format length
number of formats
keyboard layout
programmable keys
updating hyphenation dictionaries
interfaces to different types of printers and typesetters.

The limitations on job length and job numbers should not be too close to known requirements.

Another important aspect related to flexibility is the growth capacity of the system; a modular system can grow using the original system as the base module. The first system purchase will often be configured just to cope with present production. After some time more storage capacity and terminals will be needed. The reason could be that a new software module (bureau program, sports program, classified ad system, etc) will be needed or that the system will be used for new products. Thus it is important that it is first of all possible to add on to the system, and secondly, that hardware and software will not have to be exchanged just for limited growth. If it is possible to foresee a growth when purchasing a system, have the vendor document the growth possibilities and ask for an option. If more terminals are planned to be added to the system, it is often wise to install from the beginning a terminal interface which can handle the increased number.

Some time after having installed the system it may be necessary to add on new software functions to the system. When ordering, find out what it will cost to buy and install additional software later on. Some vendors may charge only the cost for the actual software module, while others may have to charge for both the software module and a complete software integration, which might be needed. Adding a new font to a typesetter requires a new width table in the system. In some systems this can be done relatively simply by the customer, while in other systems the vendor has to generate a new table.

Some vendors will offer continued system updatings as a part of a service maintenance contract. Be sure to examine carefully the conditions of such a contract.

2.3 System Developments

Before signing a contract be convinced that the vendor will support and also develop the system within the foreseeable future. Try to find out if new software is being developed for the computer which is being bought, or if a computer generation change is soon expected.

Ask the vendor about the following:

- editorial system functions (not the same as production system functions)
- classified advertising sales system (for classified ads)
- display advertising system
- pagination system
- picture processing system
- library system
- business system.

These points need not be decisive, but at least be informed on the development strategy of the vendor, since these functions will soon be of great interest for most newspaper companies.

2.4 Video Display Terminals

Can the same type of VDT be used for all different system activities or should there be different terminals for different functions?

The question of intelligent vs unintelligent terminals should not be made a major issue as long as system capabilities are the same. The allocation of computer capacity (intelligence) between the central computer and the VDTs is mainly a question of system architecture philosophy.

2.5 Speed

One of the most essential system qualities is speed. A system with long response time (i.e. the time to hyphenate and justify a story from a VDT) will definitely irritate the operators. Long response times could never be compensated by other system qualities in a newspaper system. In all systems the response times will depend on the system load in use (the number of VDTs working with H and J). Therefore the maximum response time allowed for a required system load should be specified in the contract.

Visits should be made to other installations during hours when most VDTs are being used to notice the response times. Ask the opinion of the operators on the system speed.

2.6 Documentation

Make sure there is a complete system documentation for the system you are buying. There are two kinds of documentation: User documentation and Source Program documentation.

The user documentation should be thorough and written in the user's native language. The user documentation should describe the system layout and the different hardware and software modules, the system maintenance and the daily use of the system. The operator handbook should be instructive and easily understood, making it possible for any operator readily to find the right system commands without help.

The source program documentation is necessary only for a buyer who has plans to develop his own programs to run on the production system For companies without programming needs the necessity for program documentation is not so obvious. Nevertheless, it is important that the vendor has complete documentation, making it possible in the future to correct software errors and make software changes.

2.7 The Vendor

A reliable vendor is as important as a good system. At some time or another a system will fail and help will be needed with repair and system changes. The hardware will age and have to be exchanged, and the vendor will be required when operators cannot correct minor system errors (even though the actual correction procedure may be on the first page of the documentation). In such cases the vendor must be quick to help. Existing system customers can tell you about the service quality of the vendor. Ask for a complete list of reference installations from all vendors, and talk to many customers to get an overall judgment of the vendors

Ask also for financial reports from the different vendors for the last few years and look at the sales and profit figures. If the sales figures have a downward trend, ask the vendor what measures he will take to increase the sales and how long the company could take a loss. Is there a supporting parent company behind to compensate the loss? If, on the other hand, the company is showing increasing sales figures, how fast will the service organisation adapt to the increasing service requirements for a greater number of installations?

3. SYSTEM FUNCTIONS

3.1 Editing Functions

The editing functions include all functions aimed at helping the operator to create a text or to change an existing text using a VDT. These functions are often called front-end functions.

Cursor movement The cursor is used for text entering and defining an amount of text (character, word, sentence, line, paragraph) to be changed (deleted, moved, stored, etc.).

Scrolling Moving text upwards and downward on the VDT screen to overlook and edit text.

Deleting text For editing and correcting text; deleting text blocks by mistake should be impossible.

Insert/overwrite Two different alternatives for entering text on the screen; it should be possible to go back and forth between the two on the same terminal.

Moving text Three functions are required for moving text; defining text, storing text and inserting stored text.

Search/exchange For searching text strings (words, expressions, etc.) which are to be replaced with another text string when editing.

3.2 File Managing Functions

All functions handling text communication within a system are usually called file handling or file managing functions. The files (articles, advertisements, etc.) can be divided into different queues according to different criteria (operator queues, product queues, etc.). A list of all files in a queue comprising a special selection of files from different queues is called a directory.

Frequent file managing functions are:

Creating a new file For entering new text with a unique file name or production number.

Storing a file Filing a permanent copy of the newly entered or edited text on the system disc.

Getting a stored file Reading a file from the disc to the VDT screen for editing, using the file identification (file name/number).

Outputting file/files Sending files for output on printer/typesetter or other output unit.

Reading file from input unit Text which is not written on a system VDT must be read from paper tape reader, scanner, magtape or other input unit.

Getting a directory Asking the system to output a file directory on a VDT, printer or system console.

File passing Sending a file from one VDT or queue to another VDT or queue (reporter-to-editor in an editorial system).

Linking files Linking different files which make a complete text in the final product.

Deleting files Newspaper articles that must be deleted from the system at regular intervals.

The above basic functions will be found in most systems. However, the systems may differ as to the simplicity of using the functions.

Primarily it is important that the command structure for the file managing functions is simple, logical and user oriented. The command language should be in the operator's native language and verbally associated with the actual functions (GET, STORE, etc.). For simplicity, it should be enough to write only the two first characters of the command for the system to understand the command. The command structure should be logical in the sense that all commands should be used in the same way, both regarding the command itself and the different parameters that in some cases will accompany the commands.

A system in which only English computer commands must be used is not user oriented for the average non-English speaking operator. Furthermore, this could imply that the vendor is using software or hardware which is not flexible enough to make the system user oriented or that he is not allocating resources to make the system easy to work with.

The number of keystrokes for specific functions cannot be used as an isolated criterion as to the simplicity and user orientation of the command language structure. A special key for each function would give a large and complicated keyboard layout. Frequently used file managing functions (GET, STORE, etc.) should be on separate keys, while other less frequent commands can be written or stored in programmable keys. A keyboard which allows the operator to put different commands or text strings in programmable keys can easily be adjusted to different kinds of production or operator preferences.

3.3 Hyphenation and Justification

The H and J (or composition) program is the heart of the text processing system. Following the composition commands for font choice, point size, line length, indents, etc., it will justify and if needed hyphenate the text. You should require the following of a good H and J program:

(1) Availability - the H and J program should be available from all VDTs working with editing. It should be possible to H and J a complete file or part of a file and immediately get it back on the screen.

(2) Speed - as previously stated, it should be possible to H and J files from several VDTs simultaneously without getting too lengthy response times. The response time for an average newspaper article should not exceed a couple of seconds at top load.

(3) Hyphenation quality - foremost, the hyphenation quality should be better than that of the second generation typesetter which used to handle the hyphenation.

The hyphenation quality should exceed 95% correctly hyphenated words in a standard language text. There are different philosophies as to the principle of the ideal hyphenation program - should it be based mainly on logical grammatical rules or on exception dictionaries? As long as both quality and speed are sufficient to the user the principle or philosophy is of no interest. Whatever principle, the ability for the customer to supplement the dictionary with new words is important when discussing hyphenation quality. Ask the vendors if this is possible and, if so, how this is done and what limitations are there regarding number of characters per word, number of words, effect on response time, etc.

(4) Text depth - along with the justified text, the H and J program should inform the operator on the exact length of the justified text (in picas, ciceros, millimeters or in whatever units are being used).

(5) Error indication - if there are coding errors in a text, the H and J program should clearly indicate where in the text the errors are located for simple correction.

3.4 Composition Commands

In a traditional cold typesetting system hyphenation and justification are performed by a program in the typesetter computer. The composition commands which are being punched on a paper tape or magnetic tape are unique for the typesetter program and limited to the typesetter capacity.

When a text processing system is installed, the typesetter will most often become a slave unit. That is, all hyphenation and justification decisions will be taken by the system and the typesetter will only receive justified text for typesetting. This is necessary since you want to know the H and J result before typesetting the text. In this way, all hyphenation and other errors resulting from H and J can be taken care of from a VDT, thereby making the final typeset output correct.

3.5 Format System

A number of composition commands defining a specific typography can be stored together as a typographical format. As an ordinary text file, the format will be named or numbered for reference. When that specific typography is to be used, a format call with the format name/number is written in front of the actual text. When the text is sent to the H and J program the format call will be replaced by the actual composition commands stored in the format. In this way the number of coding keystrokes can be lowered and also fewer coding errors obtained when typing long strings of composition commands.

In most typesetters it is possible to store a limited number of formats. You should demand a lot more from a text processing system, both regarding number of formats and format flexibility.

The number of formats and the allowed format length should never limit the use of formats or make the use of formats too complicated for general use. Especially in a production system, the use of formats is one of the essential system benefits.

Number of formats The need depends primarily on how many different products are being produced with different typography. Only the actual need for formats should determine the number of formats. It might be hard to guess the required number of formats when the system is first installed, but for a medium sized newspaper the user should be able to define no less than 300 formats. It should also be possible to extend the number of formats if needed later.

Format length In a format, it should at least be possible to store the composition commands for a newspaper article that logically is a unit (for example: leader, headline, text, subtitle and picture text). If the format length is limited, it should be possible to link several formats in such a way that the operator only needs to type the first format to get the complete typography.

Format definition and editing The actual format work should be as simple as text editing. It should be easy to:

(1) create new formats on screen;

(2) edit and change existing formats on screen;

(3) use the formats. Experience shows that over-complicated formats will not be used, especially not by editorial people. The simplicity of use depends both on the format system flexibility and the format structure of the newspaper (that is, the way the formats have been defined).

3.6 Classified Advertisement Program

For newspapers with large volumes of classified advertisement material, a special program/system could save a lot of routine work and at the same time simplify the advertisement handling.

In short, the use of a classified advertisement program would be as follows:

The advertisements which are taken by telephone will be written directly on a VDT. For each new advertisement, the VDT will display an advertisement form to be filled in with all the customer and advertisement information that is needed for publication, i.e. cost calculation, customer credit testing, and billing.

The advertisement text can be hyphenated and justified directly and the advertisement taker/VDT operator immediately inform the customer on the cost.

The billing information for all advertisements will then be transmitted to the business system of the newspaper, via magnetic tape or telecommunications.

Before typesetting, the system will sort the different advertisements according to classification.

The requirements for a classified advertisement program can be summarized as follows:

(1) direct advertisement input and interactive H and J with price calculation,

(2) a complete advertisement taking form on screen,

(3) data communication with business system,

(4) sorting, using the number of sort criteria presently used for advertisements,

(5) enough formats for fast advertisement input,

(6) sorted advertisement lists from printer for summary, proofing, etc.

3.7 News Bureau Program

Most text processing systems can accept newswires on-line, thereby storing all wire stories in the system. This will eliminate the tiresome handling of all the paper tape from the newswires.

First of all, the system storage capacity (disc storage) has to be large enough to store at least 24 h of bureau material in addition to other text. It should, furthermore, be easy to search for different wire stories in the system. Automatic typographical formatting of different kinds of bureau material is very useful for sports tables, radio and TV program lists, etc.

3.8 Sports Tables Program

Calculation of sports tables is a very time consuming job for many newspapers. Since this is a purely mathematical calculation, it is an ideal task for a computer. Many system vendors will offer special programs for sports tables. Look for the following:

number of different series
number of different standard formats for table typesetting
match result input on screen
program availability.

3.9 Other System Functions

In addition to the above mentioned text production functions, there is a number of system maintenance functions for the system management. The system managers will daily start and stop the system, delete old files, ask for free storage capacity and take care of system errors.

Start/stop routines Must be simple and safe enough to be carried out by persons without a special computer education. A simple typing error at system start/stop should not lead to a situation which could be hard to resolve for the average operator with adequate system instruction. It is most important that the risk of unintended text deletion when starting/stopping the system is minimized.

File deletion function Should be blocked for all operators except the systems manager who is responsible for the total systems maintenance. It should be possible to delete selections of files, for example, all files produced by a certain date, all files belonging to a specific category (sports)/edition/produced by a certain date, etc. Deleting file by file is irrational and increases the risk of accidental deletion. Automatic deletion is not recommended.

System status Information on the present system status (free storage, operator statistics, system load, output queues, etc.) should be easy to get from the system at any time, using a VDT or system console.

Error messages from the system should be:

verbal in the system operator's native language (not only error numbers)

understandable for system managers/operators

informative as to suitable action for continued system production. It should also be clear whether the system manager is to take action or if service from the vendor should be called for.

If, for example, there should be problems with the VDT-system communication, it is useful to be able to dump all text and file information on a system console or printer. The dump can be used either for re-writing on a back-up production system or for typesetting stored text. Thus it should be possible on the system console to initiate output on different output devices; console, printer, typesetter, paper tape punch, etc.

Most of the system maintenance routines which are less frequent (various system errors) should be thoroughly described in the system manager's manual and also be as fool proof as possible.

There is a great risk of operator errors in situations which occur infrequently. Often, a minor error is made worse by inexperienced operators intervening. Of course it is difficult to test a system in this respect (simplicity of infrequently used functions). Study the manager's manual and judge if the vendor has paid any attention to this problem.

4. SYSTEM HARDWARE

4.1 The Computer

Although the system computer is very important for the total system function, it should not be a high priority matter in the choice of a text processing system. Normally the overall system requirements will imply requirements on well written software as well as adequate computer capacity.

The one quality which is most telling about the system computer is speed. A slow system could be the result of bad software and/or bad hardware, while a fast system always includes an adequate computer.

A layman in computer technology should ask for the following when studying system computers:

(1) System speed - a slow system, or a system which becomes noticeably slower when heavily loaded (all VDTs at work, substantial output) implies that the computer and/or the disc drives are not adequate for the system.

(2) The reliability of the computer - ask other users with the same system configuration. A standard computer, used for many different applications, is often a safe buy (with proven technology and good service).

(3) The expandability of the computer - is it possible to add more memory to the computer? It might be hazardous for the future to buy an old computer which has been loaded to its maximum with memory, to handle the system software.

First, you will eventually want to add new software/hardware (for classified advertisements, for display advertisement terminals, for pagination, for an increased number of terminals, etc.). This will normally require more computer memory.

Secondly, the system vendor will hardly develop new software for old computers with limited memory size. Computer equipment and computer memory is becoming less expensive, which in turn allows the software people to use more computer capacity. This makes the programming faster and easier and, hopefully, the result of all this will be better software for the users.

Questions for the system vendor:

(1) What computer additions will be necessary for the software-hardware addition that could be expected to be useful to the paper within a 3-5 year period?

(2) What computer additions will be necessary for an increased number of VDTs?

(3) Is new software being developed for the computer in the system?

4.2 System Disc Drives

Many of the general system requirements (storage, speed) also apply to the disc drives. As to technical reliability, ask other users about the specific disc drives.

It is wise to buy a large storage capacity from the beginning (2-3 times larger than the calculated need).

Always discuss the effective storage capacity, which is the number of text characters (excluding system software) that the disc drive unit can be guaranteed to store. Have the vendor guarantee effective storage capacity in the sales contract.

4.3 Video Display Terminals

The VDT will be the main tool for text input, editing, H and J and text correction. It will also be used for system work like format editing, text deletion, recording of statistics, managing input/output, etc. Thus it is important to pay much attention to judgment of the VDT, both regarding technical functions and ergonomics.

Since much literature has been published on VDT ergonomics, a short summary of basic criteria for VDT comparisons will be sufficient here.

Screen characteristics:

(1) screen size (measured diagonally in inches)

(2) character matrix - the number of dots for generating screen characters (for instance 10x15 dots horizontally/vertically)

(3) character size - in points or mm

(4) amount of text on screen - characters/line, lines/screen

(5) colour - character colour and background colour

(6) refresh frequency - how often is each character regenerated on screen?

(7) stability - is the picture stable or vibrating?

(8) flicker - fast light intensity variations which might be disturbing

(9) contrast - between characters and background, should be adjustable

(10) sharpness - can focus be adjusted?

(11) light intensity - should be easily adjustable

(12) anti-reflection treatment - is essential

(13) colour of front frame around the screen - should be harmonizing with the character/background colour

(14) split screen - possible or not?

Naturally there are many possible combinations of character size, character matrix, colour, sharpness, contrast, etc, which will give a good result. What to judge is the readability as well as an overall impression of disturbing flicker, glare, contrasts, etc.

Type all characters on the VDT and judge the readability. Readability is a subjective judgment and therefore judgments should be collected from several people in the project group. Make a joint subjective judgment and note characteristics for the various VDTs to help the memory.

Keyboard characteristics:

(1) The keyboard should be separate from the screen. When entering text from a manuscript, it should be possible to place the manuscript holder in front of the keyboard and the screen to the right or left of the keyboard. Check that the connecting cable between the keyboard and the screen is long enough to allow this.

(2) It should be possible to adjust the keyboard layout according to the national standards of the country.

(3) Frequently used command keys should be positioned near the alphanumeric keys. Command keys for deleting and other crucial functions should not be too close to the alphanumeric keys, to avoid accidental deletions, etc.

(4) User definable keys are useful for storing frequently used text strings, format calls and composition commands, thus rationalizing text input. It should be easy to re-define the contents of the definable keys, adapting the keyboard to different kinds of product and VDT work. Keyboards with permanent format keys are less flexible.

(5) Aspects of the ergonomics: the keyboard should, apart from being separated from the screen, not give disturbing glare and not be too thick. Using a sculptured keyboard, the distances between each fingertip and its keys will be shorter thus making it easy to write on.

Make a total judgment of the keyboard, paying special attention to the total keyboard layout.

VDT-system communication - some supplementary questions:

(1) Maximum cable length?

(2) Total number of terminals on the system? On existing interface?

(3) Communication speed of the VDT-computer?

(4) Can a single VDT telecommunicate with the computer? How is this done technically? At what cost?

(5) Can the same VDT be used for all different tasks? (excluding pagination).

While it is often technically possible to connect a large number of terminals to a system, one should have in mind that the system speed will be affected negatively as the number of VDTs increase. In practice, the speed requirements will limit the number of VDTs for a specific computer system.

4.4 Proofing Printer

Most text processing systems will have a printer connected for paper output of stored text. In a production system, the printers are used to output text for proof reading. Proof reading typeset material is somewhat faster than proof reading printed output. But while typeset text has a better readability, it is at the same time more expensive mainly because of the photographic paper cost. Therefore, most newspapers use printed output for proof reading editorial text and classified advertisements. When proof reading complicated typography, such as display advertisements, the proof must be typeset for typography control.

Proof reading on a VDT screen is not recommended with the present screen technique.

Printer criteria to be considered:

(1) character quality

(2) character generation:

dot matrix (should not be disturbingly obvious at normal reading distance)
character types (daisy wheel, type chain, etc. often give good quality)

(3) pitch (number of characters per inch, usually 10 or 12)

(4) maximum line length

(5) fonts

(6) printer paper requirements (basis weight, quality, etc.)

(7) speed (characters/s, lines/min)

(8) carbon band (change interval, cost)

(9) multi colour carbon band? (special colour for composition commands)

(10) noise level

(11) electrical current and climate requirements

(12) service responsibility (will the system vendor service the printer?)

(13) possibility of connecting different makes of printer to the system.

The first four points are probably most important as they will decide the readability of the proofs. Compare the readability of different proof printers.

4.5 Interfacing Typesetters

The typesetter interfacing check list includes both hardware and software aspects:

(1) What different makes of typesetters can be interfaced on-line to the system? What references are there?

(2) How many (different?) typesetters can be interfaced to the system?

(3) How fast can the system drive the typesetters? At the typesetter maximum speed?

(4) Is it possible to change the font set in the typesetter from time to time (using different sets of width tables in the system)?

(5) Is it possible to buy new typesetter fonts without having to regenerate the system? Can new width tables be input by the customer?

(6) Is it possible to interface a new kind of typesetter after installation (new typesetter driver), at what cost? Are there different prices for 2nd and 3rd generation of typesetter interfaces/drivers?

(7) Can two different typesetters be interfaced to one system and run parallel? Is it possible to direct a file to a specific typesetter?

(8) Is there a backup for the connection between typesetter-computer?

(9) Will typesetter output be stopped automatically should the typesetter run out of film/paper? How is the operator warned?

(10) Can typesetter output be manually stopped and later continued?

(11) Can an operator list an output queue? Is it possible to edit an output queue (change priority, delete)?

4.6 Interfacing Other Peripherals

If you consider connecting other input/output units, like:

paper tape punches
paper tape readers
cassette tape read/write units
magnetic tape units
OCR-readers
floppy disc units,

the following questions should be answered:

(1) What different makes can be interfaced?

(2) At what cost (interface and software)?

(3) Will the system vendor service these units?

(4) What limits are there to the number of input/output units?

4.7 Telecommunications

For most small and medium size newspapers it will be necessary to communicate with their remote local bureau systems. Although local bureau systems might not be included in the first system installation, the telecommunication capabilities of the system should be investigated.

Questions to the system vendor:

(1) Is it possible to connect a single local bureau VDT to the main system using telecommunications?

(2) If (1) is possible, are you able to use the host system capabilities (software, storage, etc.) as if you were connected by cable to the system?

(3) What extra hardware/software is needed and at what cost?

(4) How many local VDTs can be connected this way?

(5) What different makes of VDT can be connected with:

  (i) one way telecommunication
  (ii) two way telecommunication with the central host system?

(6) Reference installations with different kinds of remote terminals?

(7) Can portable terminals communicate with the system? References?

(8) Can two systems communicate (local bureau system - central system)?

5. BACK-UP

There must be alternative ways of producing text should a system unit cease to work. These alternatives are called back-up. Back-up may be both spare machines and alternative methods for text production.

5.1 Hardware Back-up: alternative production method

A single computer system production is very vulnerable since a single system unit fault may stop the system. In such a system there must be a way for text to pass the faulty unit, going from input units directly to the typesetter. This will require that the input units can produce some kind of data storage medium (paper tape, floppy disc, etc.) and that the typesetter can read the same. Unless H and J can be done locally in the input units, the typesetter must be able to H and J the text before typesetting. The problem with this back-up method is that the typesetter coding is different from that of the system which is normally being used.

5.2 Hardware Back-up: redundant hardware

The most common principle for hardware back-up is to double critical system units. Critical units in a text processing system are computers, disc drives, printers, typesetters and some interfaces. In a dual computer system, the VDTs should be connected to both systems by means of a switchboard. One spare VDT for up to ten VDTs is often enough since the probability of more than one VDT failing out of ten is very small.

In a dual computer system it is important that each of the computers alone could run the production should the other break down.

There are two different dual computer back-up solutions. The hot back-up computer solution has both computers continually running and thus sharing the production load. In a back-up situation, all input/output units are switched over to the running computer and production continues.

The other back-up solution involves one running production computer and one cold spare computer. To run the spare computer in a back-up situation, it is often necessary to move some parts (memory cards, etc.) from the failing production computer to the spare computer before production may continue. This operation will take time and will be risky, since the system managers performing the switch will not likely be trained for this routine. The cost saving with a cold spare computer should carefully be weighed against the security of using a hot back-up computer, especially since production delays may be very expensive for newspaper publishers.

Of course it is very important not to depend on two computers in a dual computer system. It must be possible to carry out production on one single computer if back-up was the reason for buying the second computer.

5.3 Data Back-up

To avoid the risk of data losses, different ways of data copying are used for data back-up.

Redundant data base In many dual computer systems all text input is momentarily stored on two different discs drives. In this way, both discs will hold exactly the same information; there will be two identical data bases. If there should be a disc crash in one disc drive, no data/text would be lost since all information is also stored on the other disc.

Magnetic tape logging Copying data from the system disc onto magnetic tape can be performed in different ways. In some systems every data transfer is continually logged onto magnetic tape, while other systems only copy complete input files. A third magnetic tape back-up method is to copy manually the whole data base at regular intervals (every 4th hour, etc.). If there should be a disc crash or some other disc problem, the data base could be regenerated on a new disc from the tape back-up.

5.4 How to define back-up need

Get the proposed system hardware configurations and imagine that there are two hours left before press start. Eliminate with a pen the different hardware units in the system (one at a time) and think out a way to produce the remaining text without that unit (you cannot count on service since it is too late). Find out if you will lose any text when different units break down. Ask the vendors in detail how production should continue after break down of different units.

6. GUARANTEE

The guarantee period should start from the date when the system vendor and system buyer can agree that the system works according to contract specification. Before signing a contract, try to agree on a system acceptance test. It is important for the system buyer to learn what different functions such a test should include to be a thorough system test. Having signed an acceptance test, the buyer agrees that the system is fully delivered and he is willing to pay the last payment and that the guarantee period (usually six months) has started.

Find out the extent of the guarantee. Are travel costs and expense allowances for service engineers included in the guarantee? Will preventive maintenance be part of the guarantee? Compare different guarantee terms.

7. SERVICE

After the guarantee period, the vendor will most likely offer you a service contract, comprising preventive maintenance, repairs and a service priority. The yearly cost of a service contract will normally be 8-12% of the hardware investment. In the contract, the maximum time for a service engineer to appear at the customer site after having been called for should be specified, as should the hours of the day that service can be offered, be it repair service or telephone service. As in the case of guarantee, it should be clear whether the system buyer or the vendor should pay for the travel and allowance of the service engineer.

Before signing a contract, try to make a judgment of the quality of the service. Service quality criteria could be the number of service engineers relative to the number of installations, number of service centres with spares, reference installation opinions, etc.

8. SYSTEM TRAINING

Good system training is a prerequisite for a successful total system installation. Adequate system training should be part of the total system offer. Discuss a training plan with the system vendor, and ask that all personnel working with the system should be trained regardless of previous knowledge of text processing systems.

The training during installation should be directed at getting the production started in the new system. It is advisable to ask the vendor for follow-up training after a couple of months. In such training, minor problems and misunderstandings can be analyzed and the overall system knowledge improved.

Let all those interested get some extra education, maybe also some general computer software knowledge. They may find quick ways of using the system for production and in some cases for extending the use for other purposes. Ignorance will always be the system's worst enemy.

9. THE VENDOR

Some supplementary vendor aspects:

(1) Experience - a successful system installation will require that the vendor knows what newspaper production is all about and the specific problems pertaining to graphical production.

(2) Reference installations - they will show how the system works and how the vendor has managed system installations and system service. To avoid a mistake, buy a system seen working in a reference installation - there are always others wanting to be test pilots.

(3) Is the vendor an agent or a system manufacturer? If the vendor is an agent, try to find out if the agent is communicating well with the manufacturer.

(4) The development resources of the vendor is an important aspect to consider. The general development points towards integration of a number of sub-systems for the different newspaper production functions (editorial sub-system, local bureau systems, advertisement sub-system, library sub-system, pagination sub-system, etc.). To make this possible will require a lot from the vendor chosen for the first installation. Whether the vendor chooses to develop the different sub-systems on his own or to integrate systems from other vendors will require development resources, either way.

After the pay-off time of the system (3 - 5 years), one will probably look for a new system with all the technological features then available. If the vendor does not have this, find a new vendor, retrain personnel on a new system concept and once again work to establish good co-operation with the new vendor. The development resources of a vendor are very much interconnected with the economical stability of the vendor. A profitable operation is a necessity for successful product development and vice versa. Try to make the vendor discuss his development plans and the organisation of the development department.

AUTOMATIC CONTROL OF INKING IN WEB-OFFSET PRESS

Tapio Lehtonen [1], Kimmo Simomaa [2], Lauri Virtaperko [2], Helene Juhola [1]
1. Technical Research Centre of Finland (VTT), Graphic Arts Laboratory
2. Technical Research Centre of Finland (VTT), Instrument Laboratory

ABSTRACT

The report gives a description of the research and development project carried out at VTT [1,2], Finland. During a five-year period the project included a large number of process identification analyses, computer simulations of the printing unit and the development of new measuring and control instrumentation. The goal of this project was to develop the prototype system of an automatic inking controller for a web-offset printing press.

The control system includes the following components: remote control system to adjust and monitor the inking screws and the ink and water duct rollers and density monitor, both of which were developed at the Technical Research Centre of Finland. The master processor of the on-press densitometer also takes care of the computation of control parameters. The monitor gives the control commands to the remote control station via an in-line cable connection. This system controls the inking screws and the duct roller of a web-offset press at the Institute.

Some test trials have been carried out at the Institute. The results of these tests are reported and discussed. The future development of the control system for multi-colour production presses is also discussed.

INTRODUCTION

A great deal of time and effort have been expended to improve the high time and material waste in lithographic offset printing. Today's solution of this problem is given by the computerized control system. Figure 1 shows a block diagram of the integrated control system for an offset press. As yet, the systems in production presses are not as complete as presented in the diagram. The most advanced systems have been adapted to large newspaper printing presses: only the feedback control feature has been left out. In the case of heat-set presses the feed-forward feature, the presetting, has found few applications. The remote-control system is a common practice in new 16-page heat-set presses. Most of these constructions have been realized by means of the analog technique. This makes it difficult to expand the system with new system components such as the presetting.

Much development work has been done at VTT in Finland to design and construct system components for a control system as shown in Figure 1. The development has been reviewed [1, 2, 3, 4]. This paper describes the prototype closed-loop control system for the inking of a web-offset press and the first experience of its performance. We believe that the closed-loop inking control is the next step in the development after the remote control system and the presetting system in a heat-set offset press.

In the earlier IARIGAI reports [5, 6], we showed that the variations of the main process output variable, the print density, can be accounted for by the variations of certain state-variables of the process in lithographic printing. The possibility of giving a mathematical relation between the output and input variables of the process makes it possible to use the closed-loop system to control the variations of the output variable.

For the control system the steady-state and dynamic models of the process, an on-press densitometer for the feedback information, a control computer and remote control systems are needed.

In earlier papers [5, 7], we suggested that the control system should be realized in a two-phase controller, or a cascaded controller. The first control loop, the fast one, was formed by an ink film thickness sensor measuring the thickness on the duct roller, and a controller adjusting the ink feed according to this measurement. The second control circuit was based on the density measurement by means of an on-press densitometer. This slower second-phase controller had given the setting value to the first-phase controller. We met, however, with overwhelming difficulties in developing the ink film thickness gauge suitable for the very difficult conditions of the environment in the production presses. This is why we have not used the ink film thickness sensor in the automatic control system of inking. Still, the functioning of the ink duct has been accounted for in the controller. This has been done in the form of feed-forward control of the ink feed, especially against the effects of the speed changes of the press.

EXPERIMENTAL EQUIPMENT

A block diagram of the prototype of the inking control system is given in Figure 2. The system consists of a single-unit blanket-to-blanket printing unit, a remote control system, an on-press densitometer and a control computer. The prototype system is shown in Figure 3.

The development has been a cooperative project of the Graphic Arts Research Institute and the Instrument Laboratory of the Technical Research Centre of Finland.

THE CONTROL SYSTEM OF AN OFFSET-PRINTING PRESS

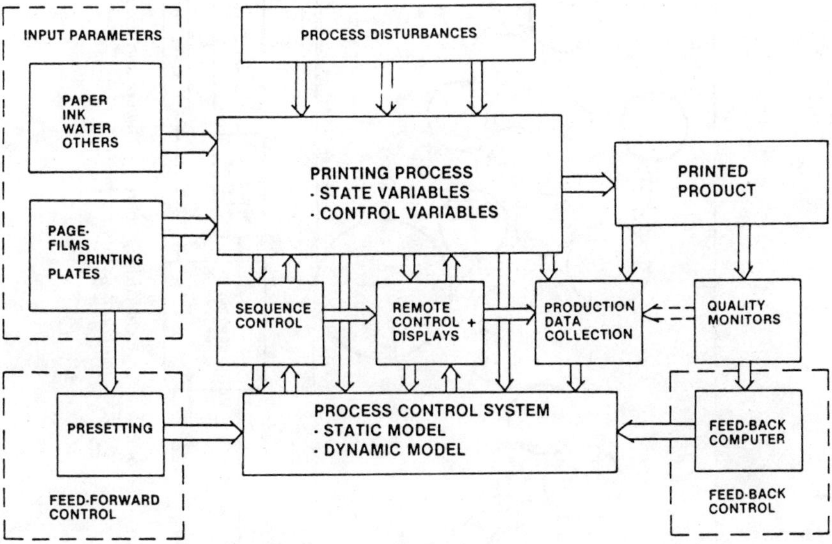

Fig. 1. Block diagram of an integrated control system of a web-offset press.

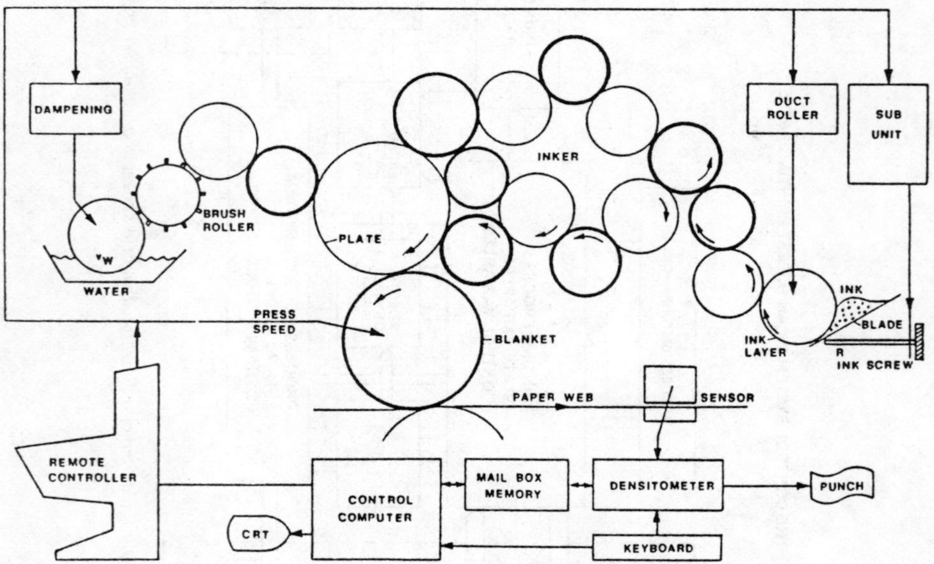

Fig. 2. A diagram of a prototype closed-loop control system of an inking process in a single-unit web-offset press.

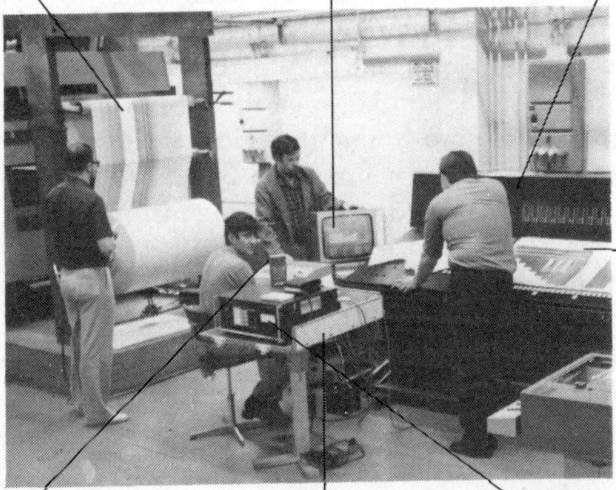

Fig. 3. An overall view of the prototype inking control system at the Graphic Arts Research Institute in Finland.

The full-scale web-offset printing unit was designed by the Finnish Manufacturer, Oy Wärtsilä Ab. The web width of the press is 990 mm and the cut-off length is 625 mm. Maximum printing speed of the press is 40,000 r/h when the printing is carried out from reel to reel. Inking is controlled by 32 duct screws and by the speed of the duct roller.

The remote control system of the press controls and displays the inking screw positions, the duct roller and fountain-roller speeds and the speed of the press. This microprocessor-controlled remote-control system was originally developed as a cooperative project of the Technical Research Centre of Finland with Oy Wärtsilä Ab for a five-unit heat-set web-offset machine having similar printing units as the test-printing press at the Institute. This remote control system includes also the controls and displays for crosswise and longitudinal register of colours, the folder register for the five printing units of the press. The control console consists of four microcomputers; a master computer, a control computer, a computer for keyboards and displays and a computer for connections to printing unit substations, and to a presetting scanner, the inking controller, etc. These processors in the console communicate mutually via a parallel data base. The data between the substations of the printing units and the control console are transmitted via a serial connection and further via a two-wire cable. The microprocessor-controlled substation takes care of the driving of the inking screws, the measurement of the screw positions and the communication with the main console. A more detailed description of the system development and of the system operation will be published later [8].

The on-press densitometer of the control system was originally a 16-channel monitoring system designed at the Institute. This monitoring system was constructed for experiments with various kinds of production printing presses. The control unit and the CRT display unit of the monitoring system are seen in Figure 3. This two-level microprocessor system has also three extra analog input channels. By way of these inputs the speed setting values of the duct roller (in manual operation), the fountain roller and the press were fed to the control system. The master computer of this measuring system also functioned as the control computer of the closed loop control system. More details of this density monitoring system and of its applications in the experiments with the production presses have been published [3]. In the inking control system only eight of the 16 measuring channels were used for the density measurements.

Figure 3 also shows the test image in the trials. The lefthand side triangle of the image was that part of it which was subjected to the closed-loop inking control. The adjacent ink screws were controlled by each of the density sensors. Thus the zonal area of 16 ink screws was controlled by means of the density sensors.

Until now the majority of the test trials have been carried out by printing on sc-newsprint with cold-set ink.

CLOSED-LOOP INKING CONTROL SYSTEM

The steady-state conditions, the dynamic conditions, the control theory behind the controller, the applied control algorithm and the realization of the controller have already been discussed in detail [9]. In this report, the system is described briefly and details of further developments are given following the earlier reported trials [9].

The steady-state model of the inking process of a lithographic press gives the operating level and some primary parameters to the controller. A well-known fact is that the relation between the print density and the ink-feed setting value is a saturating nonlinear function [5,6]. This is mainly due to the nonlinear relationship of print density to the amount of ink on the paper. The operations of the ink feeding mechanism and the ink roller train are to some extent nonlinear as well. Linear approximations of the nonlinear functions have been applied in the controller. This is why the type of paper and the level of the density setting value affect the control parameters like the gain of the controller.

The duct roller speed v_D is controlled by the closed-loop controller, so it has to be set according to the speed of the press. From the phenomena appearing in the gap between the duct roller and the duct blade [10] it follows that the gain of the process varies with the speed. For the prototype controller the gain of the differential process

$$\Delta D = \beta \Delta v_D \qquad (1)$$

the following approximation was applied

$$\beta = 10.5\, v_p^{-1} + 0.5 \qquad (2)$$

when the printing speed v_p is given in m/s.

The operation level of the water feed during the test trials was controlled manually by means of the remote control system. The manual control was, however, based on the display of the water film thickness monitor measuring the water film thickness on the non-image area on the printing plate [5].

The temperature of the inking unit was held constant by means of water circulation through the distributing rollers of the roller train.

Characteristic of the dynamic behaviour of the lithographic inking process is the deadtime delay, which in the trial press turned out to be about 15 machine revolutions, and a response characterised by many time constants. The remote-control delay time has to be accounted for as well. This raises the total delay time to d = 20 revolutions.

The approximate rise of the density output required about 100 revolutions. This result was very accurately in accordance with the results of the computer simulation carried out with the computer model of the trial press inker.

The dynamics of the printing press cannot be treated in the time domain as is customary in conventional process control dynamics.

Evidence to that effect has been given above. Instead of time units, the characteristic parameters of the dynamics of the system have been given as a number of press revolutions. This domain of revolutions instead of the time domain has already been suggested [6]. The approach calls for the application of a discrete time process. This can be given by the difference equation with an input u(k) and an output y(k) as follows:

$$y(k) + a_1 y(k-1) + \ldots, a_m y(k-m) =$$
$$b_1 u(k-d-1) + \ldots + b_m u(k-d-m) + v(k), \quad (3)$$

where k refers to a certain number of revolutions printed and the '1' is a sample interval given as the number of revolutions.

The discrete time process calls for the application of the z-transformation instead of the usual Laplace-transformation. The method of least squares was applied for the identification. This does not account for the non-linearities of the process. The noise term v(k) has been left out of the consideration.

The control model of the inking process, $G_p(z)$ is shown in Figure 4. The process output is the density D which is compared with the setting value D_{opt}. The error e is used as input of the controller $G_C(z)$. The process and the measurement are disturbed by the noises v_1 and v_2 [9].

The optimized parameters as proposed by Isermann [11] have been applied in the controller. As has been suggested [9], the control algorithm is written in the recursive form

$$\Delta R(k) = R(k) - R(k-1) = q_0 e(k) + q_1 e(k-1) + q_2 e(k-2) \quad (4)$$

where R(k) indicates the ink screw position.

In the first trials already published [9], similar control algorithms were applied for the ink screws and for the duct roller. This, however, proved unsatisfactory. For the trials described in this paper the following changes were made in the control algorithm.

The control parameters for the ink screw control $q_0 \ldots q_2$ in Equation (4) were adjusted from the values published [9]:

$$q_0 = K(1 + C_D) = 1.8\ K$$
$$q_1 = K(C_1 - 2C_D - 1) = -2.4\ K$$
$$q_2 = KC_D = 0.8\ K \quad (5)$$

The term K has the value 0.35 for the ink screw model.

The control algorithm of the duct roller rotation is based on the equation:

$$\Delta v_D\ (k) = \frac{K}{\beta}\ \{1.5E(k) - 1.4E(k-1) + 0.5E(k-2)\} \quad (6)$$

where β is given by relation (2). The factor E is the difference between the set value of the density and the average of the measured density values of the eight density sensors. The value of the parameter K depends

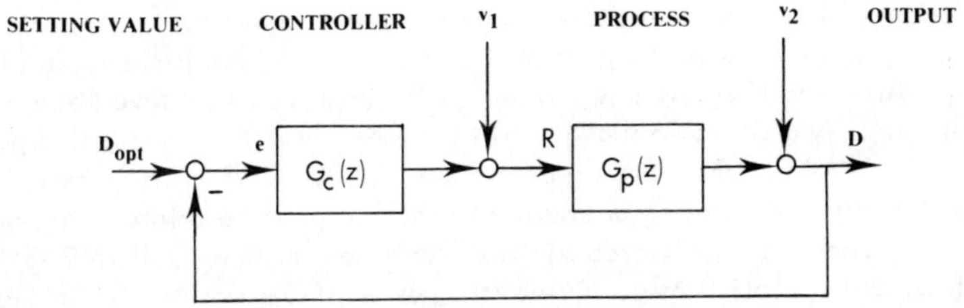

Fig. 4. The control model of an offset printing press.

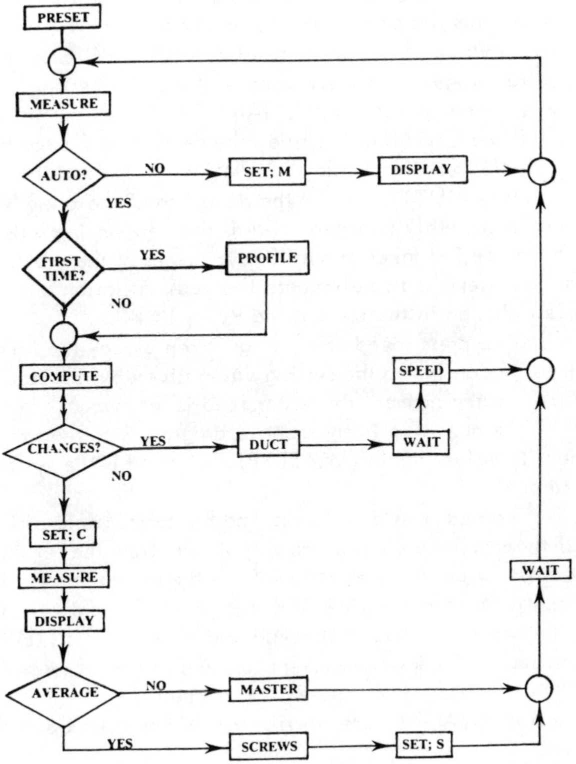

Fig. 5. A block diagram of the control program of the closed-loop inking controller.

on the direction of the control operation. When adjusting the ink feed downwards the value of K is 0.5 and when controlling upwards the value is 0.3. This is due to the well-known fact that the response of the conventional inker is much slower downwards than upwards [6].

The block diagram of the control algorithm is shown in Figure 5. After the first trials reported [9] some improvements have been made. The same symbols as in that report have been used throughout this paper.

After the start-up procedure (PRESET), which may include the presetting of the inking according to the image to be printed, the density measurements are collected via the 'mail-box' memory unit (MEASURE). If manual operation mode is true (the test AUTO? not true), the press speed, the inking and the dampening are set manually by means of potentiometers and the keyboard. The state of the process is shown in the display (DISPLAY).

If the test AUTO? is true (automatic operation mode) the first time, the ink screw profile is requested from the control console (PROFILE). After this the gain parameter β is computed according to the initial setting value of the press speed v_{po} (COMPUTE). The next step in the control program is to test whether the speed-setting value of the press has been changed (the test CHANGE?).

If the CHANGE? is true, the new value for the duct roller speed v_D is computed according to relationship (1) and this new value is set immediately (DUCT). After the deadtime of the press (WAIT), the press speed is smoothly changed towards the new setting value. The speed change period takes about 80 revolutions of the press. This procedure has been designed to eliminate the peak variation of the density after a fast change in the press speed [6].

If no press speed change has taken place, the fountain roller speed is set according to the setting value given by the potentiometer (SET; C). At the moment the water feed is set manually.

The next step in the control program is to read the newly measured density values again (MEASURE). After this the system display is accepted again.

The next test determines whether master control is necessary or not. If the measured average density differs from the set value more than the given set value of the density (5% of the set value at the moment) master control is selected (MASTER) by means of the duct roller speed. If not, the density profile is checked and controlled (SCREWS), and the new positions of the screws are sent to the control console of the remote controller (SET; S). After these control operations the controller waits 30 revolutions (WAIT) before starting a new control loop of the program.

TEST RESULTS

In the following, examples are given of test runs at the start of printing, of changing the water feed during production printing and of changing the speed of the press during printing.

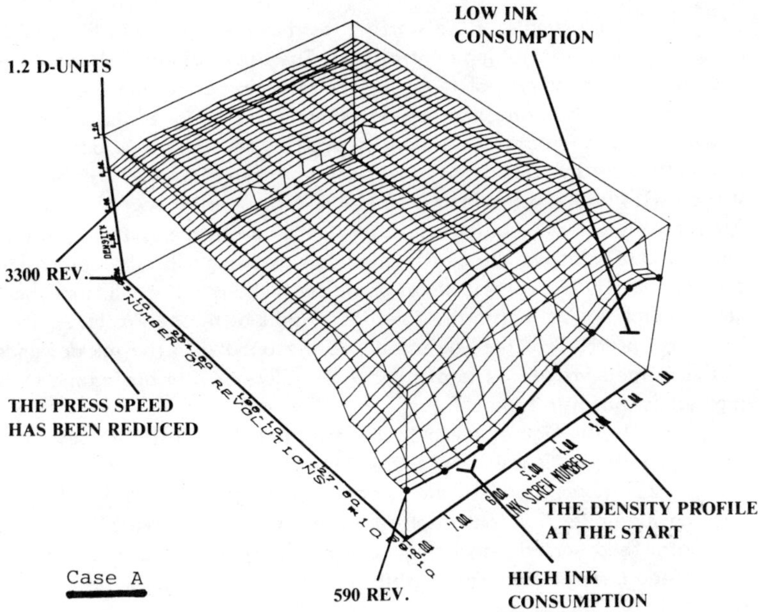

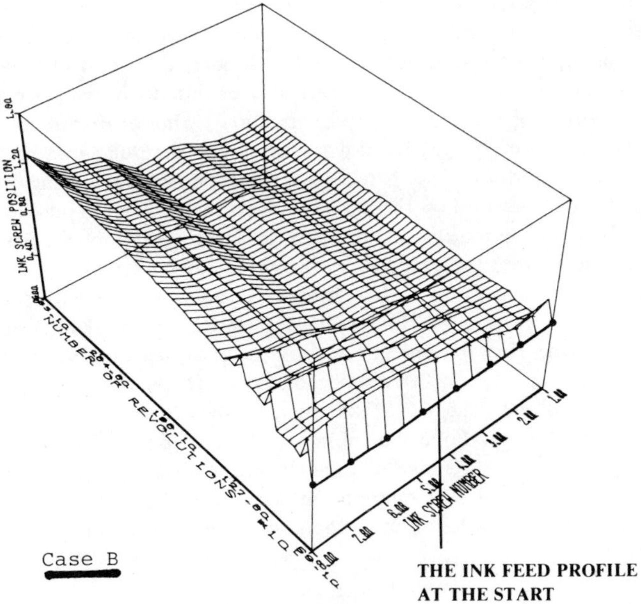

Fig. 6. A perspective graph giving an example of a starting test with the trial printing press under closed-loop inking control. Case (a): density profile as a function of the number of revolutions printed. Case (b): the ink screw position profile as a function of the number of revolutions printed.

An example of the starting test has been given in Figure 6, using the automatic inking controller. The upper diagram shows the change of the density profile during the starting period and the lower diagram shows the corresponding adjustment operations of the inking screws. The averages of the density profile, the ink screw positions and the duct roller speed settings have been given in Figure 7. Printing was started with an even ink feed profile and a duct roller speed which clearly produced a density below the set value given to the system. After about 500 revolutions the density profile was brought by the controller to a commercially acceptable level and after a thousand revolutions the density profile was within the tolerance limits of the controller. The ink screw profile was then adjusted to correspond with the ink demands of the printed image. The display of the CRT-monitor during the starting test is shown in Figure 8.

In the figures the output interval of the density values, 30 press revolutions, is the same as the control interval.

Figures 9 and 10 show, correspondingly, the results of the production run under the control of the automatic inking controller but with the water feed varied manually. During the first third of the trial, the water feed changes occurred within the tolerance limits of the ink-water balance.

By means of the controller, the density profile and the average density level could be kept fairly constant. On lowering the water feed level after this period, toning started to occur along the edge of the web. This was further stressed because of the slightly damaged surface structure of the water form roller. The result proves that such disturbances affect the operation of the closed loop controller. Similar disturbances were caused by directing additional water sprays to the dampener to open the plate more rapidly. This is a common practice in production conditions but it is not tolerated by the inking controller. After reaching the ink-water balance by feeding more water again, the edge of the web could also be taken into the control, as shown in Figure 9.

Figures 6 and 7 show the test trials when the press speed has been varied. The operation of the automatic controller has been described above: first the change of the duct roller speed and following the control interval the smooth change of the press speed in accordance with the set value of the speed. The figures show the control operations of the duct roller speed and the inking screw positions. The graphs show clearly that the press speed changes have been controlled by the automatic inking controller with acceptable accuracy: the density profile and the average density level have been kept fairly constant during the speed changes of the press.

CONCLUSIONS

The experiments described in this paper have been done on a single-unit printing press. On the whole, the prototype control system has been

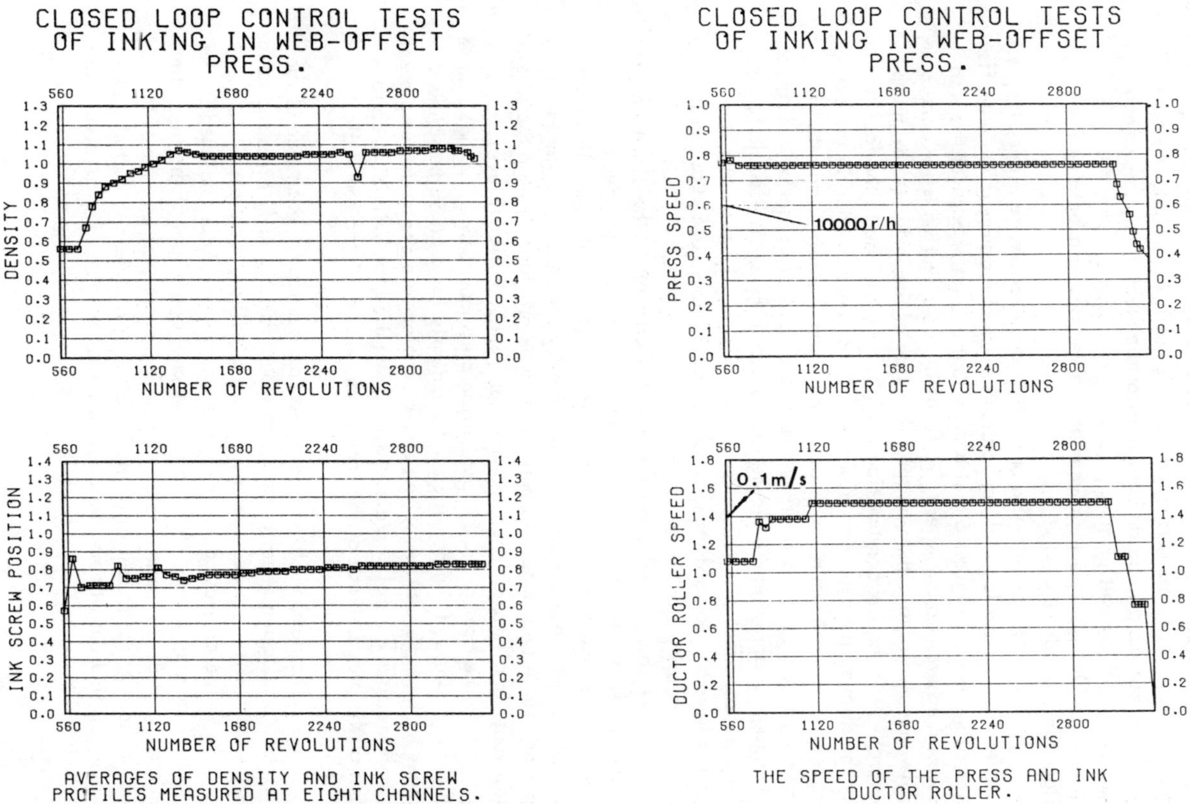

Fig. 7. The averaged densities, the ink screw positions and the duct roller speed as a function of the number of revolutions printed in the printing starting test.

functioning fairly satisfactorily. Of course further improvements to the control algorithm and the control parameters are possible and also necessary. This is especially true in the case of a multi-unit production web-offset press. More computing capacity is needed to handle the control of all the colours printed on both sides of the web and of phenomena caused by wet-on-wet printing [12].

One possible improvement is given by the adaptable control algorithm already suggested [9].

Our test results also prove that an inker pre-setting system is needed in a closed-loop inking control system of a commercial web-offset press (Figure 1). Figures 6 and 8 show that along the edge of the web (screw numbers 1 to 3), where the zonal ink consumption is low, the control works much faster than on the other side of the image (screw numbers 6 to 8) where the ink consumption is high. This is due to the fact that for the screws, the same value of the gain has been applied in the controller. The optimum result is achieved when the gain is adjusted in accordance with the ink consumption in the zone of the screw. Another advantage of applying the pre-setting is that it makes it possible to use the very fast functioning dead-beat controller [9]. Furthermore, the pre-setting lowers remarkably the number of revolutions needed at the start to make the density profile even.

In our present control system the controls of the duct roller and of the ink screws function independently: the duct roller speed is controlled if the average density shifts beyond a given tolerance limit and the ink screws are adjusted to get the density profile even. Interaction between these two modes of control has to be provided to keep the screw positions, in all conditions, near the optimum operational level.

The controller of a production press must permit manual change of the set values of the densities independently for each of the screw zones in the same way as the ink screw positions are controlled at present via the remote-control console. This is required for individual correction of the grey balance or the tone rendering of printed four-colour pictures.

The number of the density sensors needed for each side of the web of a production offset press which would satisfactorily fulfil the needs of the controller in production conditions is still an open question.

The control of the ink water balance has to be included in the closed-loop controller. All the exceptional conditions of the ink water balance outside the normal tolerance limits severely disturb the functioning of the automatic control of the inking. We have just started a new project to find out whether we could use the on-press densitometer to get an indication of the ink-water balance variations taking place during printing.

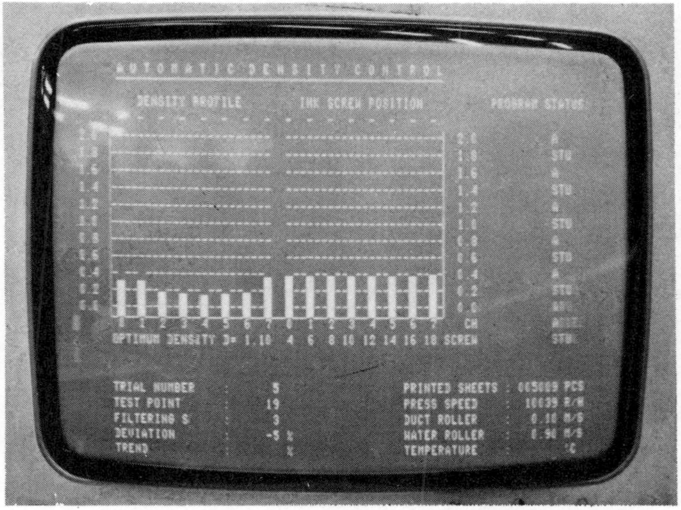

(a)

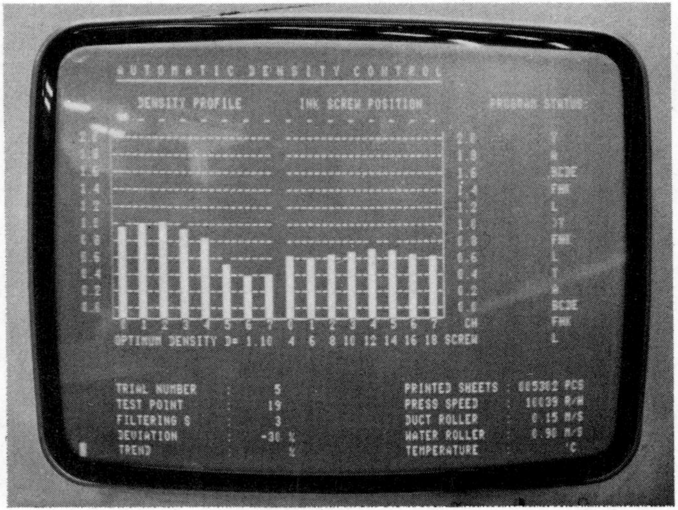

(b)

Fig. 8. *A series of photographs taken during a print starting test from CRT-display-unit of the inking controller. (a) At start, (b) after 300 revolutions, (c) after 530 revolutions, (d) after 780 revolutions, (e) after 1010 revolutions, (f) after 1530 revolutions.*

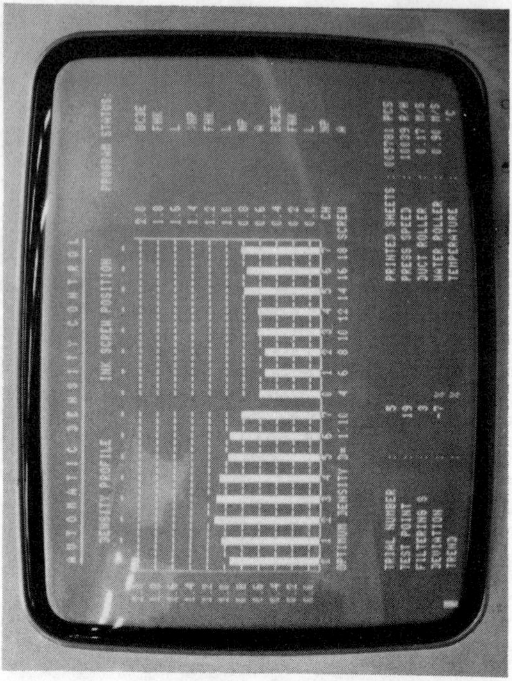
AUTOMATIC DENSITY CONTROL
DENSITY PROFILE
INK SCREEN POSITION
PROGRAM STATUS
OPTIMUM DENSITY D= 1.10
SCREEN
TRIAL NUMBER
TEST POINT
FILTERING S
DEVIATION
TREND
PRINTED SHEETS
PRESS SPEED
DUCT ROLLER
WATER ROLLER
TEMPERATURE

(d)

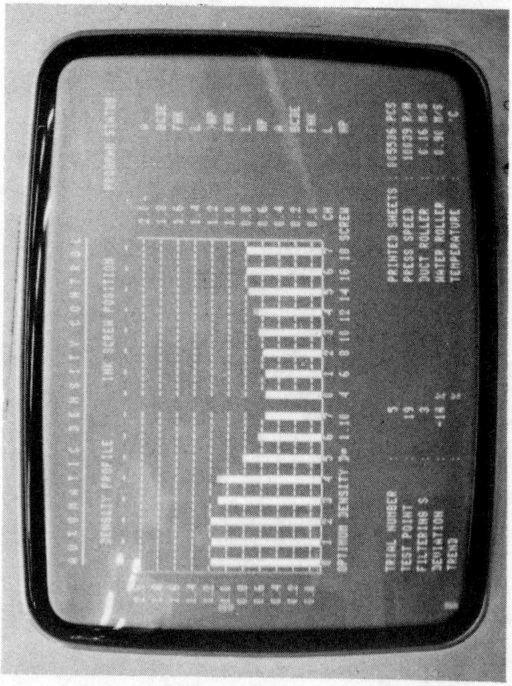
AUTOMATIC DENSITY CONTROL
DENSITY PROFILE
INK SCREEN POSITION
PROGRAM STATUS
OPTIMUM DENSITY D= 1.10
SCREEN
TRIAL NUMBER
TEST POINT
FILTERING S
DEVIATION
TREND
PRINTED SHEETS
PRESS SPEED
DUCT ROLLER
WATER ROLLER
TEMPERATURE

(c)

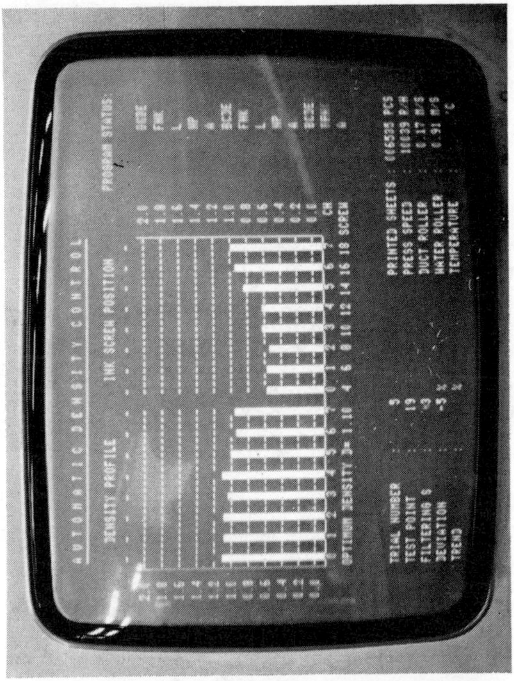
AUTOMATIC DENSITY CONTROL
DENSITY PROFILE
INK SCREEN POSITION
PROGRAM STATUS
OPTIMUM DENSITY
TRIAL NUMBER
TEST POINT
FILTERING
DEVIATION
TREND
PRINTED SHEETS
PRESS SPEED
DUCT ROLLER
WATER ROLLER
TEMPERATURE

(f)

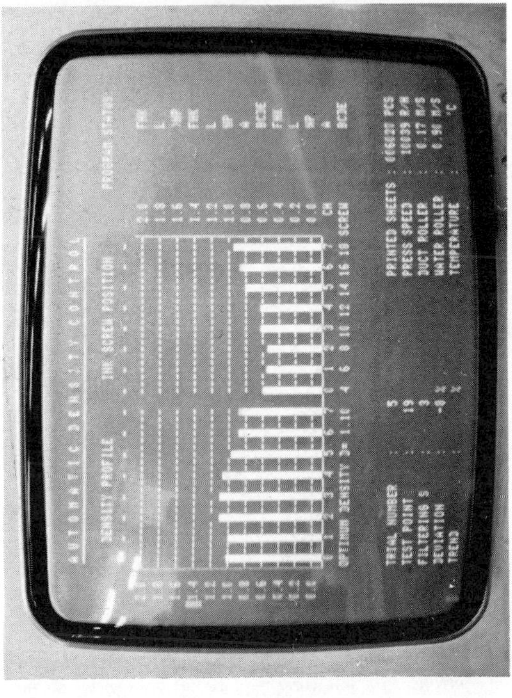
AUTOMATIC DENSITY CONTROL
DENSITY PROFILE
INK SCREEN POSITION
PROGRAM STATUS
OPTIMUM DENSITY
TRIAL NUMBER
TEST POINT
FILTERING
DEVIATION
TREND
PRINTED SHEETS
PRESS SPEED
DUCT ROLLER
WATER ROLLER
TEMPERATURE

(e)

Case A

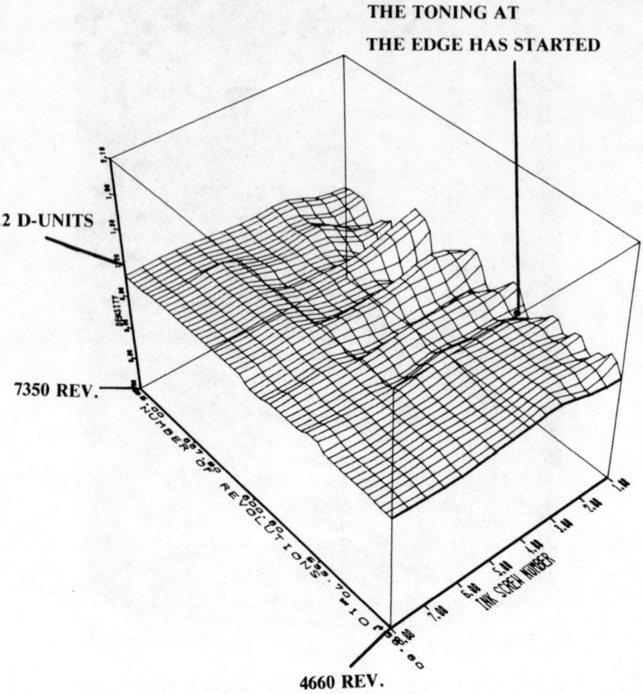

Case B

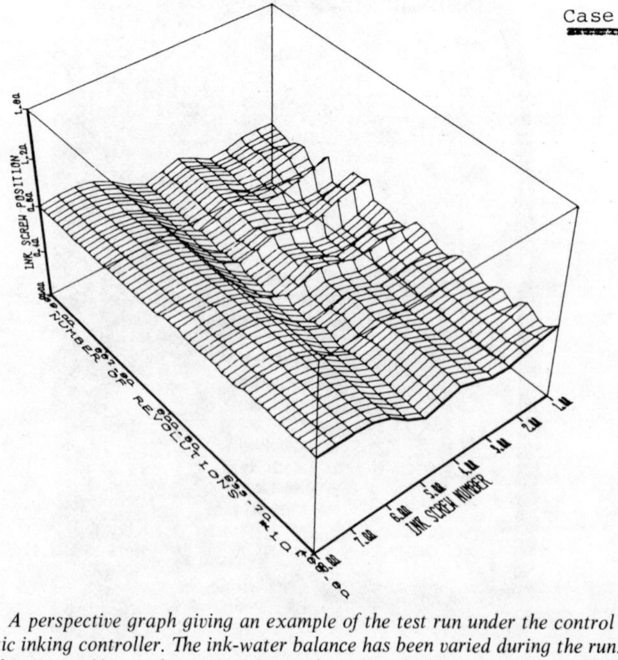

Fig. 9. A perspective graph giving an example of the test run under the control of the automatic inking controller. The ink-water balance has been varied during the run. Case (a): the density profile as a function of the number of revolutions printed. Case (b): the ink screw position profile as a function of the number of revolutions printed.

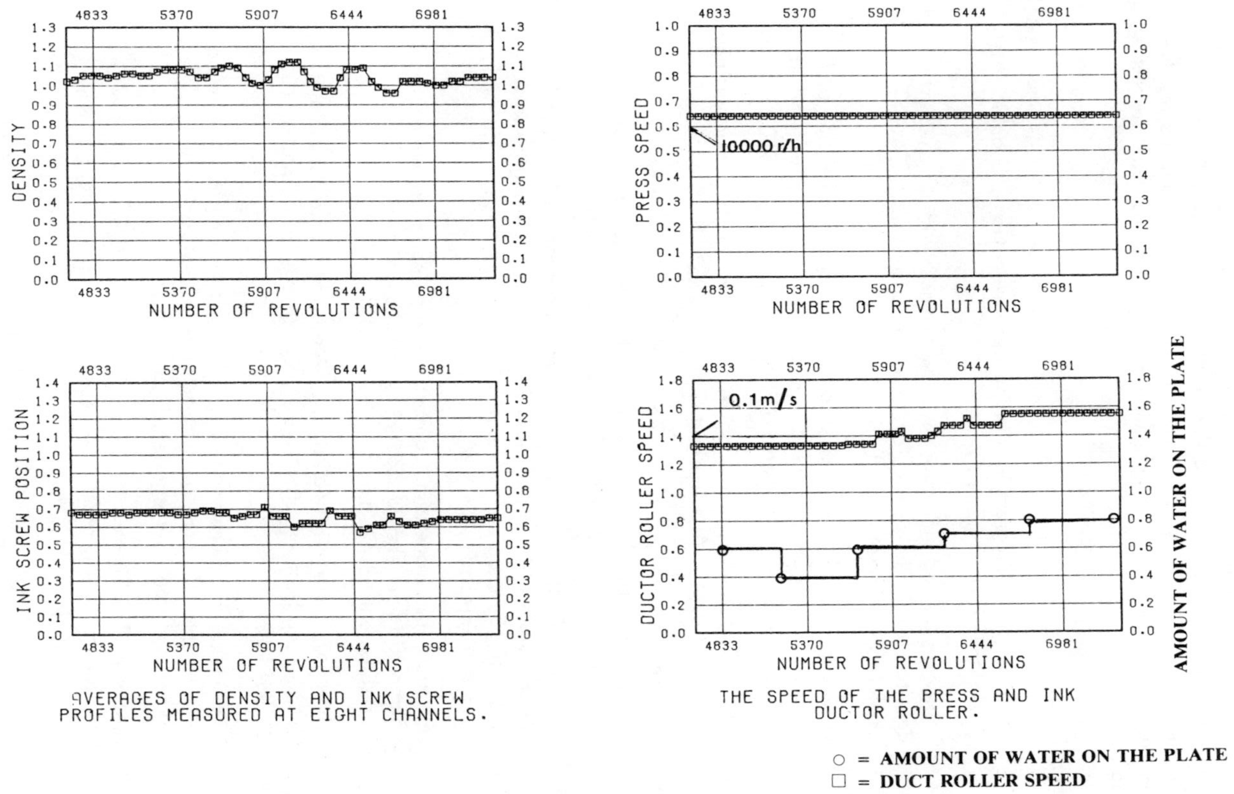

Fig. 10. The averaged densities, the ink screw positions, the duct roller speed and the water monitor readings during the test runs in which the ink-water balance was varied.

ACKNOWLEDGMENTS

The authors would like to express their gratitude to Oy WÄRTSILÅ Ab, Järvenpää Works, for providing the opportunity to implement the regulators and to Rauma-Repola Oy, who have supplied the paper for the test trials.

REFERENCES

1. Lehtonen, T., 'Progress in printing instrumentation', Graphic Arts in Finland, 8 (1979):2, p. 4-8.

2. Lehtonen, T. and Simomaa, K., 'Experiences with an on-press densitometer in the production on a web-offset press', TAGA-Conference, Kansas City, U.S.A. (1979).

3. Lindqvist, H., Lindqvist, U., Lehtonen, T., Karttunen, S. and Simomaa, K., 'On the closed loop control systems for web-offset presses', TAGA-Conference, Williamsburg, U.S.A. (1980).

4. Lehtonen, T. and Riihelä, E., 'A page-film scanner for presetting systems', Graphic Arts in Finland, 9 (1980):2.

5. Lehtonen, T., 'Effect of ink feed on the variation in density of lithographic prints', Advances in Printing Science and Technology, ed. Banks, W. H., London, Pentech Press, (1977).

6. Lehtonen, T. and Simomaa, K., 'Cross-machine variation of inking its causes and consequences in web-offset printing', Advances in Printing Science and Technology; ed. Banks, W. H., Vol. 14, London, Pentech Press (1979).

7. Lehtonen, T., Menetelmä painoalustalle painettavan painovärin määrän mittaamiseksi ja säätämiseksi laaka-painomenetelmässä, Finnish patent 58740, Int. Cl. B 41 F 31/00. (1976).

8. Lehtonen, T. and Virtaperko, L., 'A digital remote control system for five-unit heat-set offset press', Graphic Arts in Finland, 10 (1981): 1.

9. Simomaa, K. and Lehtonen, T., 'Control of inking in a web-offset press using multichannel densitometer as a feed-back device', TAGA-Conference, Rochester, U.S.A. (1981).

10. Lehtonen, T., Simomaa, K. and Riihelä, E., 'Measurement of ink-film thickness using capacitive sensors', Graphic Arts in Finland, 6 (1977): 2, p. 3-11.

11. Isermann, R., Digitale Regelsysteme, Springer Verlag, (1977).

12. Lehtonen, T., 'Effect of certain variables in offset lithography on printing wet-on-wet', 14th EUCEPA-Conference, Budapest, (1971).

DISCUSSION

A. Frøslev-Nielsen Your last point of the Conclusion concerns the future investigations of ink-water balance. Are you including an automatic control of water film thickness to minimize the number of variables?

T. Lehtonen We are trying to avoid locating delicate measuring instruments in the printing units of production presses. The solution we are trying to apply is to get an indication of ink/water balance by means of the densitometer.

J. Aspler Is your water film thickness monitor based on an infra-red detector, and is this a home-made or a commercial system?

T. Lehtonen The water-film thickness monitor is commercially available and is manufactured by an English electronics firm 'Infrared Engineering Limited'.